Advances in Intelligent Systems and Computing

Volume 519

Series editor

Janusz Kacprzyk, Polish Academy of Sciences, Warsaw, Poland
e-mail: kacprzyk@ibspan.waw.pl

About this Series

The series "Advances in Intelligent Systems and Computing" contains publications on theory, applications, and design methods of Intelligent Systems and Intelligent Computing. Virtually all disciplines such as engineering, natural sciences, computer and information science, ICT, economics, business, e-commerce, environment, healthcare, life science are covered. The list of topics spans all the areas of modern intelligent systems and computing.

The publications within "Advances in Intelligent Systems and Computing" are primarily textbooks and proceedings of important conferences, symposia and congresses. They cover significant recent developments in the field, both of a foundational and applicable character. An important characteristic feature of the series is the short publication time and world-wide distribution. This permits a rapid and broad dissemination of research results.

Advisory Board

Chairman

Nikhil R. Pal, Indian Statistical Institute, Kolkata, India
e-mail: nikhil@isical.ac.in

Members

Rafael Bello, Universidad Central "Marta Abreu" de Las Villas, Santa Clara, Cuba
e-mail: rbellop@uclv.edu.cu

Emilio S. Corchado, University of Salamanca, Salamanca, Spain
e-mail: escorchado@usal.es

Hani Hagras, University of Essex, Colchester, UK
e-mail: hani@essex.ac.uk

László T. Kóczy, Széchenyi István University, Győr, Hungary
e-mail: koczy@sze.hu

Vladik Kreinovich, University of Texas at El Paso, El Paso, USA
e-mail: vladik@utep.edu

Chin-Teng Lin, National Chiao Tung University, Hsinchu, Taiwan
e-mail: ctlin@mail.nctu.edu.tw

Jie Lu, University of Technology, Sydney, Australia
e-mail: Jie.Lu@uts.edu.au

Patricia Melin, Tijuana Institute of Technology, Tijuana, Mexico
e-mail: epmelin@hafsamx.org

Nadia Nedjah, State University of Rio de Janeiro, Rio de Janeiro, Brazil
e-mail: nadia@eng.uerj.br

Ngoc Thanh Nguyen, Wroclaw University of Technology, Wroclaw, Poland
e-mail: Ngoc-Thanh.Nguyen@pwr.edu.pl

Jun Wang, The Chinese University of Hong Kong, Shatin, Hong Kong
e-mail: jwang@mae.cuhk.edu.hk

More information about this series at http://www.springer.com/series/11156

Ryszard Jabłoński · Roman Szewczyk
Editors

Recent Global Research and Education: Technological Challenges

Proceedings of the 15th International
Conference on Global Research
and Education Inter-Academia 2016

 Springer

Editors
Ryszard Jabłoński
Institute of Metrology and Biomedical
  Engineering
Warsaw University of Technology
Warsaw
Poland

Roman Szewczyk
Institute of Metrology and Biomedical
  Engineering
Warsaw University of Technology
Warsaw
Poland

ISSN 2194-5357   ISSN 2194-5365 (electronic)
Advances in Intelligent Systems and Computing
ISBN 978-3-319-46489-3   ISBN 978-3-319-46490-9 (eBook)
DOI 10.1007/978-3-319-46490-9

Library of Congress Control Number: 2016951982

Preface

Developments in the connected fields of solid-state physics, bioengineering, mechatronics and nanometrology have had a profound effect on the emergence of modern technologies and their influence on our lives.

In all of these fields, understanding and improving the basic underlying materials is of crucial importance for the development of systems and applications.

The International Conference Inter-Academia 2016 has successfully married these fields and become a regular feature in the conference calendar. It comprises seven thematic areas in the field of material science, nanotechnology, biotechnology, plasma physics, metrology, robotics, sensors and devices.

The book Recent Global Research and Education: Technological Challenges is intended for use in academic, government and industry R&D departments, as an indispensable reference tool for the years to come. Also, we hope that the volume can serve the world community as the definitive reference source in Advances in Intelligent Systems and Computing.

This book comprises carefully selected 68 contributions presented at the 15th International Conference on Global Research and Education INTER-ACADEMIA 2016, organized by Faculty of Mechatronics, Warsaw University of Technology, during September 26–28, in Warsaw, Poland. It is the second volume in series, following the edition in 2015. It brings together the knowledge and experience of 150 leading researchers representing 13 countries.

We would like to thank all contributors and reviewers for helping us to put together this book.

Warsaw, Poland   Ryszard Jabłoński

Contents

Notations

l	Number of donors [CFU]
D_{pv}	Volume equivalent diameter [μm]
H	Height of the liquid [m]
n_t	Total molecular weight in the tank [mol]
r_c	Individual rotation radius [m]
T	Temperature [K]
y	Number of transconjugants [CFU]
z	Collision frequency [Hit/s]
Z	Collision frequency [Hit]
α	Probability of valid collision [–]
γ	Probability of inhibition [CFU/Hit]
$\aleph$	Volume number density [CFU/m^3]

Part I
Material Science and Technology, Smart Materials

Part I

The Effective Optimal Parameters of Metamaterial on the Base of Omega-Elements

Igor V. Semchenko, Sergei A. Khakhomov, Andey L. Samofalov,
Maxim A. Podalov and Qian Songsong

Abstract The properties of an artificial anisotropic structure composed of omega-elements are numerically simulated. Analytical expressions for the dielectric, magnetic and magnetoelectric susceptibilities of the structure are derived. The frequency dependence of the effective parameters of a metamaterial has been determined taking into account trajectories of the conduction electrons, the skin effect, and the electric field attenuation in metal. The relation between the effective parameters of the medium ε and μ and the tensors of magnetoelectric susceptibilities, can be determined by not only the omega-elements concentration, but also the inclusion element shape.

1 Introduction

In recent years, numerous studies of the chiral properties exhibited in the microwave band by artificial composite media have been performed. Such media can be produced on the basis of helices and Ω elements [1–11]. In this paper, the interaction of microwave electromagnetic radiation with structure consisting of Ω elements with previously calculated optimal parameters is studied. It is established that such structures can be widely used, in particular, for transformation of the polarization of electromagnetic waves of the microwave band, for example, for obtainment of a circularly polarized wave.

A circularly polarized wave is formed owing to radiation of the interrelated electric dipole moment and magnetic moment of each Ω element, which make contributions of equal absolute values to the reflected wave.

I.V. Semchenko · S.A. Khakhomov (✉) · A.L. Samofalov · M.A. Podalov
Francisk Skorina Gomel State University, Gomel, Belarus
e-mail: khakh@gsu.by

Q. Songsong
Nanjing University of Science and Technology, Nanjing, China

© Springer International Publishing AG 2017
R. Jabłoński and R. Szewczyk (eds.), *Recent Global Research and Education:
Technological Challenges*, Advances in Intelligent Systems and Computing 519,
DOI 10.1007/978-3-319-46490-9_1

3

An advantage of the application of an Ω structure for polarization transformation, in particular, for obtaining a circularly polarized wave consists in the simplicity of manufacturing and scaling. An Ω structure is plane, so it can be manufactured, for example, by the method of etching on a printed circuit board on practically any available scale. In this paper, the optimal parameters of Ω elements required for radiation of a circularly polarized wave are used [12].

2 Analytical Simulation

The interaction of the incident electromagnetic wave with an Ω element (Fig. 1) induces simultaneously matched electric dipole moment $\vec{p}$ and magnetic moment $\vec{m}$. The following relationships can be obtained for the projections on the coordinate axes for an Ω element [13]:

$$p_x = 2ql_0, \quad p_y = 0, \quad p_z = 0, \tag{1}$$

$$m_z = I\pi r^2, \quad m_x = 0, \quad m_y = 0. \tag{2}$$

We use the model of quasi-stationary current when current I does not depend on coordinate in Ω element, q is electrical charge concentrated at the edge of Ω element and $I = \frac{dq}{dt}$.

The theoretical method used in this study is based on the solution to the equation of motion of an "average" electron in the trajectory along Ω element

$$m_e\ddot{s} = -ks - \gamma\dot{s} - e\tau E_s \tag{3}$$

where s is the electron displacement along the Ω element, $\dot{s}$ is the electron velocity along the Ω element, $\ddot{s}$ is the acceleration, m_e is the electron mass, e is the elementary charge, k is the effective coefficient describing the quasi-elastic force acting on the electron in the direction opposite to its displacement, γ is the effective coefficient characterizing the scattering forces retarding electron, and τ is the coefficient of the field attenuation in the metal [14].

Fig. 1 An Ω element in the field of the incident electromagnetic wave

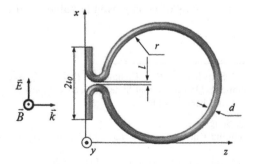

The conduction electrons in the Ω element perform harmonic oscillations that are induced by the incident electromagnetic wave. Hence, the relation

$$k = m_e \omega_0^2 = \frac{m_e \pi^2 c^2}{L^2} \tag{4}$$

holds true, where ω_0 is resonance oscillation frequency, c is light speed in vacuum, and L is the length of the wire of which the Ω element is manufactured. It is taken into account in expression (4) that, under the condition of the main resonance, the total Ω element length is approximately $\lambda_0/2$, where λ_0 is the wavelength of the electromagnetic field in free space.

The power of dissipative forces at the deceleration of conduction electrons can be calculated by the Joule–Lenz law. On this basis we arrive at the expression

$$\gamma = \rho\, e^2 N_{eff} = \rho\, e^2 N_0 N_s \tag{5}$$

where ρ is the metal resistivity, N_0 is the volumetric concentration of conduction electrons in metal, N_{eff} is the effective volumetric concentration of conduction electrons in metal,

$$N_s = \frac{2\Delta}{r_0} \tag{6}$$

is the fraction of the skin layer in the Ω element volume, r0 is the wire radius, and

$$\Delta = \sqrt{\frac{2\rho}{\mu_0 \omega}} \tag{7}$$

is the skin-layer depth [14, 15].

Expressions (5–7) show that the effective concentration of conduction electrons decreases for high frequency fields. Only the electrons in the thin surface layer contribute to the conductivity due to the skin effect. The coefficient γ characterizing the dissipative force is defined as an average value for all conduction electrons in the metal volume. When the skin effect plays an important role, the coefficient τ of the field attenuation inside the metal can be written in the form [14, 15]

$$\tau = \frac{E_{ins}}{E_0} = (1+j)\sqrt{2\varepsilon_0 \rho \omega} \tag{8}$$

where E_{ins} and E_0 are the complex amplitudes of the fields inside and outside the metal. We use the assumption that the electric and magnetic fields are monochromatic and their temporal dependence is described by the function $\exp(j\omega t)$, which is conventional in radiophysics

$$E = E_0 \exp(j\omega t) = E_x \quad s = s_0 \exp(j\omega t) \tag{9}$$

Equation (3) can be written in the following form $-m_e \omega^2 s = -ks - j\omega\gamma s - e\tau E$
and $s = \frac{-e\tau E}{m_e(\omega_0^2 - \omega^2 + j\omega\Gamma)}$, where $\Gamma = \frac{\gamma}{m_e} = \frac{\rho e^2 N_0 N_s}{m_e}$.

Then $q = -esN_{eff}\pi r_0^2$ is electrical charge concentrated at the edge of Ω element.
Nh is the concentration of Ω element the arms of which are oriented along the x
axis.

Now

$$p_x = \frac{e^2 \tau N_{eff} \pi r_0^2 \cdot 2l_0}{m_e(\omega_0^2 - \omega^2 + j\omega\Gamma)} E_x \tag{10}$$

Any Ω particle exhibits certain dielectric, magnetic, and magnetoelectric properties. Therefore, its behavior in the electromagnetic field can be described by
means of the coupling equations

$$\vec{p} = \varepsilon_0 \alpha_{ee} \vec{E} - j\sqrt{\varepsilon_0 \mu_0} \alpha_{em} \vec{H}, \quad \vec{m} = \alpha_{mm} \vec{H} + j\sqrt{\frac{\varepsilon_0}{\mu_0}} \alpha_{me} \vec{E} \tag{11}$$

Here, α_{ee} and α_{mm} are the tensor dielectric and magnetic susceptibilities and α_{em}
and α_{me} are the pseudotensors characterizing the magnetoelectric properties of Ω
elements.

From the comparison of Eqs. (10) and (11) we can see, that dielectric susceptibilities of one Ω element is

$$\alpha_{ee}^{(11)} = \frac{e^2 \tau N_{eff} \pi r_0^2 \cdot 2l_0}{\varepsilon_0 m_e(\omega_0^2 - \omega^2 + j\omega\Gamma)} \tag{12}$$

Here indexes (11) means corresponding components of tensor α_{ee}, which
describes the relationship between components p_x and E_x. We use the relationship
between the components of vectors E_x and B_y

$$E_x \cdot 2l_0 = -j\omega B_y \cdot \pi r^2 \tag{13}$$

Here field E_x or B_y belongs to incident electromagnetic wave, then second vector
B_y or E_x is induced in the Ω element. Taking into account the trajectories of the
conduction electrons, the skin effect, and the electric field attenuation in metal, one
can write the frequency dependence of the effective parameters of an medium

$$\varepsilon_r^{(11)} = 1 + N_h \alpha_{ee}^{(11)} = 1 + \frac{1}{A\varepsilon_0} \frac{1}{(\omega_0^2 - \omega^2 + j\omega\Gamma)} \tag{14}$$

$$\mu_r^{(22)} = 1 + N_h \alpha_{mm}^{(22)} = 1 + \frac{1}{A} \mu_0 B_1^2 \frac{1}{\left(\omega_0^2 - \omega^2 + j\omega\Gamma\right)} \tag{15}$$

$$\kappa^{(12)} = N_h \alpha_{em}^{(12)} = \frac{1}{A} \sqrt{\frac{\mu_0}{\varepsilon_0}} B_1 \frac{1}{\left(\omega_0^2 - \omega^2 + j\omega\Gamma\right)} \tag{16}$$

The following notation is used in Eqs. (14)–(16):

$$\frac{1}{A} = \frac{e^2 \tau N_h N_{eff} \pi r_0^2 \cdot 2l_0}{m_e}, \qquad B_1 = \frac{\pi r^2 \omega}{2l_0}.$$

Let's consider the artificial structure in which the arms of Ω elements are oriented not only along x axis as in Fig. 1, but also along y axis with the same concentration of Ω elements N_h. The sample of such structure is presented in Fig. 2. Then the relations (12), (14)–(16) can be written for the components of tensors $\alpha_{ee}^{(22)}$, $\alpha_{mm}^{(11)}$, $\alpha_{em}^{(21)}$. In the plane XOY the properties of artificial structure are isotropic and are characterized by the scalar quantities

$$\varepsilon_r = \varepsilon_r^{(11)} = \varepsilon_r^{(22)} \quad \mu_r = \mu_r^{(11)} = \mu_r^{(22)} \quad \kappa = \kappa^{(12)} = \kappa^{(21)} \tag{17}$$

It follows from the principle of symmetry of kinetic coefficients that the following relationship is valid [11]:

$$\alpha_{em} = \alpha_{me}^T \tag{18}$$

where symbol T means transposition of the tensor.

Let's assume that in the considering artificial structure for every Ω element the "pair" element exists which has the loop oriented in opposite direction. Then for the

Fig. 2 The artificial structure with compensated magnetoelectric properties

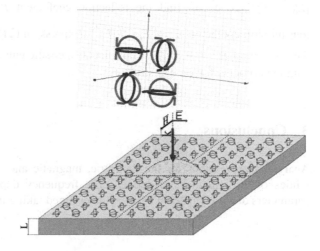

structure as whole the magnetoelectric properties are compensated and the following relation is satisfied

$$\kappa_{eff} = 0 \qquad (19)$$

The following universal relation is satisfied for optimal Ω element [12]:

$$m_z = j \frac{\omega \pi r^2}{2l_0} p_x \qquad (20)$$

To determine the wave reflection and transmission coefficients, we will consider a sample based on Ω elements and solve a boundary problem for a layer, i.e., a structure of finite thickness. Let us introduce the following designations: $\vec{E}^i$, $\vec{E}^r$, and $\vec{E}^\tau$ are, respectively, the incident, reflected, and transmitted waves and L is the thickness of the structure. We assume that the incident wave is linearly polarized and that the vector $\vec{E}^i$ oscillates along the x axis.

Using the condition of continuity of the vectors $\vec{E}$ and $\vec{H}$ at the sample boundaries, one can derive the expression for the reflected and transmitted wave amplitude:

$$E_0^r = \frac{\left(\sqrt{\frac{\varepsilon_r}{\mu_r}} - \sqrt{\frac{\mu_r}{\varepsilon_r}} \right) \left(e^{-jkL} - e^{jkL} \right) E_0^i}{\left(1 - \sqrt{\frac{\mu_r}{\varepsilon_r}} \right) \left(1 - \sqrt{\frac{\varepsilon_r}{\mu_r}} \right) e^{-jkL} + \left(1 + \sqrt{\frac{\mu_r}{\varepsilon_r}} \right) \left(1 + \sqrt{\frac{\varepsilon_r}{\mu_r}} \right) e^{jkL}}, E_0^\tau$$

$$= \frac{4 E_0^i e^{-i \frac{\omega}{c} L}}{\left(1 - \sqrt{\frac{\mu_r}{\varepsilon_r}} \right) \left(1 - \sqrt{\frac{\varepsilon_r}{\mu_r}} \right) e^{ikL} + \left(1 + \sqrt{\frac{\mu_r}{\varepsilon_r}} \right) \left(1 + \sqrt{\frac{\varepsilon_r}{\mu_r}} \right) e^{-ikL}} \qquad (21)$$

where E_0^i is the incident wave amplitude. Having calculated the squared modulus $\left| E_0^r \right|^2$, $\left| E_0^\tau \right|^2$, one can find the reflection coefficient $R = \left| E_0^r \right|^2 / \left| E_0^i \right|^2$ and the transmission coefficient $T = \left| E_0^\tau \right|^2 / \left| E_0^i \right|^2$. Expression (21) contains the relation for the wave number $k = \frac{\omega}{c} \sqrt{\varepsilon_r \mu_r}$. The calculated coefficients R and T can be compared with experiments for omega structure.

3   Conclusions

Analytical expressions for the dielectric, magnetic and magnetoelectric susceptibilities of the structure are derived. The frequency dependence of the effective parameters of a metamaterial has been determined taking into account trajectories of the conduction electrons, the skin effect, and the electric field attenuation in metal. Boundary-value problem is solved and reflection and transmission coefficients are found.

Acknowledgments This study was supported by the Belarusian Republican Foundation for Basic Research, project No. F15SO-047.

References

1. Whites, K.W., Chang, C.Y.: Composite uniaxial bianisotropic chiral materials characterization: comparison of predicted and measured scattering. JEWA **11**, 371–394 (1997)
2. Tretyakov, S.A., Sochava, A.A., Simovski, C.R.: Influence of chiral shapes of individual inclusions on the absorption in chiral composite coatings. Electromagnetics **16**, 113–127 (1996)
3. Bohren, C.F., Luebers, R., Langdon, H.S., Hunsberger, F.: Microwave-absorbing chiral composites: is chirality essential or accidental? Appl. Opt. **31**, 6403–6407 (1992)
4. Tretyakov, S.A., Sihvola, A.H., Semchenko, I.V., Khakhomov, S.A.: Reply to comment on "Reflection and transmission by a uniaxially bi-anisotropic slab under normal incidence of plane waves". J. Phys. D Appl. Phys. **32**, 2705–2706 (1999)
5. Semchenko, I.V., Khakhomov, S.A., Tretyakov, S.A., Sihvola, A.H.: Electromagnetic waves in artificial chiral structures with dielectric and magnetic properties. Electromagnetics **21**(5), 401–414 (2001)
6. Semchenko, I.V., Khakhomov, S.A., Samofalov, A.L.: Helices of optimal shape for nonreflecting covering. EPJ Appl. Phys. **49**(3), 33002-p1–33002-p5 (2010)
7. Semchenko, I.V., Khakhomov, S.A., Naumova, E.V., Prinz, V.: Ya., Golod, S.V., Kubarev, V.V.: Study of the properties of artificial anisotropic structures with high chirality. Crystallogr. Rep. **56**(3), 366–373 (2011)
8. Semchenko, I.V., Khakhomov, S.A., Tretyakov, S.A., Sihvola, A.H.: Microwave analogy of optical properties of cholesteric liquid crystals with local chirality under normal incidence of waves. J. Phys. D Appl. Phys. **32**, 3222–3226 (1999)
9. Balmakou, A., Podalov, M., Khakhomov, S., Stavenga, D., Semchenko, I.: Ground-plane-less bidirectional terahertz absorber based on omega resonators. Opt. Lett. **40**(9), 2084–2087 (2015)
10. Asadchy, V.S., Faniayeu, I.A., Ra'di, Y., Khakhomov, S.A., Semchenko, I.V., Tretyakov, S.A.: Broadband reflectionless metasheets: frequency-selective transmission and perfect absorption. Phys. Rev. X **5**(3), 031005-1–031005-10 (2015)
11. Serdyukov, A.N., Semchenko, I.V., Tretyakov, S.A., Sihvola, A.H.: Electromagnetics of bi-anisotropic materials. Gordon and Breach Science Publishers, Amsterdam (2001)
12. Semchenko, I.V., Khakhomov, S.A., Podalov, M.A., Tretyakov, S.A.: Radiation of circularly polarized microwaves by a plane periodic structure of Ω elements. J. Commun. Technol. Electron. **52**(9), 1002–1005 (2007)
13. Sivukhin, D.V.: General course of physics, vol. 3 electricity. Nauka, Moscow (1983) (in Russian)
14. Tamm, I.E.: The principles of electricity theory. Nauka, Moscow (1976)
15. Landau, L.D., Lifshitz, E.M.: Course of theoretical physics, vol. 8: electrodynamics of continuous media, 2nd ed. Nauka, Moscow (1982); Pergamon, Oxford (1984)

Impact of Ion Nitriding on Phase Composition, Structure and Properties of Carbon Films Doped with Metals

A.S. Rudenkov, D.G. Piliptsou, A.V. Rogachev, N.N. Fedosenko and Xiaohong Jiang

Abstract The main modification laws of properties, phase composition and mechanical properties of single component and metal alloyed (Cr, Ti, Cu) carbon coatings subjected to ion nitriding are defined. It is stated that ion nitriding gives rise to the fraction of sp^2-phase while reducing the cluster size in the surface layers. The formation of CN_x compounds with dominating bonds of $N-Csp^2$ type is determined at ion nitriding of carbon containing coatings. The single component carbon coatings being ion nitrided are characterized by more dispersed structure, bigger content of $N-Csp^3$ bonds in comparison with the structure and composition of coatings formed at the conditions of ion assistance or with the appearance of molecular nitrogen.

Keywords Carbon coatings · Alloying · Nitriding · Phase composition · Morphology · Hardness · Friction

1 Introduction

A wide range of properties of carbon coatings, their dependence on conditions and the modes of synthesis are caused by the variety of structural conditions of carbon, the modification of their structure and chemical activity of carbon [1–5]. While forming carbon coatings alloyed by metals and/or nitrogen, the main factors taken into consideration are their nature and concentration of the alloying elements, the way of their chemical interaction with carbon, the distribution nature on layer thickness, the existence of transitional layers. All these factors define the phase, structural condition of a carbon matrix, chemical composition of interphase layers

A.S. Rudenkov · D.G. Piliptsou (✉) · A.V. Rogachev · N.N. Fedosenko · X. Jiang
International Chinese-Belarusian Scientific Laboratory by Vacuum-Plasma Technologies, Nanjing University of Science and Technology, Xiaolingwei 200, Nanjing 210094, China
e-mail: pdg_@mail.ru

A.S. Rudenkov · D.G. Piliptsou · A.V. Rogachev · N.N. Fedosenko · X. Jiang
Gomel State University named after F. Skorina, Sovetskaya, 104, 246019 Gomel, Belarus

© Springer International Publishing AG 2017
R. Jabłoński and R. Szewczyk (eds.), *Recent Global Research and Education: Technological Challenges*, Advances in Intelligent Systems and Computing 519, DOI 10.1007/978-3-319-46490-9_2

11

and, respectively, the property of coatings [1–3]. Such layers are characterized by higher thermo- and wear resistance, a lower level of internal mechanical tension in comparison with single component coatings, while preserving high values of microhardness.

The modification of physical and chemical parameters of carbon coatings while introducing metals into their composition is explained by the chemical interaction processes with formation of carbides [1], catalytic influence of metals on synthesis processes and sp^3- and sp^2-clusters dispersion [2, 4]. The formation of CN_x compounds is defined at the alloying of carbon coatings by nitrogen carried out at a deposition stage by treating the growing layer by nitrogen ions (the ion assistance mode) [5, 6] or deposition of a coating in the environment with high concentration of molecular nitrogen [7–9], and the formation of carbonitrides and nitrides of metals is stated during the alloying process of metal-containing carbon coatings [8]. At the same time the dispersion and volume of the formed phases, the nature of their distribution along the layer thickness are substantially determined by the technological parameters of alloying, nitrogen ions energy, the degree of activation processes of chemical interaction [7, 9]. It should be noted that the given results characterize the coatings which alloying is carried out as a result of simultaneous surface deposition of carbon atoms generated in plasma, and the evaporated atoms of metal. There are no data of the nitriding influence carried out by nitrogen ions treatment of previously besieged layers on the chemical composition, structure and properties of single component and composite metal-containing carbon coatings.

In this regard, the main objective of the research is to establish the composition peculiarities, structure and phase condition, mechanical properties (hardness, wear resistance) of surface layers of single component and alloyed by metals of various nature carbon coatings subjected to ion nitriding.

2   Experimental

The research object are the single component and composite carbon coatings alloyed by Cr, Ti, Cu metals which produce various chemical activity in relation to carbon and nitrogen. The deposition of carbon coatings is produced from pulse cathodic plasma by the technique introduced in [3]. The deposition a-C:Me coatings is produced by the co-deposition of carbon atoms and metal atoms generated by magnetron dispersion (discharge power at dispersion of Cr, Ti is 400 W, at dispersion of Cu—200 W). At the same time the atomic content of metal in a coating is approximately identical 2.2 … 2.65 %.

Ionic nitriding is carried out by means of a high-energy ionic source at the following parameters: gas pressure P = 10^{-1} Pa; current—0.2 A; the accelerating voltage 2 kV.

Studying of morphology of the alloyed carbon coatings is carried out by method of atomic force microscopy (AFM) in the modes of topography and phase contrast measuring with the help of the Solver Pro device of NT-MDT (Moscow, Russia).

The determination of phase structure of carbon coatings is carried out by the analysis of the combinational dispersion spectra gathered at Senterra spectrometer with the wavelength of the exciting radiation of 532 nm, power 10 mW. The decomposition of the registered spectra at D-(~ 1400 cm^{-1}) and G-peaks (~ 1550 cm^{-1}) is produced the Gauss method. The chemical composition and structure of carbon bonds are defined by the method of x-ray photoelectronic spectroscopy (XPES). The measurements are produced by the PHI Quantera device at aluminum substance activation by Kα-radiation with quantum energy 1486.6 eV and with 250 W power.

The DM-8 microhardness tester (AFFRI, Italy) is used to measure the microhardness according Knoop. The load on a diamond pyramid is equal to 245 mN, the duration of the test is 10 s.

The tribotechnical tests are carried out according to the scheme "sphere-plane" (a ball with 5 mm radius from the tempered ShH15 steel, the coating is applied on a flat silicon substrate). The load is 0.98 N, the average speed of movement −0.0087 m/s. After rubbing, we determined the contact patch diameter using an optical microscope. The rider's wear rate was calculated as j = V/FL, where F is a load (N); L is the friction length (m); V is the worn volume of the spherical segment of the rider (m3). The friction coefficient and wear rate were calculated by averaging the data of three independent experiments.

3 Results and Discussion

It is established that at ionic nitriding the morphology of coatings considerably changes due to etching. The roughness and the amount of particular structural formations of single component and composite carbon coatings decrease after ionic nitriding (Table 1).

At the same time the most considerable roughness reduction is observed at single component carbon coating treating. Such coatings, as it is noted in [10], contain graphite microdrops, and their primary etching occurs in the ionic nitriding course, what defines roughness decrease of a coating. In metal-containing carbon layers at their nitriding the new firm phases of introduction can be formed possessing the lower etching speed in comparison with graphite, and, as a result of it, the changes of surface topography are less expressed.

Table 1 Roughness parameters of carbon coatings to/after ionic nitriding on the AFM data basis

Coatings	Average height (nm)	Rms (nm)	Grain density (grain/μm^2)	Average grain size (nm)
a-C	81/26	22.8/8.5	54/78	164/128
a-C:Cr	20/10	7.8/3.9	49/62	93/82
a-C:Cu	18/21	5.0/4.2	98/76	87/89
a-C:Ti	11/16	2.4/3.4	44/47	118/118

Table 2 The results of Raman-spectroscopy of the alloyed carbon coatings to/after ionic nitriding

Coating	D-peak		G-peak		I_D/I_G ratio
	Centre (cm^{-1})	Width (cm^{-1})	Centre (cm^1)	Width (cm^{-1})	
a-C	1438.2/1434.4	275.3/282.1	1562.8/1563.8	197.2/189.2	0.50/0.58
a-C:Cr	1395.5/1377.4	336.4/358.5	1549.1/1546.0	167.0/147.9	1.15/1.64
a-C:Cu	1406.4/1401.4	307.2/319.1	1551.8/1555.6	165.9/156.8	1.04/1.21
a-C:Ti	1404.4/1399.9	323.5/328.9	1554.8/1554.6	169.8/169.2	1.01/1.07

Ionic nitriding of carbon coatings also leads to the change of their phase state. According to the results of Raman spectroscopy presented in Table 2, after ionic nitriding of the single component and alloyed by metal carbon coatings the increase in D-peak width, decrease in G-peak width, increase of the relation of I_D/I_G are observed.

As it is stated in research [3], narrowing of G-peak width is caused by the order increase of sp^2-clusters, the increase in a ratio of I_D/I_G can be caused by the reduction of their size and growth of the number of sp3 bonds. The broadening of D-peak and its shift to the area of low wave numbers can demonstrate the disorder and reduction of sp^3-clusters number. Such changes can be caused by the destruction of grains borders and substrate heating under the influence of the subsequent ionic bombing that correlates with AFM data and the results of experiments on formation of carbon coatings in the conditions of ionic bombing stated in research [10].

The formation of carbon coatings in the conditions of ionic assistance (irrespective of the assisting ions energy) has more significant effect on their phase structure [11]. At assisting the I_D/I_G ratio increases from 0.5 to 1.38, and G-peak width decreases to 166.6 cm^{-1}. After ionic nitriding the I_D/I_G ratio equals to 0.58, and G-peak width equals to 189.2 cm^{-1}. Such changes are explained by various mechanisms of phase transitions $sp^3 \rightarrow sp^2$ of carbon coatings. At ionic assistance, the main cause for the change of phase structure are inelastic impacts of ions of nitrogen and carbon in plasma, while the subsequent ionic nitriding has a superficial character, the processes of etching, implantation and local heating of the top coat layers prevail.

It should be noted that the introduction of metal in a carbon coating a little reduces graphitization degree, and in the case of an alloying copper ionic nitriding even reduces the surface area of the substrate occupied by sp^2-clusters (Fig. 1).

The change of a chemical composition of the coatings subjected to ionic nitriding is defined according to the change of location and the N1s form of peak of XPS of the spectrum (Fig. 2).

It is stated that both the single component and alloyed by metals carbon coatings contain chemically bonded atoms of nitrogen in a surface layer. The decomposition of N1s peak of single component carbon coatings allows to allocate N–Csp^3 bonds with ~398.2 eV energy, N–Csp^2 (399.9 eV) and N–O (401.7 eV). At N1s peak of the coverings alloyed by titan there is a component with 397.3 eV energy, corresponding to titan nitride N–Ti [12]. N1s peak of the carbon coatings alloyed by

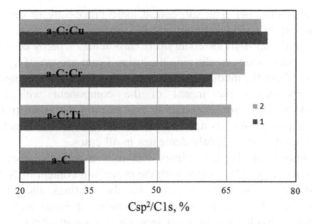

Fig. 1 The content of a sp²-phase in a coating before (*1*) and after (*2*) ionic nitriding

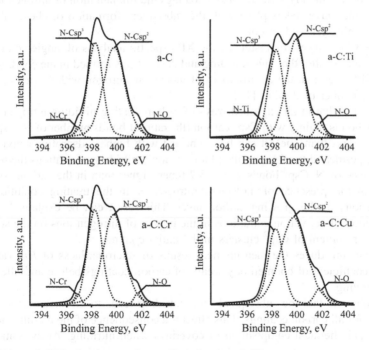

Fig. 2 N1s-peak XPS of carbon coatings subjected to ion nitriding

chrome is displayed on the following components: N–Cr with binding energy 396.7 eV [13], N–Csp³ with binding energy $\sim$398.3 eV, N–C sp² (399.8 eV) and N–O (401.7 eV). It is difficult to reveal the presence of copper nitride by the XPS method as the binding energy of copper nitride equals to $\sim$932.7 eV for Cu2p, that is very close to Cu–O component with spin-orbit orbital modification $2p_{3/2}$ with a

priori higher intensity [14]. The subsequent ionic processing of the single component and alloyed by carbide-forming metals carbon coatings leads to the growth
of the contents of the sp^2-hybridized atoms of carbon that is partly confirmed by the
data of Raman-spectroscopy confirming the insignificant growth of I_D/I_G ratio. It is
difficult to judge about the modification of a sp^3-phase as at the decomposition Csp^3
and C–N, C=N bonds are meant of the component with the binding
energy ~ 285.4 eV. However, even taking into account the intensity contribution
of the given component of C=N and C–N bonds it is possible to conclude that the
share of the integrated area slightly decreases in all cases.

From the analysis of N1s it becomes obvious that at ionic processing of the
deposited carbon coating CN_x bonds with the prevailing content of N–Csp^2 bonds
are formed. Besides, after ionic nitriding for the coatings alloyed by carbide
forming metals, the decrease in content of carbides of metal and formation of
nitrides are fixed. Also the insignificant increase in content of the sp^2-hybridized
atoms of carbon is observed. The above mentioned modifications prove that at
nitriding of the already formed carbon coatings the implantation of nitrogen ions in
a near-surface layer takes place with the subsequent formation of chemical compounds of nitrogen and carbon.

Table 3 introduces the results of the XPS spectra analysis of single component
carbon coatings after the ionic nitriding and the coatings formed in the conditions of
ionic assistance or in the environment of molecular nitrogen with its pressure in a
vacuum chamber of 0.1 Pas [11].

It is stated that in all exploded ways of carbon coating nitriding nitrogen forms
chemical bonds mainly with Csp^2 carbon (the ratio of bonds number of N–Csp^3 and
N–Csp^2 is less than one), it can be connected with higher adsorptive activity and
diffusive permeability of a graphite phase. At ionic nitriding of coatings the relative
maintenance of N–Csp^3 bonds is 3.5–3.7 times higher than in the carbon coatings
formed in the presence of molecular nitrogen or in the treating conditions by
nitrogen ions of the growing carbon layer. This result can be explained by the
modification of activation conditions at the impact of nitrogen ions on the surface,
the smaller content of Csp^2-clusters in the initial coating.

Table 4 introduces the measurement results of microhardness of $H\kappa$, value of
volume coefficient of counterbody wear j of carbon coatings before and after their
ionic nitriding.

It is obvious that the subsequent ionic nitriding leads to the increase in microhardness for all carbon coatings. This effect is a consequence of the modifications of
a phase and chemical composition of coverings when nitriding. At the same time

Table 3 The results of XPS of carbon coatings subjected to nitriding by different methods

Alloying means	Csp^2/C1s (%)	N–C sp^3/N–C sp^2
No alloying	33.98	–
Ion nitriding	50.51	0.68
Deposition in N2 environment	56.00	0.18
Ion assisting	59.00	0.19

Table 4 The influence of ionic nitriding on the hardness of Hk and wear coefficient of counterbody j

Coating	Hk (GPa)		j ($\times 10^{-18}$ m^3/(N m))	
	Before	After	Before	After
a-C	10.74 ± 0.22	11.21 ± 0.15	419.3 ± 4.6	345.6 ± 2.9
a-C:Cr	12.45 ± 0.23	12.81 ± 0.18	138.1 ± 2.7	172.5 ± 3.3
a-C:Cu	10.23 ± 0.26	10.79 ± 0.21	120.1 ± 4.2	165.2 ± 3.8
a-C:Ti	11.91 ± 0.25	12.62 ± 0.14	128.9 ± 5.0	191.0 ± 4.2

considering that the hardness increase is observed also when nitriding single component carbon coatings, apparently, the greatest contribution to the hardness increase is produced by formation processes of CN_x bonds.

The decrease in volume coefficient of wear of a counterbody at nitriding of non-alloyed carbon coatings (Table 4) is explained by the significant increase in content of a graphite sp^2-phase (by 1.5 times) which being on surfaces produces a lubricant effect at contact interaction. At friction subjected to nitriding of the carbon coatings alloyed by metal the value of wear volume coefficient of a counterbody increases that is caused by the formation in a surface layer of firm heat-resistant bonds (nitrides, carbonitrides), capable to abrasive destruction of the contacting surface.

4 Conclusion

The main modification laws of properties and phase structure of the single component and Cr, Ti, Cu alloyed carbon coatings subjected to ionic nitriding are defined. The increase in a sp^2-phase content, concentration decrease in near-surface layers of Me$_y$C$_x$ bonds, formation of Me$_y$N$_x$ and CNx bonds with the prevailing N–Csp^2 content of bonds are established. After ionic nitriding of single component and composite coatings the increase in microhardness is observed at 0.5–0.7 GPa and in 1.2–1.5 wear volume coefficients of a steel counterbody at friction with a coating.

It is defined that at ionic nitriding of single component carbon coatings the relative content of N–Csp^3 bonds is 3.5–3.7 times higher in comparison with carbon coatings, formed in the molecular nitrogen presence or in the conditions of simultaneous deposition and treatment of the surface by nitrogen ions.

Acknowledgments This work was supported by the Belarusian Republican Foundation for Fundamental Research with the Project (T16 M-015) for 2016-2018 and National Natural Science Foundation of China (51373077), 2015–2016 Intergovernmental Cooperation Projects in Science and Technology of the Ministry of Science and Technology of PRC (No. CB11-09 & CB11-11).

References

1. Рогачев, А.В.: Триботехнические свойства композиционных покрытий, осаждаемых вакуумно-плазменными методами. Трение и износ **29**(3), 285–592 (2008)
2. Donnet, C., Erdemir, A.: Tribology of diamond-like carbon films: fundamentals and applications. Springer Science & Business Media, p. 680 (2007)
3. Руденков, А.С.: Влияние концентрации металла на фазовый состав, структуру и свойства углерод-металлических покрытий. Проблемы физики, математики и техники **3** (24), 26–32 (2015)
4. Robertson, J.: Diamond like carbon. Pure Appl. Chem. **66**, 1789–1796 (1994)
5. Lifshitz, Y.: Diamond like carbon—present status. Diam. Relat. Mater. **8**, 1659–1676 (1999)
6. Khurshudov, A.G., Kato, K.: Tribological properties of carbon nitride overcoat for thin-film magnetic rigid disks. Surf. Coat. Technol. **9**, 537–542 (1996)
7. Koskinen, J., et al.: Tribological characterization of carbon-nitrogen coatings deposited by using vacuum arc discharge. Diam. Relat. Mater. **5**, 669–673 (1996)
8. Bauer, C., et al.: Mechanical properties and performance of magnetron-sputtered graded diamond-like carbon films with and without metal additions. Diam. Relat. Mater. **11**, 1139–1142 (2002)
9. Broitman, E., et al.: Mechanical and tribological properties of CN_x films deposited by reactive magnetron sputtering. Wear **248**, 55–64 (2001)
10. Пилипцов, Д.Г., Руденков, А.С., Бекаревич, Р.В.: Морфология композиционных покрытий на основе углерода, подвергнутых обработке ионами азота. Проблемы физики, математики и техники **3**(4), 31–34 (2010)
11. Чжоу, Б.: Легированные азотом алмазоподобные покрытия: структура и свойства. Материалы, технологии, инструменты **18**(3), 16–21 (2013)
12. Haasch, R.T., et al.: Epitaxial TiN (001) grown and analyzed in situ by XPS and UPS. I. Analysis of as-deposited layers. Surf. Sci. Spectra **7**, 193–203 (2000)
13. Haasch, R.T., et al.: Epitaxial CrN (001) grown and analyzed in situ by XPS and UPS. I. Analysis of as-deposited layers. Surf. Sci. Spectra **7**, 250–261 (2000)
14. Navío, C., et al.: Thermal stability of Cu and Fe nitrides and their applications for writing locally spin valves. Appl. Phys. Lett. **94**, 263112-1–263112-3 (2009)

Effect of Shungite Nanocarbon Deposition on the Luminescent Properties of ZnS:Cu Particles

M.M. Sychov, S.V. Mjakin, K.A. Ogurtsov, N.N. Rozhkova,
P.V. Matveychikova, V.V. Belyaev, F.I. Vysikailo and Y. Nakanishi

Abstract Modification of surface of solid state materials allow to control on the
one hand interface interactions and on the other hand physical properties of modified semiconductor or dielectric. Especially this is true in case of carbon
nanoparticles due to their electron acceptor properties. In this paper modification of
a ZnS:Cu electroluminescent phosphor by the deposition of shungite carbon nan-
oclusters is described. Modification resulted in increased intensity of "blue" band in
its luminescence spectra and oscillating change of EL brightness. Effect is discussed
in terms of possible mechanisms relating to electric field concentration at the areas
of nanocarbon adsorption on specific surface sites connected with "blue" luminescent centers. This effect is promising for an adjustable control over electroluminescence spectra and brightness as well as control of interface interactions.

Keywords Phosphor · Electroluminescence · Zinc sulfide · Surface · Interface ·
Carbon · Shungite · Nanoparticles

M.M. Sychov (✉) · S.V. Mjakin · K.A. Ogurtsov · P.V. Matveychikova
St. Petersburg State Institute of Technology (Technical University), St. Petersburg, Russia
e-mail: msychov@yahoo.com

M.M. Sychov
Institute of Silicate Chemistry of the RAS, St. Petersburg, Russia

N.N. Rozhkova
Institute of Geology, Karelian Research Centre of the RAS, Petrozavodsk, Russia

V.V. Belyaev
Moscow Regional State University, Moscow, Russia

F.I. Vysikailo
Moscow Radiotechnical Institute of the RAS, Moscow, Russia

Y. Nakanishi
Research Institute of Electronics, Shizuoka University, Hamamatsu, Japan

© Springer International Publishing AG 2017

R. Jabłoński and R. Szewczyk (eds.), *Recent Global Research and Education:*
Technological Challenges, Advances in Intelligent Systems and Computing 519,
DOI 10.1007/978-3-319-46490-9_3

19

1   Introduction

Luminescent phosphors based on $A^{II}B^{VI}$ are widely used in various electronic devices including displays, electroluminescent panels, etc. [1]. Their efficient and durable exploration in combination with adjustable control over the luminescence spectra, brightness and stability can be achieved due to specific doping and modification in addition to conventional activators and additives. In a series of earlier studies [1–3] spectral and brightness performances of ZnS based phosphors were shown to strongly depend on their surface functional composition, particularly on the presence of Brønsted and Lewis with a certain composition and donor-acceptor (acid-base) properties involved in the formation of luminescence centers and capable of activation or suppression of luminescent electron transitions. Furthermore, a possibility for a significant increase of their luminescent brightness due to a specific functionalization of their surface using various physicochemical techniques was demonstrated [1–3].

To further develop the considered approach, in this study the surface layer of an electroluminescent phosphor was modified by the deposition of shungite carbon (SC) nanoparticles featured with a high activity and ability to change the donor-acceptor properties and electronic structure of the surface layer of materials even upon addition in micro-quantities [4]. The deposition of these conductive nanoparticles was expected to promote an increase in the electroluminescence brightness of the modified phosphor due to electric field concentration on its particles [5, 6] taking into consideration a fractal structure of SC additive involving voids acting as quantum resonators for electrons [5, 6] and promoting the capture of free electrons and polarization or crystallites (Fig. 1) [7–9].

2   Experimental

A ZnS:Cu electroluminescent phosphor was modified by the deposition of SC nanoparticles from aqueous dispersions containing 0.06 and 0.1 mg/L of 54 ± 25 nm (according to dynamic light scattering) sized SC clusters prepared as

Fig. 1 Formation of internal electric fields in a nanocrystal (NC) upon the addition of allotropic carbon. NC and the dopant (acting as an "electron trap") are charged positively and negatively accordingly [7–9]

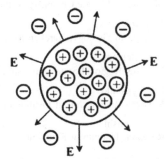

described in [4]. The ratio between SC and modified phosphor varied from 0.79 to 1.25 mg/g corresponding to SC volume content $(1.74–2.76) \times 10^{-3}$.

The modified phosphors were characterized by Raman spectroscopy using a Nicolet Almega XP (Thermo Scientific) spectrometer, adsorption of acid-base indicators to study the surface functional composition [1, 2] and measuring the luminescence brightness (radiometer IL1700) and spectra (spectrofluorimeter AvaSpec-3648).

3 Results and Discussion

Raman spectroscopy data confirm SC binding onto the phosphor particles as confirmed by the appearance of specific peaks at 1330 cm^{-1} (D-band) and 1580 cm^{-1} (G-band) (Fig. 2).

The indicator adsorption analysis indicates that SC deposition results in a significant change of the phosphor surface functional composition, particularly in a decrease of the content of surface Brønsted acidic centers with pK$_a$ 2.5 (according to [3] corresponding to Cu$_x$S-H groups) clearly (with the coefficient of −0.99) correlating with the increase of the content of centers with pK$_a$ 2.1 intrinsic to SC surface. This result suggests that hydrophobic SC binds to relatively hydrophobic

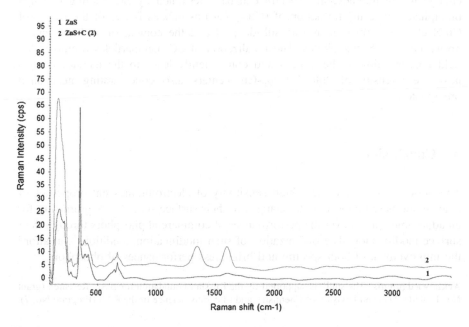

Fig. 2 Raman spectra of the initial (*1*) and SC-modified (0.79 mg/g) ZnS:Cu phosphors (*2*)

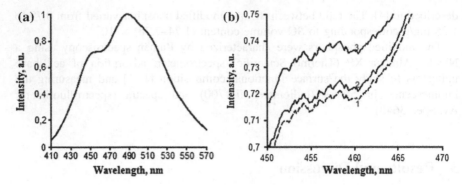

Fig. 3 Electroluminescence spectrum of a non-modified ZnS:Cu phosphor (**a**) and fragment of this spectrum (**b**) before (*1*) and after SC deposition in the content of 0.79 (*2*) and 1.25 mg/g (*3*)

areas of the phosphor comprising Cu_xS phase precipitates shielding the corresponding groups and forming SC centers on the surface.

Although SC deposition did not provide a growth in the overall brightness of the studied phosphor, it afforded a certain increase in the relative intensity of the "blue" luminescence band at 455 nm (Fig. 3) corresponding to the associates of Cu ions at Zn sites with the interstitial copper (Cu_{Zn}' $Cu_i^{\cdot}$) compared with the "green" band at 490 nm corresponding to (Cu_{Zn}' $Cl_S^{\cdot}$) donor-acceptor pairs [10]. The considered change in the luminescence spectra can be determined by SC binding to Cu_xS precipitates since the formation of (Cu_{Zn}' $Cu_i^{\cdot}$) associates is one of the stages of Cu_xS clusters formation in zinc sulfide [11] and the content of such associates grows near Cu_xS precipitates. The localization of SC nanoparticles promotes the field concentration in these areas and consequently leads to the increased luminescence intensity of "blue" Cu_{Zn}-Cu_i centers also concentrating near such precipitates.

4   Conclusion

The presented data indicate a high sensitivity of electroluminescent properties of zinc sulfide based phosphors to changes of their surface state that is promising for an adjustable control over the performances of commercial phosphors through their surface modification. The optimization of such modification conditions can afford the improvement of both spectral and brightness performances of phosphors.

Acknowledgments The study is supported by the Russian Foundation for Basic Research (grant No. 14-07-00277) and Division of Chemistry and Materials Science of the RAS (Program No. 7).

References

1. Sychov, M.M., Mjakin, S.V., Nakanishi, Y., Korsakov, V.G., Vasiljeva, I.V., Bakhmetyev, V. V., Solovjeva, O.V., Komarov, E.V.: Study of active surface centers in electroluminescent ZnS:Cu,Cl phosphors. Appl. Surf. Sci. **244**(1–4), 461–464 (2005)
2. Mjakin, S.V., Sychov, M.M., Vasiljeva, I.V. (eds.): Electron beam modification of solids: mechanisms, common features and promising applications. Nova Science Publishers, Inc. (2009)
3. Bakhmet'ev, V.V., Sychov, M.M., Korsakov, V.G.: A model of active acid-base surface sites for zinc sulfide electroluminescent phosphors. Russ. J. Appl. Chem. **83**(11), 1903–1910 (2010)
4. Rozhkova, N.N.: Shungite nanocarbon. Karelian Research Centre of RAS, Petrozavodsk (2011)
5. Kim, J.-Y., Kim, H., Jung, D., SeGi, Y.: Enhanced electroluminescence performances by controlling the position of carbon nanotubes. J. Appl. Phys. **112**, 104515 (2012)
6. Sheka, E.F., Rozhkova, N.N.: Shungite as the natural pantry of nanoscale reduced graphene oxide. Int. J. Smart Nanomater. **5**(1), 1–16 (2014)
7. Vysikaylo, P.I.: Physical fundamentals of hardening of materials by space charge layers. Surf. Eng. Appl. Electrochem. **46**(4), 291–298 (2010)
8. Vysikaylo, P.I.: Cumulation of de Broglie waves of electrons, endoions and endoelectrons of fullerenes, and resonances in the properties of nanocomposite materials with spatial charge layers. Surf. Eng. Appl. Electrochem. **46**(6), 547–557 (2010)
9. Vysikaylo, P.I.: Cumulative quantum mechanics (CQM). Part II. Application of cumulative quantum mechanics in describing the Vysikaylo polarization quantum—size effects. Surf. Eng. Appl. Electrochem. **48**(5), 395–411 (2012)
10. Shionoya, S., Yen, W.M. (eds.): Phosphor handbook. CRC Press, LLC, New York (2006)
11. Miloslavsky, A.G., Suntsov, N.B.: Defect structure and luminescence centers of zinc sulfide phosphors. High Press. Phys. **7**(2), 94–103 (1997)

Nano-Sized Calcium Phosphates: Synthesis Technique and Their Potential in Biomedicine

Linda Vecbiskena

Abstract Nowadays, there is particular interest in nano-sized inorganic crystals for use as calcium phosphate bone substitutes, especially tricalcium phosphate (TCP) and hydroxyapatite (HA). This research takes on the challenge to produce nano-sized TCP and HA by heating amorphous calcium phosphate (ACP) precursors above the crystallization temperature. ACP precursors were synthesized by modified wet-chemical precipitation method. ACP treated with ethanol, favoured the formation of pure α-TCP, as well as others calcium phosphates—β-TCP and HA.

Keywords Wet-chemical precipitation · Amorphous calcium phosphate · Tricalcium phosphate · Hydroxyapatite · Bone substitutes

1 Introduction

Calcium phosphate (CaP) biomaterials have long been of interest in biomaterials science since they are found in biological organisms, and form the main inorganic component of the human bones and teeth [1–3]. These materials have been used increasingly in the dental and orthopaedic applications since 1970s [4–6]. There has been a growing demand for better life and improved lifestyle, the need for bone tissue repair and regeneration includes patients with damaged tissue from trauma or cancer, and cases where a spinal or bone deformity must be corrected. Less critical cases include reconstruction for aesthetic reasons, but corrections to bone tissue are more important, since this forms the basis for accommodating implants. A choice from different calcium phosphates that requires exposure of the defect for filling, the granules, blocks and pastes cover the larger segment of the biomaterials market [5].

L. Vecbiskena (✉)
Faculty of Materials Science and Applied Chemistry, Riga Technical University,
3 Paula Valdena St., Riga 1048, Latvia
e-mail: linda.vecbiskena@gmail.com

© Springer International Publishing AG 2017
R. Jabłoński and R. Szewczyk (eds.), *Recent Global Research and Education: Technological Challenges*, Advances in Intelligent Systems and Computing 519,
DOI 10.1007/978-3-319-46490-9_4

Fig. 1 The global market of
biomaterials for the forecast
period of 2015–2020

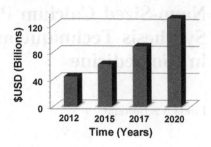

Additionally, the global market of biomaterials was estimated to be $44.0 billion in 2012 growing from 2012 to 2017 to reach $88.4 billion by 2017. The last completed report for the period of 2015–2020 shows that it is expected to reach $130.6 billion by 2020 from $62.1 billion in 2015 [7] (Fig. 1).

Lately, nano-sized calcium phosphates (10–100 nm), especially tricalcium phosphate (TCP) and hydroxyapatite (HA), have received much attention as promising biomaterials for applications in different fields of biomedicine: bone regeneration, tissue engineering scaffolds, teeth implants, self-setting bone cements, drug/gene delivery systems, vaccine adjuvants, contrast agents for imaging and multi-modal imaging, nanosystems for photodynamic therapy and antifungal/antibacterial agents [8]. Nano-sized CaP based biomaterials can achieve significantly better bone regeneration in vivo than conventional micro-sized CaP based biomaterials, but further studies are needed to understand the bone tissue regeneration mechanisms via nano-sized calcium phosphates [9].

Different synthesis methods have been reported for the production of calcium phosphate powders [10–12]. Generally, these powders can be produced using wet- and dry-based methods. The wet-chemical precipitation from aqueous supersaturated solutions containing calcium and phosphate ions is the most commonly used. However, special emphasis has been placed on scaling up the suggested methods; therefore, it is important to choose the simplest and cost-effective method for the production of nano-sized calcium phosphates. The wet-chemical precipitation from aqueous supersaturated solutions containing calcium and phosphate ions is most commonly used, especially the precipitation reaction between $Ca(NO_3)_2 \cdot 4H_2O$ and $(NH_4)_2HPO_4$ [13, 14]. The amorphous calcium phosphate precursors can be obtained at ambient temperature. *Rajesh* et al. reported the reaction between $Ca(NO_3)_2 \cdot 4H_2O$ and $NH_4H_2PO_4$ to produce micro-sized HA and TCP powders for calcium phosphate coatings without attention to the phase transformation from ACP [15].

Although many studies have been published about the development of bioactive bone substitute biomaterials, for example, CaP bioceramics, bone cements and coatings, there is still potential for development and improvement. The central concept of this research is the development of bone substitutes through a modified wet-chemical precipitation method to obtain nano-sized calcium phosphates from amorphous calcium phosphate precursors.

2 Materials and Methods

2.1 Synthesis of Calcium Phosphates

Nano-sized calcium phosphates—amorphous calcium phosphate, α-tricalcium phosphate (α-TCP) and β-tricalcium phosphate (β-TCP), and hydroxyapatite—were synthesized by the modified wet-chemical precipitation method. The general experimental scheme is illustrated in Fig. 2. A calcium nitrate solution made by

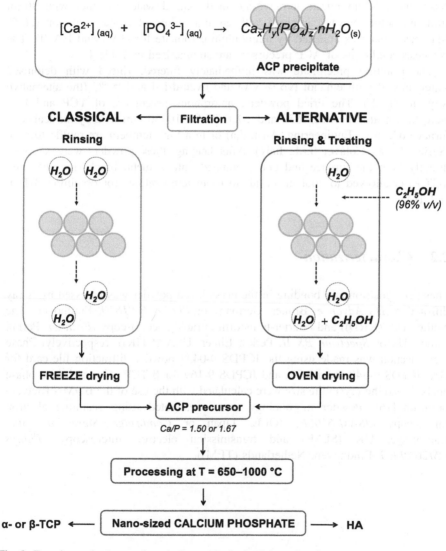

Fig. 2 Experimental scheme of producing nano-sized calcium phosphates

Table 1 Synthesis conditions of amorphous calcium phosphate precursors

	Molar ratio (Ca/P)	$c([Ca^{2+}])^a$ (M)	$c([HPO_4^{2-}])^b$ (M)	$c([H_2PO_4^-])^c$ (M)	pH ($[Ca^{2+}] + [PO_4^{3-}])^e$
TCP	Near 1.50	0.30	0.24	–	~10.00
HA I	Near 1.67	0.30	0.22^d	–	~10.00
HA II		0.30	–	0.25^d	~10.00

[a]$Ca(NO_3)_2 \cdot 4H_2O$ (Sigma-Aldrich, analytical grade)
[b]$(NH_4)_2HPO_4$ (Sigma-Aldrich or abcr GmbH, analytical grade)
[c]$NH_4H_2PO_4$ (Fluka, analytical grade)
[d]$(NH_4)_2CO_3$, (Sigma-Aldrich, analytical grade)
[e]26 % NH_4OH solution (Sigma-Aldrich, analytical grade)

dissolving calcium nitrate tetrahydrate in deionised water together with 30 ml ammonia solution was combined with an ammonium phosphate solution (26 % NH_3 basis, pH was adjusted by ammonium carbonate for HA I and HA II). The synthesis conditions of ACP precursors are summarised in Table 1.

The resulting precipitate was immediately filtered, rinsed with deionised water, treated with ethanol (96 % v/v) and oven-dried at 105 °C (the alternative way in Fig. 2.). The dried powders, amorphous precursors of TCP and HA, were then heated between 650 and 1000 °C for 10–60 min in a cylindrical tube furnace (Ceramic Engineering, Australia) or in a high temperature muffle furnace (Series Z1200, Colaver S.r.l., Italy). After heating, these powders were removed directly from the furnace and either emptied onto a metal block (for obtaining α-TCP) or allowed to cool in the air to room temperature (for obtaining β-TCP and HA).

2.2   Characterization

Phase composition and bonding in the crystallized powder was assessed by X-ray diffraction (*D8 Advance*, Bruker, Germany or *Philips PW1830*, Eindhoven, The Netherlands (XRD)) and Fourier-transform infrared spectroscopy (*Frontier*, Perkin Elmer, US or *Spectrum BX II*, Perkin Elmer, USA (FTIR)), respectively. Phase identification was made using the JCPDS 9-0432 powder diffraction file card for HA, JCPDS 9-348 for α-TCP, and JCPDS 9-169 for β-TCP. A quantitative phase analysis and the crystallite size were calculated with the use of the BGMN Rietveld program [16]. Powder morphology was investigated using scanning electron microscopy (*JSAM-6500F*, JOEL, Japan or *Cambridge Stereoscan 360*, Cambridge, UK (SEM)) and transmission electron microscopy (*Philips CM200/FEG*, Eindhoven, Netherlands (TEM)).

3 Results and Discussion

The pathway for producing nano-sized calcium phosphates consists of four main steps: (1) chemical precipitation, (2) precipitate treatment with ethanol, (3) processing at temperatures between 650 and 1000 °C, and (4) final product (mainly analysed by XRD, FTIR and SEM). The final products were investigated in the potential applications: nano-sized α-TCP as solid phase for producing calcium phosphate bone cements, and nano-sized β-TCP and HA—for increasing the bioactivity in biocomposite coating.

The broad XRD pattern suggested an amorphous calcium phosphate precursors, a broad maximum at approximately 30° and a weaker—at about 45° [17]. XRD patterns of powders heated at temperatures 675–800 °C demonstrated pure α-TCP and heated at 850 °C—β-TCP and apatite (Fig. 3). All peaks correspond to the standard XRD pattern of HA (JCPDS standard 9-432), α-TCP (JCPDS standard 9-348) and β-TCP (JCPDS standard 9-169). Freeze-dried amorphous calcium phosphate (Fig. 3c, black line) after crystallization showed α-TCP and β-TCP but oven-dried amorphous calcium phosphate (Fig. 3c, red line)—pure α-TCP. In addition, nano-sized HA can be synthesized using an alternative phosphorus source, ammonium dihydrogen phosphate.

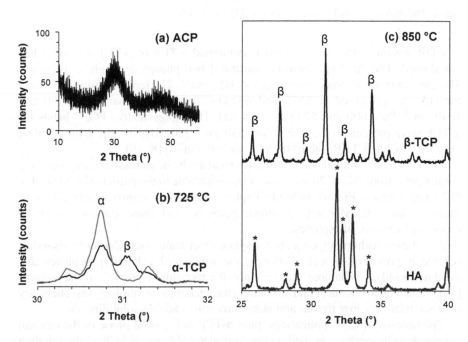

Fig. 3 a XRD pattern of an amorphous calcium phosphate (the same pattern was seen in ACP from all synthesis conditions); ACP heated at **b** 725 °C (α-TCP peak intensity was 3663 counts) and **c** 850 °C. Legend: α = α-TCP, β = β-TCP and * = HA

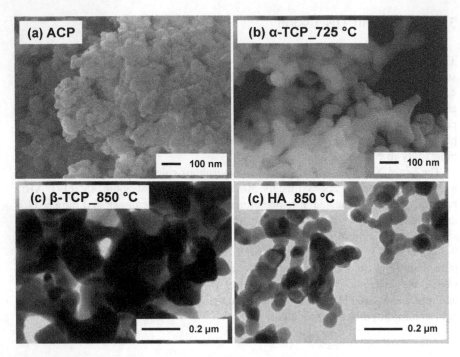

Fig. 4 Morphology of ACP and processed at 725 and 850 °C

FTIR spectra of the heated powders confirmed α-TCP or β-TCP and HA (data not shown). The ACP spectrum demonstrated two phosphate bands at 1050 and 570 cm^{-1}, and an adsorbed water band at 3450 cm^{-1}. The representative OH^{-} band for HA was presented at 3575 and 632 cm^{-1}, and PO_4^{3-} bands at 1090 and 1040 cm^{-1}, 963, 600 and 570 cm^{-1} [18–21]. The representative PO_4^{3-} bands for α-TCP were present at 1050, 950 and 550 cm^{-1} [20, 21], but the representative PO_4^{3-} bands for β-TCP—at 1048, 970 and 945 cm^{-1} [18, 19].

The crystallite size, calculated from Rietveld analysis, increased with processing temperature (from 50 to 120 nm), and was comparable to the particle size viewed in SEM micrographs (Fig. 4). Individual spherical particles, approximately 50 nm in diameter, have transformed to longer particles, and these connect to create nano-sized calcium phosphates.

An efficient technology concept for synthesis of nano-sized CaP in biomaterials science is providing a real challenge; the nano-sized calcium phosphates that exhibit "bio-like" behaviour can enhance the recognised possible disadvantages [22]: variations in chemical and structural characteristics (technology and chemistry related), relatively low tensile and shear strengths, variable solubility, etc.

The nano-sized CaP applications, pure α-TCP as the solid phase of the calcium phosphate bone cement, as well as the nano-sized HA and β-TCP as the building blocks into the biopolymer matrix have been studied. The results demonstrate a new technique of converting an amorphous calcium phosphate to nano-sized pure α-TCP, and this finding provides a faster pathway to CaP bone cements. The effect of both phases, nano-sized HA and β-TCP, included into the chitosan matrix via electrochemical/electrophoretic deposition, has not been studied yet. The results demonstrate a potential design of the functional biocomposite coating on titanium, focusing on improving the coating in vitro bioactivity and biocompatibility.

4 Conclusions

The efficient synthesis technology towards nano-sized tricalcium phosphate and hydroxyapatite has been developed and verified. Nano-sized calcium phosphates, TCP and HA, were synthesized from an amorphous calcium phosphate, treated in ethanol and then crystallized at 650–1000 °C.

Acknowledgments This research has been partly supported by the European Social Fund within the project "Support for the implementation of doctoral studies at Riga Technical University" and European Union student exchange programme "ERASMUS Student Mobility for Placements".

References

1. Dorozhkin, S.V., Epple, M.: Biological and medical significance of calcium phosphates. Angew. Chem. Int. Ed. **41**, 3130–3146 (2002)
2. Dorozhkin, S.V.: Calcium phosphates and human beings. J. Chem. Educ. **83**(5), 713–719 (2006)
3. Dorozhkin, S.V.: A detailed history of calcium orthophosphates from 1770s till 1950. Mater. Sci. Eng. C **33**(6), 3085–3110 (2013)
4. Garrido, C.A., Lobo, S.E., Turibio, F.M., LeGeros, R.Z.: Biphasic calcium phosphate bioceramics for orthopaedic reconstructions: clinical outcomes. Int. J. Biomater. **2011** (129727), 1–9 (2011)
5. Bohner, M., Tadier, S., Van Garderen, N., De Gasparo, A., Dobelin, N., Baroud, G.: Synthesis of spherical calcium phosphate particles for dental and orthopedic applications. Biomatter. **3**(2), 1–15 (2013)
6. Al-Sanabani, J.S., Madfa, A.A., Al-Sanabani, F.A.: Application of calcium phosphate materials in dentistry. Int. J. Biomater. **2013**, 1–12 (2013)
7. Markets and markets: biomaterials market worth 130.57 billion USD by 2020. http://www.marketsandmarkets.com/PressReleases/global-biomaterials.asp
8. Loomba, L., Sekhon, B.S.: Calcium phosphate nanoparticles and their biomedical potential. J. Nanomater. Mol. Nanotechnol. **4**(1), 1–12 (2015)
9. Wang, P., Zhao, L., Liu, J., Weir, M.D., Zhou, X., Xu, H.H.K.: Bone tissue engineering via nanostructured calcium phosphate biomaterials and stem cells. Bone Res. **2**, 1–13 (2014)
10. Rey, C., Combes, C., Drouet, C., Somrani, S.: Tricalcium phosphate-based ceramics, bioceramics and their clinical applications. In: Cambridge. 1st ed., Woodhead Publishing Limited, pp. 326–358 (2008)
11. Brunner, T.J., Grass, R.N., Bohner, M., Stark, W.J.: Effect of particle size, crystal phase and crystallinity on the reactivity of tricalcium phosphate cements for bone reconstruction. J. Mater. Chem. **17**, 4072–4078 (2007)

12. Sadat-Shojai, M., Khorasani, M., Dinpanah-Khoshdargi, E., Jamshidi, A.: Synthesis methods for nanosized hydroxyapatite with diverse structures. Acta Biomater. **9**(8), 7591–7621 (2013)
13. Somrani, S., Rey, C., Jemal, M.: Thermal evolution of amorphous tricalcium phosphate. J. Mater. Chem. **13**, 888–892 (2003)
14. Bernache-Assollant, D., Ababou, A., Champion, E., Heughebaert, M.: Sintering of calcium phosphate hydroxyapatite $Ca_{10}(PO_4)_6(OH)_2$ I. Calcination and particle growth. J. Eur. Ceram. Soc. **23**(2), 229–241 (2003)
15. Rajesh, P., Muraleedharan, C.V., Sureshbabu, S., Komath, M., Varma, H.: Preparation and analysis of chemically gradient functional bioceramic coating formed by pulsed laser deposition. J. Mater. Sci. Mater. Med. **23**(2), 339–348 (2012)
16. Taut, T., Kleeberg, R., Bergmann, J.: The new seifert Rietveld program BGMD and its application to quantitative phase analysis. Mater. Struct. **5**, 57–66 (1998)
17. Sun, L., Chow, L.C., Frukhtbeyn, S.A., Bonevich, J.E.: Preparation and properties of nanoparticles of calcium phosphates with various Ca/P ratios. J. Res. Natl. Inst. Stand. Technol. **115**(4), 243–255 (2010)
18. Carrodeguas, R.G., De Aza, S.: α-tricalcium phosphate: synthesis, properties and biomedical applications. Acta Biomater. **7**, 3536–3546 (2011)
19. Jilavenkatesa, A., Condrate, R.A.: The infrared and raman spectra of beta- and alpha-tricalcium phosphate $(Ca_3(PO_4)_2)$. Spectrosc. Lett. **31**, 1619–1634 (1998)
20. Alobeedallah, H., Ellis, J.L., Rohanizadeh, R., Coster, H., Dehghani, F.: Preparation of nanostructured hydroxyapatite in organic solvents for clinical applications. Trends Biomater. Artif. Organs **25**, 12–19 (2011)
21. Gandolfi, M.G., Taddei, P., Tinti, A., De Stefano Dorigo, E., Prati, c: Alpha-TCP improves the apatite-formation ability of calcium-silicate hydraulic cement soaked in phosphate solutions. Mater. Sci. Eng. C **31**, 1412–1422 (2011)
22. Lemon, J.E., Misch-Dietsh, F., McCracken, M.S.: Biomaterials for dental implants, dental implant prosthetics. In: Missouri. Elsevier Inc., pp. 66–94 (2008)

Frequency Resolution and Accuracy Improvement of a GaP CW THz Spectrometer

Tetsuo Sasaki, Tadao Tanabe and Jun-ichi Nishizawa

Abstract High resolution terahertz (THz) spectroscopy is available for sensitive gas detection and also useful for vibrational mode analysis of organic molecules. We developed a widely frequency tunable (0.1–7.5 THz), monochromatic coherent THz-wave Signal Generator on the basis of difference frequency generation (DFG) in a GaP crystal via excitation of phonon-polariton mode under small-angle non-collinear phase matching condition. The pump beams for DFG were supplied from semiconductor lasers as seed beams after amplification by fiber amplifiers. Then we applied it as a light source for a high resolution and high accuracy THz spectrometer. Combining with a superconducting Transition Edge Sensor (TES) bolometer cooled by a pulse tube refrigerator, a non-stop spectrometer system was realized. The system had advantages of high stability, small size, easy operation and maintenance free in the lifetime of the pump lasers. As frequency resolution and accuracy were limited by those of pump laser beams, we have newly developed a frequency control system for two beams at the same time. Frequency resolution of 8 MHz constant in whole the frequency tunable range and absolute accuracy of better than 3 MHz were evaluated by the measurements of water vapor absorptions in vacuum.

T. Sasaki (✉)
Research Institute of Electronics, Shizuoka University, Shizuoka, Japan
e-mail: tsasaki@rie.shizuoka.ac.jp

T. Tanabe
Institute of Multidisciplinary Research for Advanced Materials,
Tohoku University, Sendai, Japan

J. Nishizawa
Nishizawa Memorial Center, Tohoku University, Sendai, Japan

© Springer International Publishing AG 2017
R. Jabłoński and R. Szewczyk (eds.), *Recent Global Research and Education:
Technological Challenges*, Advances in Intelligent Systems and Computing 519,
DOI 10.1007/978-3-319-46490-9_5

33

1 Introduction

Terahertz (THz) spectroscopy is an important tool for detection and evaluation of materials and crystals, since unique molecular vibrations of organic materials can be found in this frequency range. We have realized widely frequency-tunable (0.1–7.5 THz) monochromatic THz Signal Generator based on difference frequency generation (DFG) via excitation of the phonon-polariton mode in semiconductor Gallium Phosphide (GaP) crystals. Then we applied it as a light source for high accurate THz spectrometer that must be called as THz Laser spectrometer. We claimed such high accurate THz spectroscopy can detect defects in organic molecules or crystals [1, 2]. In this paper, we describe recent progress of our GaP continuous wave (CW) THz Spectrometer especially in frequency resolution and accuracy.

2 GaP CW THz Spectrometer

For THz wave DFG, two CW infrared (IR) beams were incident on a GaP crystal with small angle to satisfy phase-matching condition. When the energy conservation and momentum conservation laws were fulfilled at the same time, THz wave could be generated. Frequency could be varied by tuning wavelength and angle of two IR beams to satisfy the phase-matching condition at each frequency. Figure 1 shows a frequency resolution and accuracy improved spectrometer system. Two IR beams were delivered from a distributed feedback (DFB, Toptica DL-DFB) laser and a mode-hop-free external-cavity laser diode (ECLD, Spectra Quest Lab. λ-master 1040). The wavelengths of the ECLD and the DFB laser were tunable from 1051 to 1075 nm (278.9–285.2 THz) and from 1071.7 to 1074.4 nm (279.0–279.7 THz), respectively. THz wave from 0 to 6.2 THz could be obtained continuously. We applied two independent frequency feedback control. Two wavelengths were monitored by a two-channel frequency meter (HighFinesse WS-7) and the ECDL laser was feedback controlled in this loop. For the DFB laser, a part of the beam was introduced to a Fabry-Perot interferometer and the wavelength of the beam was frequency-locked to the resonance frequency depending on the cavity length. Each beam was amplified to constant power 3 W at any wavelength by polarization-maintained ytterbium (Yb)-doped optical fiber amplifiers (YbFA, FITEL HPU60217). The DFB laser beam was chopped at around 1 kHz, which enabled high sensitive lock-in detection of low power THz-wave. Application of a Niobium Transition Edge Superconductor (TES) bolometer cooled by a low vibration pulse tube cooler (QMC instruments Ltd, QNbB/PTC) as a detector, the spectrometer is now working as liquid helium-free, non-stop system. Due to realization of the high speed feedback control, frequency deviations of both laser beams were decreased.

We have confirmed frequency resolution and accuracy by measuring the Full Width Half Maximum (FWHM) and center frequencies of water vapor absorption

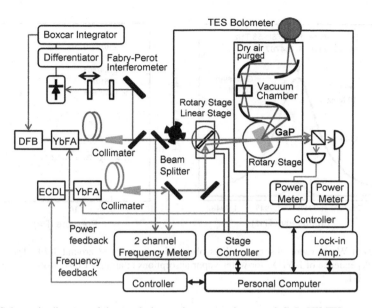

Fig. 1 Schematic diagram of the resolution and accuracy improved GaP CW THz spectrometer

in vacuum. Figure 2 (Left) shows water vapor THz absorbance spectra at 4.0 and 110 Pa pressure at around 1.795 THz. Absorbance spectrum at higher and lower pressure can be well fitted by a Lorentzian and a Gaussian curve, respectively. This is because the FWHM value was dominated by water molecular collisions at higher pressure and the Doppler effect became dominant at lower pressure. The other absorption spectra at around 1.097, 1.661, 2.774 and 5.107 THz showed similar tendencies. The results were plotted in Fig. 2 (Right). The FWHM values decreased with decreasing pressure and reaches to constant at lower pressure less than 10 Pa. For example, convergent value for the absorption at 1.0973663 THz is 8.5 MHz. Doppler linewidth Δf_D [3] caused by the velocity dispersion of molecules given by the Maxwell distribution can be described as,

$$\Delta f_D = \frac{2}{c}\sqrt{\frac{2kT \ln 2}{m}} \cdot f_a. \tag{1}$$

Here, k: Boltzmann constant, T: temperature, m: molecular weight, c: speed of light, and f_a: absorption frequency. The Doppler widths at 295 K for 1.0973663 THz is 3.21 MHz. Comparing with the measured FWHM and the calculated doppler width, the resolution of the system could be estimated to be 7.87 MHz. Such estimated values from the other absorptions are listed in Table 1. All the results suggested the system resolution to be almost 8 MHz. Also we evaluated the frequency accuracy by comparing with our measurements of center peak absorption values with JPL database [4]. The accuracy must be within a few MHz, as the maximum difference was 2.65 MHz as listed in Table 1.

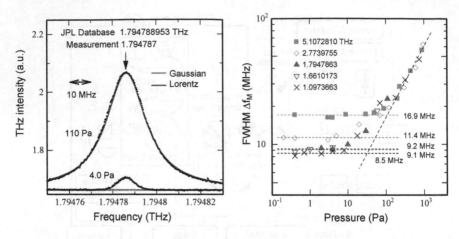

Fig. 2 (*Left*) Water vapor absorption lines observed in vacuum at around 1.795 THz. (*Right*) Pressure dependence of FWHM for water vapor absorptions at around 1.097, 1.661, 1.795, 2.774 and 5.107 THz

Table 1 Frequencies and FWHM of measured water vapor absorption lines. Frequency values from JPL and calculated doppler widths are listed

H_2O vapor absorption					
Measured f_A (THz)	JPL data f_P (THz)	Deviation Δ_f (MHz)	Doppler width Δf_D (MHz)	Measured FWHM $\Delta f_{M\,sat.}$ (MHz)	Estimated linewidth $\sqrt{\Delta f_{M\,sat.}^2 - \Delta f_D^2}$ (MHz)
1.0973663	1.0973648	−1.51	3.21	8.5	7.87
1.6610073	1.6610076	0.34	4.85	9.1	7.70
1.7947863	1.7947890	2.65	5.25	9.2	7.55
2.7739755	2.7739766	1.09	8.11	11.4	8.01
5.1072810	5.1072811	0.10	14.93	16.9	7.92

3 High Resolution THz Spectra in Middle Molecular Weight Pharmaceuticals

One of our main targets to apply high accurate and high resolution spectrometer is crystallinity evaluation of organic crystals such as pharmaceuticals especially in middle weight molecules (molecular weights (MW) in about 500–2,000). The middle molecular weight pharmaceuticals have small side effects due to high target specificity and selectivity compared with small molecular pharmaceuticals, and are effective even inside of cells unlike heavyweight pharmaceuticals (antibody drugs).

As an example of THz spectra for middle weight molecules, Fig. 3 shows THz absorption spectra of α-Cyclodextrin ($C_{36}H_{60}O_{30}$, MW = 972.85) 10 wt% polyethylene pellet (t = 1.0 mm) measured at 10, 70 and 300 K. α-Cyclodextrin is sometimes used to improve solubility of poorly water-soluble drugs [5]. While only

Fig. 3 THz absorption
spectra of α-Cyclodextrin at
10, 70, 300 K

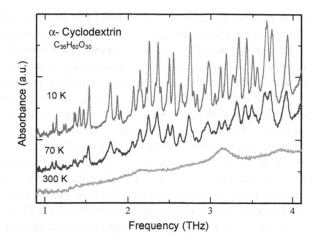

a slight dull peaks could be seen in the spectrum at 300 K, a lot of sharp peaks
became clearly observable at low temperature. These peaks were still not resolved
at 70 K, but well resolved at 10 K. The Lorentz FWHM of the absorption peak at
1.0974 THz was obtained to be 7.05 GHz by the least squares fitting method. Such
sharp peaks could be evaluated correctly with using high resolution spectrometer,
and only at enough low temperature where atomic movements in crystal were
suppressed.

Figure 4 (Left) graph shows THz absorption spectra of α-Cyclodextrin supplied
from four different product makers. Sharpness of absorption peaks (peak height and
FWHM) measured at 10 K were different sample to sample. Figure 4 (Right) graph
shows the comparison of absorption peak heights at 1.14, 1.54, 2.75 and 3.94 THz

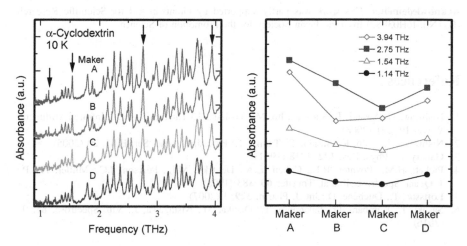

Fig. 4 (*Left*) THz absorption spectra of α-Cyclodextrin supplied from four different product
maker. (*Right*) Peak height comparison of absorptions at 1.14, 1.54, 2.75 and 3.94 THz

(marked with allow in the left graph). The sharpness of the absorption peaks are thought to be affected by the crystallinity of the sample which must be quantifiable as the sum of "defect effects" in the crystal. Here, the effects of point defects such as impurities, vacancies, etc., those of line defects such as dislocation, and those of planar defects including surface, grain boundary, etc. should be taken into account. As all the defects would interrupt normal mode vibrations, absorption peaks must become dull with increasing defect amounts. So, the sample provided from Maker A seems to have good quality crystal.

For example, a defect named the "surface" treated as crystal habit is known as an important parameter to control crystal dissolution speed or long term storage stability of pharmaceuticals. Crystal orientation of the dominant surface critical for material property can be detected by THz spectroscopy, because each absorption peaks can be assigned to each direction [6]. So we believe high accurate terahertz spectroscopy at low temperature can predict the dissolution speed and/or the long term stability of the pharmaceutical crystal powders and it becomes an inspection tool for higher quality control in future.

4 Summary

We have applied high speed frequency feedback control of pump lasers and successfully obtained 8.0 MHz resolution and ~ 3 MHz accuracy in the whole measurement frequency range (0.5–6.2 THz). We have demonstrated that high resolution spectroscopy of middle weight molecule at low temperature could show a lot of sharp absorption peaks. Such spectra must be available for crystallinity evaluation of organic crystals like as pharmaceuticals developed in future.

Acknowledgments This work was partly supported by Grants-in-Aid for Scientific Research (B) (No. 16H03882) from the Japan Society for the Promotion of Science.

References

1. Nishizawa, J., Sasaki, T., Suto, K., Ito, M., Yoshida, T., Tanabe, T.: Int. J. Infrared Millimeter Waves **19**, 291 (2008)
2. Nishizawa, J., Suto, K., Sasaki, T., Tanno, T.: Proc. Jpn. Acad. Ser. B **81**, 20 (2005)
3. Galatry, L.: Phys. Rev. **122**, 1218 (1961)
4. Pickett, H.M., Poynter, R.L., Cohen, E.A., Delitsky, M.L., Pearson, J.C., Muller, H.S.P.: J. Quant. Spectrosc. Radiat. Transfer **60**, 883 (1998)
5. Loftsson, T., Duchêne, D.: Int. J. Pharm. **329**, 1 (2007)
6. Sasaki, T., Kambara, O., Sakamoto, T., Otsuka, M., Nishizawa, J.: Vib. Spectrosc. **85**, 91 (2016)

Study on the Magnetizing Frequency Dependence of Magnetic Characteristics and Power Losses in the Ferromagnetic Materials

Maciej Kachniarz and Dorota Jackiewicz

Abstract The following paper presents results of investigation of the magnetic properties of several ferromagnetic cores influenced by the magnetizing field of different frequencies with particular consideration of power losses. The description of investigated materials and developed measurement setup is included in the paper. Obtained results are presented and discussed. The conclusion about magnetizing frequency dependence of magnetic properties of investigated materials is formulated in the last section of the paper.

Keywords Magnetism · Ferromagnetic materials · Frequency dependence · Power loss

1 Introduction

In modern energy and electronic industry energy saving is a very important matter. All electronic devices and end elements of energy grids are developed in a way allowing to minimize energy consumption. In both energy and electronic industry, inductive components with magnetic cores made of magnetic materials are widely utilized. For energy grid inductive elements, steel is very important material used for magnetic cores of transformers, while ferrite and nanocrystalline cores are mainly elements of many inductive components like filters, chokes, resonant circuits and many other inductive components found in electronic devices [1].

M. Kachniarz (✉)
Institute of Metrology and Biomedical Engineering, Warsaw University
of Technology, Sw. Andrzeja Boboli 8, 02-525 Warsaw, Poland
e-mail: m.kachniarz@mchtr.pw.edu.pl

D. Jackiewicz
Industrial Research Institute for Automation and Measurements,
Al. Jerozolimskie 202, 02-486 Warsaw, Poland
e-mail: d.jackiewicz@mchtr.pw.edu.pl

© Springer International Publishing AG 2017
R. Jabłoński and R. Szewczyk (eds.), *Recent Global Research and Education:
Technological Challenges*, Advances in Intelligent Systems and Computing 519,
DOI 10.1007/978-3-319-46490-9_6

Inductive components are often working with alternating current, which results in alternating magnetizing filed acting on the magnetic core. Magnetic properties of ferromagnetic materials, like steels, ferrites or nanocrystalline alloys, are strongly correlated with frequency of the magnetizing field [2]. Especially important parameter, influenced by the frequency of magnetizing field, is power loss, which determines how much energy is lost during single magnetizing cycle. Taking into account energy saving trends in modern industry, it is very important to study the dependences between magnetizing filed frequency and magnetic parameters of the material with special attention paid to the power loss, which has great importance for energy efficiency.

The following paper is dedicated to the studies on magnetizing frequency dependence of magnetic parameters of several magnetic cores representing the most important groups of ferromagnetic materials used in modern industry: steels, ferrites and nanocrystalline alloys.

2 Investigated Ferromagnetic Materials

For the performed investigation, three different ferromagnetic materials were selected, each one representing different group of ferromagnetic materials utilized in modern industry. Each of them has different chemical composition and exhibits unique physical properties.

- **F-3001** is a manganese zinc (Mn-Zn) ferrite material. Ferrites are ceramic materials chemically composed of iron oxide Fe_2O_3 and one or more metallic elements [3, 4]. They are most frequently utilized in chokes, filters and transformers. Chemical composition of Mn-Zn ferrites is described with general formula $Mn_{1-x}Zn_xFe_2O_3$. For investigated F-3001 ferrite value of x parameter is within the range $0.8 \div 1.0$. Magnetic parameters of the material declared by the manufacturer are: saturation magnetic flux density $B_m = 0.37$ T, coercive field $H_c = 15$ A/m and initial relative magnetic permeability $\mu = 3500$. Material is utilized in cores of wideband transformers.
- **M-450** is rapidly quenched nanocrystalline alloy based on iron [5], which is commercially available as NANOPERM material (chemical composition $Fe_{73.5}Cu_1Nb_3Si_{15.5}B_7$). Nanocrystalline alloys are relatively new group of ferromagnetic materials, intensively developed due to their extraordinary soft magnetic properties [6, 7]. Magnetic properties of the M-450 alloy are: saturation magnetic flux density B_m: 1.2 T, coercive field $H_c = 1.2$ A/m and initial relative magnetic permeability $\mu = 8000$. Material is mostly utilized in EMC (electromagnetic compatibility) applications and current transformers [8].
- **13CrMo4-5** is heat resistant alloy steel [9, 10] with addition of chromium and molybdenum. It is utilized mostly in energy industry. 13CrMo4-5 steel is material exhibiting much harder magnetic properties than ferrites and nanocrystalline alloys, but is still classified as soft magnetic material (coercive

Table 1 Geometric parameters and numbers of magnetizing (n_m) and sensing (n_s) windings of investigated materials

Material	F-3001	M-450	13CrMo4-5
l_e (mm)	96.4	78.0	126.0
S_e (mm^2)	116.0	38.0	20.0
V_e (mm^3)	11182.4	2964.0	2520.0
n_m	10	10	350
n_s	3	50	100

field lower than 1000). Experimentally determined magnetic parameters of the material are: saturation magnetic flux density $B_m = 1.5$ T, coercive field $H_c = 580$ A/m and initial relative magnetic permeability $\mu = 250$.

All investigated materials were formed into the shape of toroidal magnetic core. F-3001 and 13CrMo4-5 were bulk materials while M-450 was thin nanocrystalline ribbon formed into multilayer ring core. On the basis of geometrical dimensions (outer diameter, inner diameter and thickness of the core) three parameters necessary for investigations of magnetic properties of the materials were calculated: average flow path of magnetic flux within the core l_e, cross-sectional area of the core S_e and volume of the core V_e. Values of this parameters for each investigated material are presented in Table 1.

On each sample two sets of windings were made: magnetizing winding and sensing winding. For each material different numbers of turns in both windings were necessary due to varied magnetic properties of investigated materials. Numbers of turns in magnetizing and sensing windings for investigated samples are given in Table 1.

3 Experimental Setup

In the Fig. 1 experimental setup for measurements of magnetic properties of soft magnetic materials is shown. Measurements of hysteresis loops are performed using a hysteresisgraph HB-PL30. It contains current waveforms generator and fluxmeter. The work of the hysteresisgraph is controlled by computer. Current generator is used to generate magnetizing and demagnetizing field H according to the equation:

$$H = \frac{n_m I}{l_e} \tag{1}$$

where I is magnetizing current. The magnetizing waveform goes to magnetizing winding. Magnetizing field acts on the sample.

Sensing winding is connected to the fluxmeter. On the basis of voltage u_i induced in the sensing winding, value of magnetic flux density B within the sample is determined:

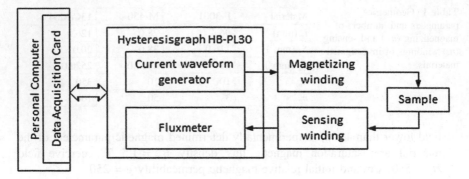

Fig. 1 Schematic block diagram of the experimental setup

$$B = \frac{K_c}{n_s S_e} \int u_i \mathrm{d}t \qquad (2)$$

where K_c is constant of the integrator circuit.

For each measured hysteresis loop utilized measurement system calculated value of power loss P_s according to the well-known equation [1]:

$$P_s = f V_e \int_0^{B_m} H \mathrm{d}B \qquad (3)$$

where f is magnetizing field frequency and B_m is saturation magnetic flux density. On the basis of this calculation, characteristics of dependence between power loss and magnetizing frequency. Integral operation was performed numerically.

The magnetizing frequency dependence of magnetic characteristics was tested for the values of magnetizing field H_m above values of saturation on all three samples. The measurements were made for values of frequency within the range $0.01 \div 1000$ Hz. Settings of measurement system for each investigated sample are shown in Table 2. The settings were adjusted for each sample individually.

Table 2 Settings of the measurement system for investigated materials

Material	F-3001	M-450	13CrMo4-5
Amplitude of magnetizing field H_m (A/m)	200	250	5000
Minimum magnetizing frequency f_{min} (Hz)	0.01	0.01	0.01
Maximum magnetizing frequency f_{max} (Hz)	1000	1000	1000
Amplitude of demagnetizing field (A/m)	200	250	5550
Frequency of demagnetizing field (Hz)	24	24	7
Damping coefficient of demagnetizing field	1.04	1.03	1.04

4 Experimental Results

For all investigated materials measurements were performed on the developed measurement setup according to presented methodology. As a result, family of hysteresis loops for constant value of magnetizing filed amplitude and magnetizing frequency within the range from 0.01 to 1000 Hz was obtained for each material. The results for investigated materials are presented in Figs. 2, 3 and 4. All characteristics increases monotonically. The curves for the F-3001 ferrite and M-450 nanocrystalline alloy exhibit similar shape. The least power loss was measured for M-450 material. Values for F-3001 ferrite are significantly higher. The curve for the 13CrMo4-5 steel has a different character. Steel is much harder magnetic material than ferrite and nanocrystalline alloy, so increase of power loss is observed for much lower frequencies, than in softer materials. However, after reaching value of $P_s = 4.0$ W, the increase of power loss with the magnetizing frequency is lower. The maximum value of the power losses for 13CrMo4-5 steel is order of magnitude higher than for other investigated materials.

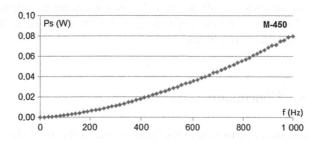

Fig. 2 Characteristic of power loss for a sample made of M-450 nanocrystalline alloy

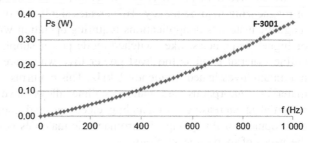

Fig. 3 Characteristic of power loss for a sample made of F-3001 ferrite

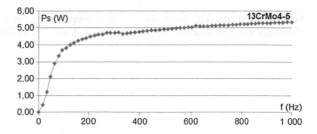

Fig. 4 Characteristic of power loss for a sample made of 13CrMo4-5 steel

Fig. 5 Characteristics of maximum flux density for the investigated materials: F-3001 ferrite, M-450 nanocrystalline alloy and 13CrMo4-5 steel

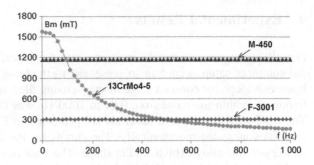

Figure 5 presents characteristics of maximum flux density for the three tested materials. The value of maximum flux density doesn't change throughout whole frequency range for the F-3001 ferrite and M-450 nanocrystalline. In turn for 13CrMo4-5 steel the value of maximum flux density decreases up to ten times. This is the result of higher magnetic hardness of steel. Due to magnetic properties of the material, for higher frequency magnetic domain structure of the steel is not fully organized according to magnetizing field. Changes of magnetizing field are too fast for the domain walls to fully change their magnetic orientation, which is the reason of lower maximum magnetic flux density observed in 13CrMo4-5 steel for high magnetizing frequencies.

5 Conclusion

Results of performed investigation presented in the following paper indicates that in all investigated materials there is distinct increase of the power loss with the growth of magnetizing field frequency. This creates significant obstacle to the utilization of magnetic materials in applications requiring operation with AC magnetizing fields of higher frequencies, like switched-mode power supplies. The M-450 nanocrystalline material exhibits the least power loss, which are relatively small even for maximum investigated frequency 1 kHz. This confirms that modern ferromagnetic materials—amorphous and nanocrystalline alloys have the greatest application potential in increasingly demanding electronic and energy industry. Yet further development of this group of ferromagnetic materials is necessary in order to face the needs of modern technology.

Acknowledgments This work was partially supported by the statutory founds of Institute of Metrology and Biomedical Engineering, Warsaw University of Technology (Poland).

References

1. O'Handley, R.: Modern magnetic materials—principles and applications. John Wiley & Sons, New York (2000)
2. Derebasi, N., Rygal, R., Moses, A.J., Fox, D.: Variation of magnetic properties of toroidal cores with magnetizing frequency. J. Magn. Magn. Mater. **215–216**, 611–613 (2000)
3. Carter, C.B., Norton, M.G.: Ceramic materials: science and engineering. Springer, NY (2007)
4. Sridhar, C.S.L.N., Lakshmi, C.S., Govindraj, G., Bangarraju, S., Satyanarayana, L., Potukuchi, D.M.: Structural, morphological, magnetic and dielectric characterization of nano-phased antimony doped manganese zinc ferrites. J. Phys. Chem. Solids **92**, 70–84 (2016)
5. Zheng, R., Chen, J., Xiao, W., Ma, C.: Microstructure and tensile properties of nanocrystalline $(FeNiCoCu)_{1-x}Ti_xAl_x$ high entropy alloys processed by high pressure torsion. Intermetallics **74**, 38–45 (2016)
6. Salach, J., Bieńkowski, A., Szewczyk, R.: Magnetoelastic properties of hitperm—type alloys from torque moment. Acta Phys. Pol. A **113**, 115–118 (2008)
7. Bieńkowski, A., Szewczyk, R., Kulik, T., Ferenc, J., Salach, J.: Magnetoelastic properties of HITPERM-type Fe(41,5)Co(41,5)Cu(1)Nb(3)B(13) nanocrystalline alloy. J. Magn. Magn. Mater. **304**, e624–e626 (2006)
8. Petruk, O., Szewczyk, R., Salach, J., Nowicki, M.: Digitally controlled current transformer with hall sensor. Adv. Intel. Syst. Comput. **267**, 641–647 (2014)
9. Abang, R., Findeisen, A., Krautz, H.J.: Corrosion behavior of selected power plant materials under oxyfuel combustion conditions. Górnictwo i Geoinżynieria **35**, 23–42 (2011)
10. Berns, H., Theisen, W.: Ferrous materials: steel and cast iron. Springer, NY (2008)

Synthesis and Study of Luminescent Materials on the Basis of Mixed Phosphates

Vitalii V. Malygin, Lev A. Lebedev, Vadim V. Bakhmetyev,
Mariia V. Keskinova, Maxim M. Sychov, Sergey V. Mjakin
and Yoichiro Nakanishi

Abstract Highly dispersed $NaBaPO_4:Eu^{3+}$ and $NaBaPO_4:Eu^{2+}$ phosphors were prepared using sol-gel and self-propagating high temperature synthesis routes. The effects of thermal treatment and dopant concentration on phase composition, crystallite size, lattice parameters and luminescent properties of the synthetized phosphors are studied. Using excitation and emission spectra, energy-level positions of luminescent centers were determined. The obtained phosphors provide an effective luminescence upon either UV or X-ray excitation that makes them promising for the application as components of white LEDs and medicine.

Keywords Phosphate phosphors · Luminescence · Europium · Eu^{2+}

1 Introduction

$NaBaPO_4:Eu$ phosphors providing an effective luminescence upon UV or X-ray excitation are useful as components of white LEDs and pharmaceutical drugs for photodynamic therapy of cancer. The latter medical application requires finely dispersed (particle size 30…220 nm, preferably 70…100 nm) phosphors [1] obtainable using such approaches as sol-gel synthesis and self-propagating high temperature synthesis (SHS).

The luminescence spectra of the considered $NaBaPO_4:Eu$ phosphors depend on the activator oxidation state, with the activation by Eu^{3+} and Eu^{2+} providing red and blue luminescence respectively. Furthermore, the luminescence spectra of Eu^{2+}-activated phosphors also significantly depend on Eu^{2+} ions environment in the

V.V. Malygin · L.A. Lebedev · V.V. Bakhmetyev (✉) · M.V. Keskinova ·
M.M. Sychov · S.V. Mjakin
Saint-Petersburg Institute of Technology (Technical University),
Saint Petersburg, Russia
e-mail: vadim_bakhmetyev@mail.ru

Y. Nakanishi
Research Institute of Electronics, Shizuoka University, Shizuoka, Japan

© Springer International Publishing AG 2017
R. Jabłoński and R. Szewczyk (eds.), *Recent Global Research and Education:
Technological Challenges*, Advances in Intelligent Systems and Computing 519,
DOI 10.1007/978-3-319-46490-9_7

crystal lattice. Therefore, the study of luminescence centers in these phosphors is very important in order to adjust their luminescence performances.

In our previous research [2–5] finely dispersed $Zn_3(PO_4)_2:Mn^{2+}$, $Ba_3(PO_4)_2:Eu^{2+}$, $NaBaPO_4:Eu^{3+}$ and $NaBaPO_4:Eu^{2+}$ phosphors were prepared using sol-gel and SHS methods. Although the effect of the activator concentration upon their phase composition, crystallite size and luminescent properties was characterized in these studies, most of the synthesized $NaBaPO_4:Eu^{2+}$ samples were multiphase including the admixtures of such phases as $Ba_3(PO_4)_2$ and $Ba_2P_2O_7$ complicating the study of their luminescent properties since the phase composition affects Eu^{2+} environment and consequently the luminescence spectrum.

This aim of this study was the synthesis of single-phase $NaBaPO_4:Eu^{2+}$ phosphors and characterization of their luminescence performances.

2 Experimental

$NaBaPO_4:Eu^{2+}$ phosphor was synthetized via SHS and sol-gel methods, Eu content was 5 mol% for both samples. Solution-combustion type of SHS method that similar to reported in [6] used in current work, but instead of urea b-alanine was used as fuel. The starting materials were analytical grade $Ba(NO_3)_2$, NaH_2PO_4, Eu_2O_3, HNO_3, NH_4NO_3 and b-alanine. Firstly, appropriate amount of Eu_2O_3 and $Ba(NO_3)_2$ were placed in thin wall quartz beaker with concentrated nitric acid to form nitrate solution. Than solution of NaH_2PO_4, NH_4NO_3 and b-alanine in minimum amount of distilled water was added. The amount of b-alanine was doubled from stoichiometric and NH_4NO_3 used to increased temperature of reaction. The precursor solution was placed on a hot plate to evaporate water and form viscous mass that self ignites and burn producing large amount of gases and form white sponge like product. To produce $NaBaPO_4:Eu^{2+}$ phosphor, product was milled in mortar and placed in corundum boats and introduced in tube furnace sintered under reductive atmosphere (5 %H_2 + 95 %N_2) at 1050 °C for two hours.

To produce sodium-barium phosphate via sol-gel route, required amount of europium oxide was dissolved in 35 % nitric acid in a glass beaker under magnetic stirring to form clear solution. Barium nitrate solution was mixed in the beaker with europium nitrate and sodium phosphate solution was added dropwise to form translucent sol. The resulting product was filtered and washed with distilled water on Buchner funnel under the vacuum of water jet pump and dried at 120 °C at vacuum drying cabinet. The xerogel was milled in mortar placed in corundum crucible and introduced in muffle furnace for two hours at 600 °C. The reduction procedure was the same as described above.

XRD characterization of the synthesized phosphors was carried out using a RINT 2200 diffractometer. Photoluminescence (PL) and photo-excitation spectra were also studied for all the prepared samples. PL was stimulated by xenon lamp using a monochromatic system. PL spectra were measured in the range 350... 600 nm upon excitation with the wavelengths 270, 300 and 335 nm. The excitation spectra were measured in the range 250...400 nm and recorded at 420, 460 and 480 nm. The selection of excitation wavelengths and luminescence wavelengths for the measurements was based on the reference data on the positions of bands in the luminescence and excitation spectra [7], as well as on our earlier studies [5].

3 Results and Discussion

XRD spectra of the synthesized phosphors presented in Fig. 1 indicate that their phase composition is identical to PDF 33–1210 card and corresponds to NaBaPO$_4$ with a prominent crystal structure free of admixture phases. The crystallite size calculated according to Scherrer equation is 86.2 and 80.0 nm or the samples synthesized using sol-gel and SHS procedures accordingly.

PL spectra (stimulated by 270, 300 and 335 nm UV radiation) of the phosphors synthesized by sol-gel and SHS methods are shown in Fig. 2 and their excitation

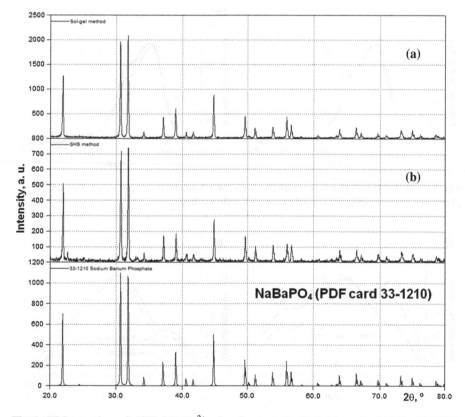

Fig. 1 XRD spectra of NaBaPO$_4$:Eu^{2+} phosphors synthesized via sol-gel (**a**) and SHS (**b**) procedures

spectra measured by detectors of 420, 460 and 480 nm light are illustrated in Fig. 3. The shapes of luminescence and excitation spectra markedly vary depending on the excitation and detection wavelength correspondingly, with this variation being more prominent for the sample prepared via sol-gel route.

The above luminescence and excitation spectra were deconvoluted into separate Gaussian bands. The exemplary deconvolutions of luminescence and excitation spectra are shown in Figs. 4 and 5 correspondingly.

According to [7], the luminescence and excitation spectra of $NaBaPO_4:Eu^{2+}$ phosphors are expected to include two bands corresponding to the luminescence and excitation of Eu^{2+} ions present in $NaBaPO_4$ lattice in the locations with coordination numbers (CN) 10 and 12. However, the spectra shown in Figs. 4 and 5 indicate more than two Gaussian bands in the luminescence and excitation spectra. The luminescence spectra involve four Gaussian bands at 399 ± 8, 423 ± 3, 458 ± 2 and 476 ± 2 nm. The excitation spectra also include four Gaussian bands at 274 ± 4, 322 ± 3, 345 ± 3 and 372 ± 5 nm. In view of the single phase nature of the samples the presence of additional bands cannot be accounted for admixture

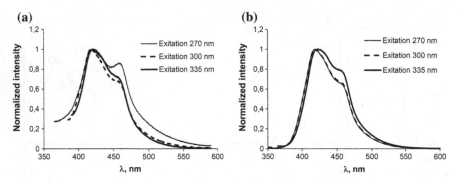

Fig. 2 270, 300 and 335 nm wavelengths stimulated luminescence spectra of $NaBaPO_4:Eu^{2+}$ phosphors synthesized using sol-gel (**a**) and SHS (**b**) methods

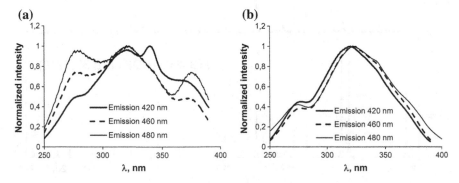

Fig. 3 Excitation spectra of $NaBaPO_4:Eu^{2+}$ phosphors synthesized using sol-gel (**a**) and SHS (**b**) methods measured at detector emission wavelengths 420, 460 and 480 nm

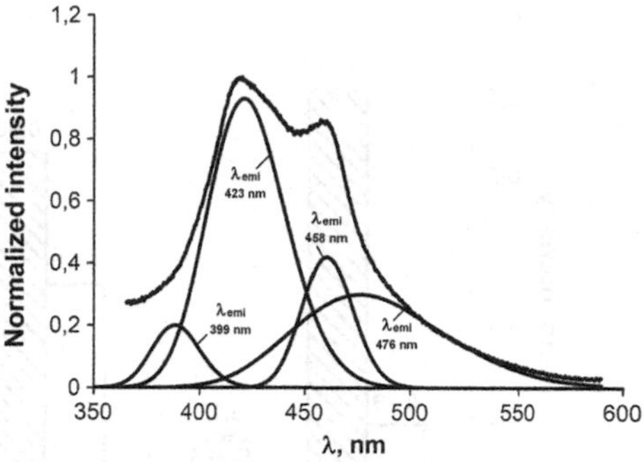

Fig. 4 Deconvolution of
sol-gel synthesized NaBaPO$_4$:
Eu^{2+} phosphor luminescence
spectrum into Gaussian bands

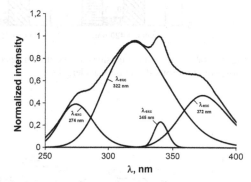

Fig. 5 Deconvolution of
sol-gel synthesized NaBaPO$_4$:
Eu^{2+} phosphor excitation
spectrum into Gaussian bands

phases. Moreover, neither of the considered bands in both luminescence and excitation spectra relates to Eu^{3+} ions. Therefore, all these bands can only correspond to Eu^{2+} ions in different spatial environment in NaBaPO$_4$ crystal lattice.

In order to identify the above Gaussian bands the intensity of excitation spectrum bands was plotted and analyzed as a function of emission wavelengths at which the spectra were measured (Fig. 6).

As shown in Fig. 6, the band at 274 nm in the excitation spectrum is mostly intensive at the emission wavelength 480 nm, i.e. near the luminescence band with the peak at 476 nm suggesting that these excitation and luminescence bands are interrelated and correspond to the same luminescence center (because NaBaPO$_4$: Eu^{2+} is featured with an intra-center luminescence [7]). The highest intensity of 322 nm band in the excitation spectrum is observed at the detection wavelength 420 nm located near the luminescence band at 423 nm also suggesting that these bands relate to a certain luminescence center.

As considered above Eu^{2+} can occupy two positions with CN 10 and 12 in NaBaPO$_4$ crystal lattice. Under the effect of crystal field 4f^{6}5d^1 level of Eu^{2+} ion is cleaved into several sub-levels separated by the gap growing with the increase of the crystal field intensity [8]. Since the position with CN 12 is featured with a higher field intensity compared with CN 10, the excitation band at λ_{exc} = 274 nm and the corresponding luminescence band at λ_{emi} = 476 nm relate to Eu^{2+} ions occupying the position with CN 12 (with a higher gap between sublevels), whereas

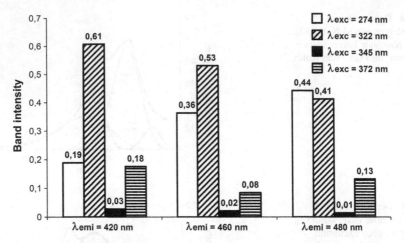

Fig. 6 Ratios of band intensities in excitation spectra of sol-gel synthesized $NaBaPO_4:Eu^{2+}$ phosphor at various detection wavelengths

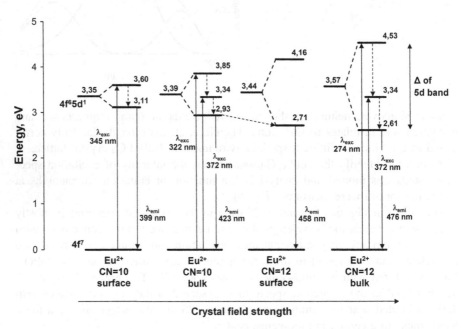

Fig. 7 Supposed structure of luminescence centers in $NaBaPO_4:Eu^{2+}$ phosphors

the excitation and emission bands at $\lambda_{exc} = 322$ nm and $\lambda_{emi} = 423$ nm probably correspond to Eu^{2+} ions in CN 10 (with lower gap between sublevels). This conclusion is illustrated in Fig. 7.

The initial (non-cleaved) $4f^6 5d^1$ level of Eu^{2+} ion can be determined from the excitation and emission levels (i.e. the corresponding wavelengths and energies) as their arithmetic mean. The corresponding calculations illustrated in Fig. 7 indicate that Eu^{2+} locate in the position with CN 12 is featured with a higher $4f^6 5d^1$ level energy compared with Eu^{2+} ion in CN 10 position. This can be accounted for the nephelauxetic effect of electron cloud growth upon overlapping with orbital of the neighboring atoms [9]. From this viewpoint, CN 12 position provides a more prominent electron cloud increase and correspondingly higher $4f^6 5d^1$ non-cleaved level energy compared with CN 10 position, as supported by the above data.

From the energy of $4f^6 5d^1$ level, it can be defined what other bands in luminescence and excitation spectra correspond to levels symmetrical to this non-cleaved one. According to Fig. 7, such bands include λ_{exc} at 345 nm and λ_{emi} at 399 nm. These weak bands likely related to Eu^{2+} ions located on the surface of $NaBaPO_4$ crystal. A weak crystal field and consequently low level cleavage in the surface layer determine their low intensity (Figs. 4, 5 and 6).

The excitation spectra also involve a band at λ_{exc} = 372 nm. As shown in Fig. 6, its intensity is minimal at the detector wavelength 460 nm and grows at emission wavelengths 420 and 480 nm corresponding to Eu^{2+} ions in CN 10 and CN 12 locations. Probably, europium ions involve the corresponding energy levels, possibly intermediate sublevels resulting from $4f^6 5d^1$ level cleavage (Fig. 7).

The luminescence band λ_{emi} = 458 nm is mostly difficult to identify. No bands with intensities correlating with this one are observed in excitation spectra. As shown in Fig. 6, the maximum intensity of this band is achieved upon excitation with the wavelength λ_{exc} = 322 nm corresponding to a luminescence center connected with Eu^{2+} (CN 10) and suggesting a possible energy transfer probably from Eu^{2+} ion (CN 10) to this center. According to the phosphor composition, the luminescence center with λ_{emi} = 458 nm also relates to Eu^{2+} ion, supposedly located on the surface. In this case, the luminescence center with λ_{exc} = 345 nm and λ_{emi} = 399 nm can correspond to the surface Eu^{2+} with CN 10 whereas the center with λ_{emi} = 458 nm associates with the surface Eu^{2+} ion with CN 12 (Fig. 7). Since the lowest energy level of this ion resulting from $4f^6 5d^1$ level cleavage is intermediate between the lowest levels for Eu^{2+} (CN 10) and Eu^{2+} (CN 12), the highest energy level of this center can be also intermediate between the corresponding highest levels, i.e. about 4.16 eV (Fig. 7). Therefore, the initial non-cleaved $4f^6 5d^1$ of this center should also be intermediate between the energies for Eu^{2+} (CN 10) and Eu^{2+} (CN 12) in accordance with the nephelauxetic effect, as confirmed by the observed data.

Thus the obtained data allowed us to define an energy scheme of luminescence centers determined in $NaBaPO_4:Eu^{2+}$ phosphors with the indication of energy transitions corresponding to specific bands in their luminescence and excitation spectra.

4 Conclusion

Finely dispersed single phase $NaBaPO_4:Eu^{2+}$ phosphors are synthesized and studies in respect of their luminescence and excitation spectra with the identification of all the determined Gaussian bands. In addition to already known luminescence centers represented by Eu^{2+} ions occupying positions with coordination numbers 10 and 12 in $NaBaPO_4$ crystal lattice, these phosphors comprise two other types of luminescence centers probably corresponding to europium ions located in the surface with the same coordination numbers.

Acknowledgment The reported study was funded by RFBR according to the research project No. 16-33-00998 мол_а.

References

1. Vorob'ev, S.I.: Colloidal-chemical characteristics of perfluorocarbon emulsions. Pharm. Chem. J. **41**(11), 614–619 (2007)
2. Bakhmetyev, V., Sychov, M., Orlova, A., Potanina, E., Sovestnov, A., Kulvelis, Yu.: Nanophosphors for Roentgen photodynamic therapy of oncological diseases. Nanoindustry **8**, 46–50 (2013)
3. Minakova, T.S., Sychov, M.M., Bakhmetyev, V.V., Eremina, N.S., Bogdanov, S.P., Zyatikov, I.A., Minakova, LYu.: The Influence of $Zn_3(PO_4)_2$: Mn—Luminophores Synthesis Conditions on their Surface and Luminescent Features. Adv. Mater. Res. **872**, 106–111 (2013)
4. Mjakin, S.V., Minakova, T.S., Bakhmetyev, V.V., Sychov, M.M.: Effect of the Surfaces of $Zn_3(PO_4)_2:Mn^{2+}$ Phosphors on their Luminescent Properties. Russ. J. Phys. Chem. A **90**(1), 240–245 (2016)
5. Bakhmetyev, V.V., Lebedev, L.A., Malygin, V.V., Podsypanina, N.S., Sychov, M.M., Belyaev, V.V.: Effect of composition and synthesis route on structure and luminescence of $NaBaPO_4:Eu^{2+}$ and $ZnAl_2O_4:Eu^{3+}$. JJAP Conf. Proc. **4**, 011104-1–011104-6 (2016)
6. Jiayue, Sun, Xiangyan, Zhang, Haiyan, Du: Combustion synthesis and luminescence properties of blue $NaBaPO_4:Eu^{2+}$ phosphor. J. Rare Earths **30**(2), 118–122 (2012)
7. Zhang, S., Huang, Y., Nakai, Y., Tsuboi, T., Seo, H.J.: The luminescence characterization and thermal stability of Eu^{2+} ions-doped $NaBaPO_4$ phosphor. J. Am. Ceram. Soc. **94**(9), 2987–2992 (2011)
8. Huang, C.-H., Wu, P.-J., Lee, J.-F., Chen, T.-M.: (Ca, Mg, Sr)9Y(PO4)7:Eu^{2+}, Mn^{2+}: Phosphors for white-light near-UV LEDs through crystal field tuning and energy transfer. J. Mater. Chem. **21**, 10489–10495 (2011)
9. Jang, H.S., Won, Y.-H., Vaidyanathan, S., Kim, D.H., Jeon, D.Y.: Emission band change of $(Sr_{1-x})_3SiO_5:Eu^{2+}$ (M = Ca, Ba) phosphor for white light sources using blue/near-ultraviolet LEDs. J. Electrochem. Soc. **156**(6), J138–J142 (2009)

The Effect of Cutting Edge Sharpness on Cutting Characteristic of Polycarbonate

Yuki Kurita, Katsuhiko Sakai and Hiroo Shizuka

Abstract Polycarbonate is used in various fields because it exhibits high-impact resistance and light-weight properties. Products requiring high flatness are made using an end mill. When the cutting edge of an end mill becomes dull, the surface quality worsens. This is a serious problem. In this experiment, sharp and worn tools were prepared to investigate the effects of an edge's sharpness on the cutting characteristics of polycarbonate. The cutting force was measured, and the finished surface was observed. When the sharpness of the cutting edge was changed from 1 to 12 μm, the cutting force increased 20 % and the quality of the finished surface greatly worsens. In conclusion, small wear of a cutting edge, e.g. within 11 μm, degrades a surface's finish and transparency.

Keywords Cutting · Polycarbonate · Cutting edge sharpness

1 Introduction

Various kinds of resin are widely used as industrial materials because of their high chemical resistance, high corrosion resistance and light-weight properties. In comparison to other resins, polycarbonate exhibits impact resistance, heat resistance, flame-retardancy and transparency. Therefore, polycarbonate is used in optical components, automobile parts and casings for household electric appliances.

Y. Kurita (✉) · K. Sakai · H. Shizuka
Shizuoka University, 3-5-1 Johoku Naka-ku, Hamamatsu, Japan
e-mail: kurita.yuhki.15@shizuoka.ac.jp

K. Sakai
e-mail: tksakai@ipc.shizuoka.ac.jp

H. Shizuka
e-mail: tkshizu@ipc.shizuoka.ac.jp

© Springer International Publishing AG 2017
R. Jabłoński and R. Szewczyk (eds.), *Recent Global Research and Education: Technological Challenges*, Advances in Intelligent Systems and Computing 519, DOI 10.1007/978-3-319-46490-9_8

Generally, these products are made by injection molding; however, it is impossible to obtain high flatness in thin components because the heat required by the process causes deformations in those components. Consequently, products requiring high flatness, e.g. protection films for smart phones and displays in car navigation systems, are made by a trimming process using an end mill from a thin plate. Polycarbonate is a softer material than metal; thus, it is easy to cut, and the tool wear rate is very slow. However, polycarbonate's melting point is very low (220 °C [1]); consequently, facile melting of the workpiece makes it difficult to obtain a good surface finish, and cracks form on the finished surface. To establish a precision machining technology using resin, the cutting mechanisms of resin must first be understood, and the appropriate cutting conditions and geometric designs for the cutting tools must be determined. To date, only a few studies have been conducted on the cutting of resin materials [2–4]. Currently, the cutting of resin at manufacturing sites is only performed on a trial-and-error basis. In this study, the orthogonal cutting of polycarbonate was performed; the effects of the sharpness of the cutting edge and the cutting speed on cutting characteristics were investigated. In addition, the basic cutting phenomenon of polycarbonate was described.

2 Experimental Setup and Work Material

Figure 1 shows the experimental apparatus. In this study, a general-purpose lathe equipped with a continuously variable transmission was used for varying the cutting speed and performing orthogonal cutting. A polycarbonate disk was mounted on a main spindle using a fixing jig. Uncoated sintered carbide tools were used in the experiments. The cutting force was measured with a dynamometer (Kistler 9227)

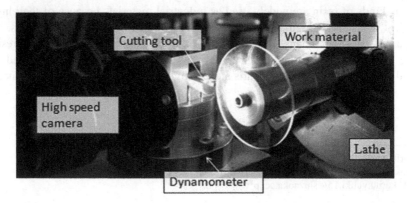

Fig. 1 Experimental setup

and analysed by a personal computer with software (Kistler Dyno Ware). The finished surfaces and chips were observed with a video microscope and the chip formation during cutting was recorded with a high-speed video camera.

3 Effect of the Cutting Edge's Sharpness

3.1 Experimental Purpose

Since the hardness of polycarbonate is lower than that of general metal materials, a very sharp cutting tool was used (roundness of edge was approximately 1 μm). However, when the cutting edge became dull from tool wear, the surface quality worsened because of melting of work piece due to cutting temperature rising and deflection of cutting tool due to increase in cutting force. Therefore, orthogonal cutting of the polycarbonate material was performed to investigate how the sharpness of the cutting edge affected the cutting force and surface finish; cutting was performed with both sharp and worn tools. Furthermore, decreasing the sharpness of the cutting edge was expected to change the internal stress distribution around the cutting point during cutting. A photo-elastic experiment was performed to observe how the sharpness of the cutting edge affected the internal stress distribution around the cutting point.

3.2 Experimental Method

Worn tools used for this experiment were made using a tool grinding machine (Matsuzawa Factory MZ-3BG) and a wrapping machine (Nano Factor FACT-200). The shape of the cutting edge was measured with a laser microscope (KEYENCE VK-X200). The experimental conditions are listed in Table 1, and these conditions were derived from the practical conditions employed at production sites.

Table 1 Cutting conditions

Cutting speed (m/min)	100, 200, 400, 800
Feed rate (mm/rev)	0.1
Rake angle (deg)	20
Relief angle (deg)	15
Roundness of cutting edge (μm)	1, 12, 22, 50
Workpiece	Polycarbonate
Thickness of workpiece (mm)	3
Cutting tool	Carbide

4 Experimental Results

4.1 Finished Surface Characteristics

When the cutting edge became dull by tool wear, the surface quality worsened. Therefore, the effects of the cutting edge's sharpness on the surface finish were investigated using different cutting edge tools with variable sharpness. Figure 2 shows the finished surfaces for cutting edges of variable sharpness; the cutting speed used in these experiments was 100 m/min. The figure illustrates that the finished surface worsened as the sharpness of the cutting edge decreased. This result is probably caused by the deformation of work material caused by lost cutting quality and an increase in cutting force.

4.2 Cutting Force

The experiment described in the preceding section verified that the finished surface worsened as the sharpness of the cutting edge decreased. It is also likely that the

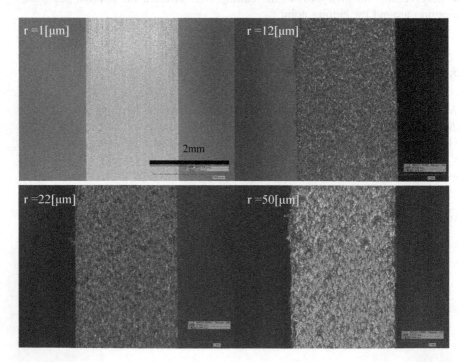

Fig. 2 Finished surfaces of cutting edges(r) having variable sharpness; cutting speed was 100 m/min

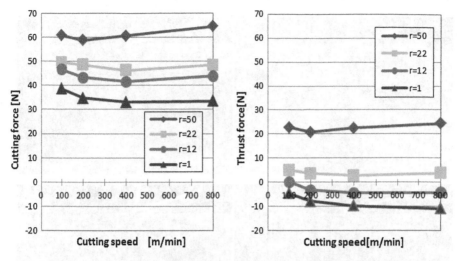

Fig. 3 Relationship between the sharpness of the cutting edge and the cutting resistance

cutting force will be affected by the sharpness of the cutting edge. Therefore, the effect of the sharpness of a cutting edge on the cutting force was investigated. Figure 3 shows the relationship between the sharpness of a cutting edge and cutting resistance at different cutting speeds. The graph illustrates that the cutting force and thrust force increased as the sharpness of the cutting edge decreased. This was probably caused by deformation of the work material and an increased friction due to an increase in the contact area between the tool and the workpiece. Also, the cutting force and thrust force significantly changed even with small changes in the sharpness of the cutting edge. Therefore, a small degree of tool wear results in the degradation of the surface finish due to tool vibration and bending.

4.3 Photo-Elastic Experiment

When the sharpness of the cutting edge changed, the cutting force changed significantly. The internal stress is likely affected by the sharpness of the cutting edge. Therefore, the internal stress was observed with a photo-elastic experiment. Figure 4 shows an isochromatic line picture for cutting edges with variable sharpness; a cutting speed of 100 m/min was employed. The figure shows that the fringes in front of the tool are apt to spread in a deeper part of the workpiece, and the stressed area expands as the sharpness of the cutting edge increases. This causes degradation of the surface finish due to an increase in the workpiece's internal strain. Also, the density of fringes at the tip of the cutting tool increased as the sharpness of the cutting edge decreased. This causes cracking of the machined surface due to a higher stress concentration at the tool's tip.

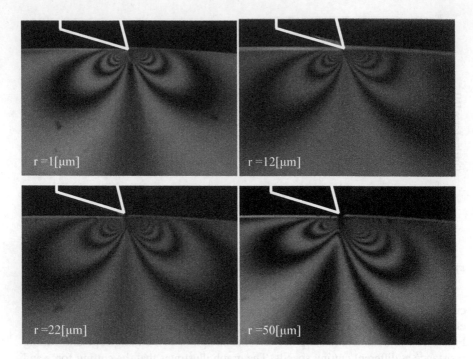

Fig. 4 Isochromatic line picture of cutting speed 100 m/min

5 Summary

In this paper, the effects of the sharpness of a cutting edge on cutting characteristics was studied using orthogonal cutting of polycarbonate with cutting edge tools of variable sharpness. The findings of the experiment are presented herein.

1. Even a slight decrease in the cutting edge's sharpness greatly changed the cutting resistance. When the sharpness of the cutting edge was changed from 1 to 12 μm, the cutting force increased by 20 %.
2. A finished surface is greatly affected by the sharpness of a cutting edge. Therefore, slight tool wear causes degradation of the finished surface.
3. A decrease in a cutting edge's sharpness causes higher stress at the tool tip and an expansion of the internal stress area.

References

1. Akitoshi, H.: Plastic properties introduction to the molding technician. Nikkan Kogyo Shimbun
2. Akio, K.: Study on cutting of plastic. JSPE J. **27**(322), 726–732 (1961)

3. Akira, K.: Machinability of plastics. Soc. Polym. Sci. **14**(161), 660–669 (1965)
4. Katsumasa, S.: Mechanism of chip formation in plastics. Trans. JSME **31**(224) (322), 657–671 (1969)

Investigation of the Magnetoelastic Villari Effect in Steel Truss

Dorota Jackiewicz, Maciej Kachniarz and Adam Bieńkowski

Abstract The following paper presents results of investigation of the magnetoelastic Villari effect in steel truss. Construction of the specially developed steel truss allowed to investigate three identical steel elements at the same time, where one element was subjected to the tensile stress and two elements were influenced by the compressive stress. For all three elements the magnetic characteristics for different values of applied mechanical stress were measured. Purpose of the performed investigation was experimental validation of the theory of magnetoelastic Villari effect for steel. For tensile stress initial increase of the magnetic flux density was observed up to the Villari reversal point and then values of magnetic flux density started to decrease. For compressive stress only decrease of the magnetic flux density was observed. Obtained experimental results confirmed theoretical description of the Villari magnetoelastic effect.

Keywords Magnetism · Magnetoelastic villari effect · Villari reversal point · Constructional steel · Steel truss

1 Introduction

Magnetomechanical effects are group of physical phenomenon connecting magnetic and mechanical properties of the solid matter. They can be applied in both actuators (e.g. magnetostriction) [1, 2] and sensors (e.g. magnetoelastic effect) [3, 4] tech-

D. Jackiewicz (✉) · A. Bieńkowski
Institute of Metrology and Biomedical Engineering, Warsaw University
of Technology, Sw. Andrzeja Boboli 8, 02-525 Warsaw, Poland
e-mail: d.jackiewicz@mchtr.pw.edu.pl

M. Kachniarz
Industrial Research Institute for Automation and Measurements,
Al. Jerozolimskie 202, 02-486 Warsaw, Poland
e-mail: mkachniarz@piap.pl

© Springer International Publishing AG 2017
R. Jabłoński and R. Szewczyk (eds.), *Recent Global Research and Education:
Technological Challenges*, Advances in Intelligent Systems and Computing 519,
DOI 10.1007/978-3-319-46490-9_9

nology. Therefore magnetomechanical effects are intensively investigated and new applications are constantly developed.

The most fundamental magnetomechanical effect is magnetostriction known also as Joule effect, which involves change of dimensions of material under the influence of magnetic field [5]. Effect thermodynamically opposite to the magnetostriction is magnetoelastic Villari effect called also inverse magnetostrictive effect. Villari effect is phenomenon of changing magnetic state under the influence of the mechanical stress [6]. The magnetoelastic effect creates the possibility of measuring mechanical force or stress by measuring magnetic parameters of the material.

Magnetoelastic Villari effect was intensively investigated in ferrites [7, 8] and amorphous alloys [9, 10], where theoretical description of the phenomenon was confirmed by the experimental results. However, for steel materials, which are very important for the modern industry, there is still little research results on their magnetoelastic properties. In this paper, special steel truss is presented, which allows to investigate magnetoelastic properties of the steel elements for both tensile and compressive stress at the same time. Elements made of 13CrMo4-5 constructional steel were investigated in order to validate correctness of theoretical description of Villari effect for steel materials.

2 Magnetoelastic Villari Effect

Magnetoelastic effect was discovered by Italian physicist Emilio Villari in 1865. He observed change of magnetic flux density B within the ferromagnetic material under the influence of mechanical stress. Typical characteristics of dependence of magnetic flux density B on the mechanical stress σ are presented in Fig. 1.

As it was previously reported, the character of changes of magnetic flux density B under the influence of mechanical stress σ is determined by the sign of factor $\lambda_s \sigma$,

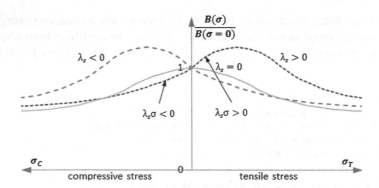

Fig. 1 Typical characteristics of dependence of magnetic flux density B on the mechanical stress σ for the material with positive $(\lambda_s > 0)$, negative $(\lambda_s < 0)$ and zero $(\lambda_s = 0)$ saturation magnetostriction coefficient λ_s [11]

where λ_s is saturation magnetostriction coefficient [11], which results from the equation:

$$\left(\frac{\partial B}{\partial \sigma}\right)_H = \left(\frac{\partial \lambda}{\partial H}\right)_\sigma \tag{1}$$

where H is magnetizing field. For the positive value of $\lambda_s \sigma$ factor magnetic flux density B increases with the increase of applied stress σ. For example in material with positive saturation magnetostriction λ_s magnetic flux density B will increase for tensile stress σ_T, which is commonly considered as a positive one as opposed to the negative compressive stress σ_C. The same process will occur for material with negative λ_s under compressive stress σ_C. When the sign of the factor $\lambda_s \sigma$ is negative, decrease in value of the magnetic flux density B is observed. Each of characteristics presented in Fig. 1 exhibits maximum for certain value of applied stress σ. This point, where $(\partial B/\partial \sigma)_H = 0$, is known as Villari reversal point [7]. After reaching this point material reacts for further increase of stress σ like the one with opposite sign of saturation magnetostriction λ_s (material with positive λ_s reacts like the one with negative λ_s and vice versa).

3 Developed Steel Truss

For the purpose of performed investigation special steel truss with three replaceable identical elements was developed, presented in Fig. 2. Three central elements K1, K2 and K3 are investigated steel elements. They have significantly smaller cross-sectional area than rest of the elements in order to allow performing investigation without damaging entire truss. K1 element is influenced by tensile stress σ_T while K2 and K3 elements are subjected to the identical compressive stress σ_C.

Fig. 2 3D model of the developed steel truss, K1, K2, K3—investigated steel elements with magnetizing and sensing windings on them, F—force applied to the truss

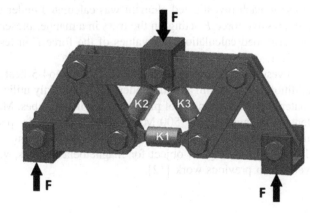

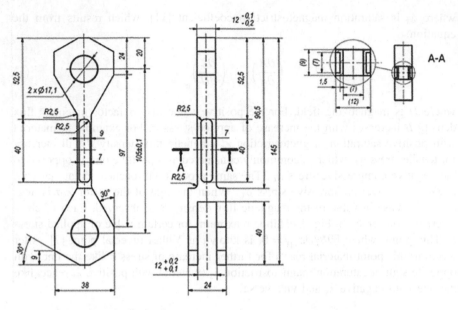

Fig. 3 Design of the single investigated element of the developed steel truss

Single investigated element of the truss was frame-shaped magnetic sample presented in Fig. 3. Such shape allowed to obtain nearly uniform distribution of applied stress σ in the columns of the sample as well as closed magnetic circuit within the element, which is important due to reduction of the demagnetization effect. On the columns of each investigated element two sets of windings were made: 260 turns of magnetizing winding and 100 turns of sensing winding, which allowed to investigate magnetic characteristics of the elements for given value of applied stress σ.

On the basis of truss geometry and dimensions of the investigated elements, stress in each investigated element was calculated under the influence of specified compressive force F acting on the truss in a manner presented in Fig. 1. The results of performed calculations for values of the force F in test points are presented in Table 1.

Investigated elements were made of 13CrMo4-5 heat resistant alloy steel with addition of chromium and molybdenum. It is mostly utilized in energy industry as a material for development of pipe lines and heater tubes. Material is characterized by tensile strength about 500 MPa, relative magnetic permeability $\mu = 250$ and magnetostriction coefficient about $\lambda_s = +25$ μm/m. Such properties make 13CrMo4-5 steel good object for magnetoelastic tests, which was experimentally verified in previous work [12].

4   Experimental Setup

The schematic block diagram of the experimental setup developed for the purpose of the performed experiment is presented in Fig. 4. Main component of the experimental setup is Automated System for Testing Ferromagnetic Materials developed in Institute of Metrology and Biomedical Engineering [13]. The system is controlled by PC with National Instruments Data Acquisition Card (DAQ) installed. Special software, developed in NI LabVIEW environment is controlling the measurement equipment and collecting measurement data. Magnetizing voltage waveform is generated from DAQ and converted to the current waveform with voltage-current converter KEPCO BOP 36-6 M. Current waveform is driving magnetizing winding, so the magnetizing field acts on the steel sample. As a result voltage is induced in the sensing winding, which is measured and converted to the magnetic flux density B with the fluxmeter Lakeshore 480.

Compressive force F acting on the truss was generated with oil hydraulic press. Value of the force was measured with strain gauge force sensor and calculated to the value of stress for each investigated element according to Table 1.

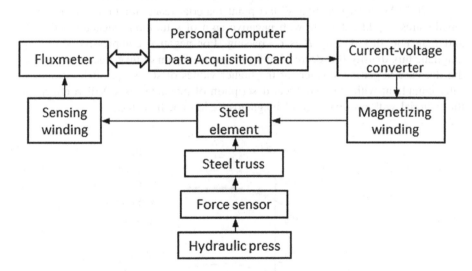

Fig. 4 Schematic block diagram of the experimental setup

Table 1 Tensile (σ_T) and compressive (σ_C) mechanical stress applied to the investigated elements of the truss for the compressive force F acting on the truss

F (kN)	0.6	1.1	1.7	2.2	2.8	3.3	3.9	4.4	5.0	5.5	6.9	8.3	9.7	11.1	12.5
σ_T (MPa)	20	40	60	80	100	120	140	160	180	200	250	300	350	400	450
σ_C (MPa)	13	27	40	53	67	80	93	107	120	133	167	200	233	267	300

5 Experimental Results

During the performed investigation, developed steel truss was subjected to the compressive force F of value within the range from 0 to 12.5 kN, which resulted in values of tensile stress σ_T acting on element K1 from 0 to 450 MPa and compressive stress σ_C acting on elements K2 and K3 from 0 to 300 MPa. For each value of applied force F magnetic characteristics of the K1, K2 and K3 elements were measured for several values of magnetizing field amplitude H_m. As a result, family of $B(\sigma)_H$ magnetoelastic characteristics was obtained for each investigated element, which are presented in Fig. 5. For positive stress values (tensile stress σ_T) results were obtained on K1 element. Results for negative compressive stress σ_C obtained on the elements K2 and K3 were identical as their positions are symmetrical to the direction of force F vector, so results for only one of this elements are presented in Fig. 5.

Obtained magnetoelastic $B(\sigma)_H$ characteristics are consistent with the theoretical description of the Villari effect. Investigated 13CrMo4-5 steel is characterized by positive value of the magnetostriction coefficient ($\lambda_s = + 25$ µm/m), so for the positive tensile stress σ_T, increase of the magnetic flux density B of the material can be observed up to the Villari reversal point, which for investigated steel is about $\sigma_V = 120$ MPa. After reaching Villari point material reacts for further increase of tensile stress σ_T like the one with negative sign of saturation magnetostriction λ_s and magnetic flux density B is decreasing. For negative compressive stress σ_C magnetic flux density B is decreasing for the entire region of investigated stress, but the decrease is getting smaller for the higher values of stress σ_C. Such results are fully compliant with the theoretical description of magnetoelastic Villari effect for the material with positive sign of magnetostriction coefficient λ_s.

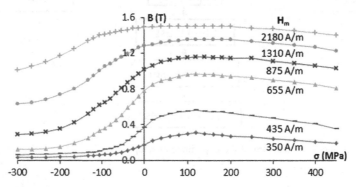

Fig. 5 Magnetoelastic $B(\sigma)_H$ characteristics for compressive and tensile stress of investigated steel elements made of 13CrMo4-5 constructional steel

6 Conclusion

Experimental results presented in the following paper indicates, that theoretical description of the magnetoelastic Villari effect is correct for investigated 13CrMo4-5 steel characterized by positive magnetostriction coefficient. For tensile stress initial increase of the magnetic flux density was observed up to the Villari reversal point and then values of the parameter started to decrease. For compressive stress only decrease of the magnetic flux density was observed.

Developed steel truss with replaceable elements presented in the paper is very important contribution to the investigations on magnetoelastic properties of steel materials. It allows to test elements made of any kind of steel for the influence of both tensile and compressive stress at the same time. Moreover, performing the investigation does not lead to destruction of the entire truss. Only investigated elements are damaged, while rest of the construction is unaffected.

Results of the performed experiment indicates the possibility of application of the magnetoelastic Villari effect in the Non-Destructive Testing (NDT) of the elements made of constructional steel materials. Magnetoelastic-based methodology of NDT could be useful instrument for stress assessment in steel construction elements.

Acknowledgments This work was partially supported by the statutory founds of Institute of Metrology and Biomedical Engineering, Warsaw University of Technology (Poland).

References

1. Zhang, T., Jiang, T., Zhang, H., Xu, H.: Giant magnetostrictive actuators for active vibration control. Smart Mater. Struct. **13**, 473–477 (2004)
2. Braghin, F., Cinquemani, S., Resta, F.: A model of magnetostrictive actuators for active vibration control. Sens. Actuator A Phys. **165**, 342–350 (2011)
3. Salach, J., Bieńkowski, A., Szewczyk, R.: The ring-shaped magnetoelastic torque sensors utilizing soft amorphous magnetic materials. J. Magn. Magn. Mater. **316**, E607–E609 (2007)
4. Bieńkowski, A., Szewczyk, R., Salach, J.: Industrial application of magnetoelastic force and torque sensor. Acta Phys. Pol. A **118**, 1008–1009 (2010)
5. Tumański, S.: Handbook of magnetic measurements. CRC Press, New York (2011)
6. Bozorth, R.: Ferromagnetism. Van Nostrand, New York (1951)
7. Bieńkowski, A.: magnetoelastic villari effect in Mn-Zn ferrites. J. Magn. Magn. Mater. **215–216**, 231–233 (2000)
8. Szewczyk, R.: Modelling the magnetic and magnetostrictive properties of high-permeability Mn-Zn ferrites. PRAMANA. J. Phys. **67**, 1165–1171 (2006)
9. Bieńkowski, A., Szewczyk, R., Kulik, T., Ferenc, J., Salach, J.: Magnetoelastic properties of HITPERM-type $Fe_{41.5}Co_{41.5}Cu_1Nb_3B_{13}$ nanocrystalline alloy. J. Magn. Magn. Mater. **304**, E624–E626 (2006)
10. Meydan, T., Oduncu, H.: Enhancement of magnetostrictive properties of amorphous ribbons for a biomedical application. Sens. Actuator A Phys. **59**, 192–196 (1997)

11. Švec Sr., P., Szewczyk, R., Salach, J., Jackiewicz, J., Švec, P., Bieńkowski, A., Hoško, J.: Magnetoelastic properties of selected amorphous systems tailored by thermomagnetic treatment. J. Electr. Eng. **65**, 259–261 (2014)
12. Jackiewicz, D., Kachniarz, M., Rożniatowski, K., Dworecka, J., Szewczyk, R., Salach, J., Bieńkowski, A., Winiarski, W.: Temperature resistance of magnetoelastic characteristics of 13CrMo4-5 constructional steel. Acta Phys. Pol., A **127**, 614–616 (2015)
13. Urbański, M., Charubin, T., Rozum, P., Nowicki, M., Szewczyk, R.: Automated system for testing ferromagnetic materials. Adv. Intell. Syst. Comput. **440**, 817–825 (2016)

Atomic Force Microscopy Study of Contamination Process of Glass Surface Exposed to Oleic Acid Vapors

F. Samoila, A. Besleaga and L. Sirghi

Abstract In this study, we exposed glass substrates to acid oleic vapor (vapor pressure 9×10^{-5} Torr at 78 °C) for different periods of time (10–60 min). Before the exposure to oleic acid vapor, the glass substrates were cleaned and hydroxylated by low-pressure water vapor plasma. The contaminated surfaces were examined by atomic force microscopy, which revealed formation and growing of oleic acid nanodroplets on glass surfaces.

Keywords Vapor contamination · Oleic acid · Nanodroplet

1 Introduction

Surface contamination can occurs in many forms and may be present in a variety of states on the surfaces. Some examples of contamination sources are: machining oils and greases, hydraulic and cleaning fluids, adhesives, waxes, human contamination, and particulates [1]. Oleic acid is seldom used as a model contaminant material for study of surface contamination with hydrophobic oil molecules [2]. Oleic acid is a surface modifying agent of which molecule has a carboxyl group, a long alkyl chain with an unsaturated bond. In the process of contamination of silica surface, the carboxyl group may react with active groups on the surface (hydroxyl) until the alkyl chain and unsaturated bond provide a hydrophobic character of the surface [2].

In this work we investigate by atomic force microscopy the process dynamics of oleic acid vapor condensation on glass substrates kept at a temperature lower than the vapor temperature. This study has a double importance: theoretical, it gives physical insight on the process dynamics of silica surface wetting, and

F. Samoila (✉) · A. Besleaga · L. Sirghi
Faculty of Physics, Iasi Plasma Advanced Research Center (IPARC), Alexandru Ioan Cuza University of Iasi, Blvd. Carol I nr. 11, 700506 Iași, Romania
e-mail: samoila.f@yahoo.com

© Springer International Publishing AG 2017
R. Jabłoński and R. Szewczyk (eds.), *Recent Global Research and Education: Technological Challenges*, Advances in Intelligent Systems and Computing 519,
DOI 10.1007/978-3-319-46490-9_10

71

practical, because in many regions the pollution with organic species is an important problem. Garland et al. [3] studied the vapor deposition of oleic acid onto different types of surfaces (flat and spherical, hydrophilic and hydrophobic) and observed that the oleic acid forms nanodroplets rather than uniform coverage of surface. They have found that nanodroplets formation may depend on particle size, the amount of coating, referring to monolayer or multilayer deposition, and also on the mixture of the oleic acid and other compounds present under typical atmospheric conditions [3]. Katrib et al. [4] obtained layers of oleic acid from 2 to 30 nm in thickness for covering polystyrene latex particles of 101 nm.

2   Materials and Methods

Glass substrates (coverglasses from Agar scientific, 13 mm in diameter) were cleaned with ethanol, acetone and deionized water (each step for 15 min) in a sonication bath. In order to clean and activate their surfaces (generation of hydroxyl group on the surface) the substrates were exposed for 10 min to a d c discharge plasma in water vapor at low pressure [5]. Finally, the substrates were placed upside-down on the top of a Petri dish that contains oleic acid heated at 78 °C (Fig. 1).

After exposure to the oleic acid vapor for various time durations, the samples were loaded on the atomic force microscope for investigation of the surface morphology of contaminated surfaces. The AFM topography images were obtained in non contact tapping mode using a commercial atomic force microscope (XE70 from Park Systems, Korea). The AFM probe used was from μmasch HQ: NSC35/NoAl with a silicon tip with nominal curvature radius of 10 nm. The scanning parameters were: set-point amplitude, 4.23 nm, free amplitude, 20.22 nm, and frequency, 388.11 kHz.

Fig. 1 Schematic representation of the experimental setup

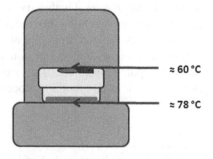

≈ 60 °C

≈ 78 °C

3　Results and Discussion

The AFM images acquired show that oleic acid forms nanodroplets on the hydrophilic glass surfaces and, with the increase of the exposure time, the number of the nanodroplets and their size increase (Fig. 2). Figure 3 presents the histograms of droplet height and diameter values obtained from analysis of 150 nanodroplets on the topography images presented in Fig. 2. After 10 min of exposure, we could not observe formation of droplets or theirs dimensions were smaller than the resolution power of the microscope. For 20 min, it can be seen the formation of nanodroplets (most probable height and diameter values of 5 nm and 190 nm, respectively). After 30 min of exposure, we can observe an increase in the number of nanodroplets, a noticeable increase of droplet height (most probable value of 15 nm) and a slight increase of droplet diameter (200 nm) (Fig. 3). For 60 min of exposure, we can see that the amount of oleic acid adsorbed onto the surface increased. The droplet diameter increased to 325 nm (most probable value), while the droplet height increased to 27.5 nm (most probable value).

These results are different from those reported by Garland et al. They observed also formation of acid oleic nanodroplets on flat hydrophilic surfaces of silica, but the dynamic of the contamination process was different. By the increase of the exposure time, the size of nanodroplets remained constant and only the number of nanodroplets increased. This difference may be caused by different experimental conditions. In our experiments, a difference between the substrate temperature and oil vapor temperature of about 20 °C was maintained constant during the contamination process, while in the experiments of Garland et al. the substrate and vapor had the same temperature. The difference of temperature between the substrate and vapors favored the condensation and growth of the nanodroplets. Also, in the present experiments the glass surfaces were cleaned and hydroxylated by plasma treatment, which increased their oleic acid contact angle values (macroscopic measurements, from 28 to 34°). A larger contact angle favored formation and growth of oleic acid nanodroplets during exposure of surfaces to oleic acid vapors.

| 10min | 20min | 30min | 60min |

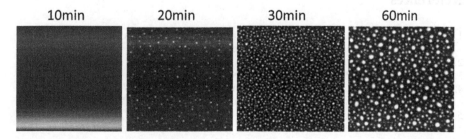

Fig. 2 AFM topography images ($10 \times 10 \ \mu m^2$) of oleic acid condensed on glass substrates for different duration time of exposure to oleic acid vapors

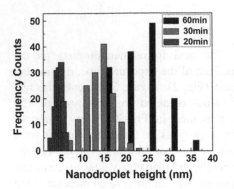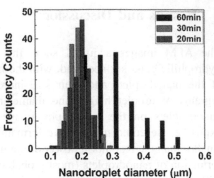

Fig. 3 Histograms of nanodroplet heights and diameter for different duration time of exposure to oleic acid vapors (statistics took into account a number of 150 nanodroplets)

4 Conclusion

The dynamics of the contamination of cleaned and hydrophilic glass surfaces exposed to oleic acid vapor was studied by atomic force microscopy investigation of the contaminated surfaces. Topography AFM images revealed that the oleic acid vapor condensed on the hydrophilic glass substrates and formed nanodroplets. The amount of oleic acid adsorbed on the substrates, i.e. the size and number of oleic acid nanodroplets, increased with the exposure time of substrate to the oleic acid vapor. This result differs from the results reported previously in literature for silica surface contamination with oleic acid, i.e. the size of droplets remained constant and only the number of droplets increased with the exposure time. The difference can be caused by the plasma pretreatment of glass surfaces and a difference between substrate and vapor temperatures in our experiment.

Acknowledgments This work was supported by CNCSIS, IDEI Research Program of Romanian Research, Development and Integration National Plan II, Grant no. 267/2011.

References

1. Kohli, R., Mittal, K.L. (eds.): Developments in surface contamination and cleaning—vol. 6: methods of cleaning and cleanliness verification (2013)
2. Chin. J. Chem. E **14**(6), 814 (2006)
3. Phys. Chem. Chem. Phys. **10**, 3156–3161 (2008)
4. Katrib, Y., Martin, S.T., Rudich, Y., Davidovits, P., Jayne, J.T., Worsnop, D.R.: Density changes of aerosol particles as a result of chemical reaction. Atmos. Chem. Phys. (2005)
5. Sirghi, L.: Rom. J. Phys. **56**, 144 (2011)

Thin Film Formation of the Polyvinylpyrrolidone-Added Europium Tetrakis (Dibenzoylmethide)-Triethylammonium and Its Mechanoluminescent Properties

R.A.D.M. Ranashinghe, Masayuki Okuya, Masaru Shimomura and Kenji Murakami

Abstract Mechanoluminescence (ML) is a light emission from materials induced by any mechanical stress. We have synthesized europium tetrakis (dibenzoylmethide)-triethylammonium (EuD$_4$TEA) and investigated its mechanoluminescent properties. The study has revealed that the ML intensity of EuD$_4$TEA is enhanced by an addition of polyvinylpyrrolidone (PVP). Recently we have succeeded to form a thin film of the PVP-added EuD$_4$TEA. The thin films with different amounts of the PVP addition are formed on an Al$_2$O$_3$ buffer layer prepared on Ni substrate. The present study has investigated a relationship between both the ML and the photoluminescence properties of thin films and the addition amount. The results suggest that the properties are strongly related to a crystallinity of the thin films.

Keywords Mechanoluminescence · Europium · Organic materials · Thin film

1 Introduction

Mechanoluminescence (ML) is a phenomenon that light emission induced by a mechanical action on a solid [1]. There are several forms of ML materials such as elastico-, plastic- and fracto-luminescence, usually denoted by a deformation of materials induced by mechanical stress [2]. It is suggested that a fracture of materials leads to break the crystal bonds along the planes with opposite charge,

R.A.D.M. Ranashinghe
Graduate School of Science and Technology, Shizuoka University, Johoku, Naka-Ku, Hamamatsu 432-8561, Japan

M. Okuya · M. Shimomura · K. Murakami (✉)
Graduate School of Integrated Science and Technology, Shizuoka University, Johoku, Naka-Ku, Hamamatsu 432-8561, Japan
e-mail: murakami.kenji@shizuoka.ac.jp

© Springer International Publishing AG 2017
R. Jabłoński and R. Szewczyk (eds.), *Recent Global Research and Education: Technological Challenges*, Advances in Intelligent Systems and Computing 519,
DOI 10.1007/978-3-319-46490-9_11

and a recombination of the charges emits a light [3]. The ML can be also observed by peeling the tape in a vacuum. [4].

Many ML materials have been synthesized based on the inorganic materials with different dopants such as $SrAl_2O_4$:Eu [5], $BaAl_2O_4$:Eu [6], $SrAl_2O_4$:Eu+3Dy+3 [7], ZnS:Mn+2 [8]. Almost all the inorganic ML materials have been synthesized by the very high temperature over 1000 °C. Recently, Fontenot et al. has proposed a europium tetrakis dibenzoylmethide triethylammonium (EuD4TEA) as an organic ML material [3, 9]. The first practical EuD4TEA ML material was synthesized by Hurt et al. [10].

From the point of view of practical applications, thin film formation of the ML materials is strongly required. Conventional ML thin films have been fabricated combining with binder or resin [11] which suppresses a direct transmission of the mechanical stresses.

In the present study, we have formed the organic ML thin film without any binder or resin by using the spray pyrolysis deposition technique. The organic ML material is fine-grained prior to the formation of thin film. High surface to volume ratio of the fine-grains could enhance a solid-solid reaction.

2 Experiment

Firstly, 3 mmol dibenzoylmethane(1,3-diphenylpropane-1,3dione) ($C_{15}H_{12}O_2$) (98.0 %, Wako) and 1.5 ml trimethylamine (99.0 %, Wako) were completely dissolved in 80 ml of ethanol (99.5 %, Wako) under vigorously stirring on the hot plate at 70 °C. Then 0.3 mol europium(iii) nitratehexahydrate (99.9 %, Wako) was added into the stirred solution and further stirred for 20 min until the solution became clear. The solution contained beaker was capped tightly and kept in the thermos to cool down slowly for overnight. Finally, the precipitated crystal (EuD$_4$TEA) was filtered under reduced pressure and washed several times with ethanol. In order to enhance a ML intensity of EuD$_4$TEA, different concentrations of polyvinylpirrolidone (pvp) (4, 7, 9, 11, 13 and 15 %w/w) were added in 40 ml ethanol solutions with 0.1 g synthesized EuD$_4$TEWA and sonicated for 5 min.

Thin films were formed as follows: 0.101 g Al_2O_3 was dissolved in 40 ml of ethanol and sprayed on Ni substrate heated at 400 °C by using spray gun. Then the synthesized EuD$_4$TEA with different amount of pvp was sprayed again after the Ni substrate with Al_2O_3 buffer layer was cooled down to 130 °C.

The formed thin films were characterized by using the X-Ray diffractometer (XRD, Rigaku RINT UltimaIII) with Cukα ($\lambda = 0.15418$ nm), the scanning electron microscopy (SEM, JEOL JSM-6320F) and the transmission electron microscopy (TEM, JEOL JEM-2100F). Mechanoluminescence and photoluminescence of the thin films were observed by using the multichannel spectroscope (Hamamatsu Photonics, PMA-12).

3 Results and Discussion

Figure 1 shows TEM image of the synthesized material. The image reveals that the material is well crystalized and composed of nanoparticles with a size of 2–5 nm in diameter. Inserted diffraction pattern gives a polycrystalline one.

Amount of the added pvp changes an intensity of photoluminescence of the synthesized material exited by the light of 357 nm as shown in Figs. 2 and 3. Spectra correspond to the emission from europium, Eu^{3+} with 592.7 nm ($^5D_0-^7F_1$), 612 nm ($^5D_0-^7F_2$), 651.2 nm ($^5D_0-^7F_3$) and 701.3 nm ($^5D_0-^7F_4$). The intensity increases as the pvp amount increases from 4 to 9 %w/w and keep constant to 11 % w/w followed by gradually decrease to 15 %w/w. The results suggest that the Eu^{+3} is coordinated with pvp molecules and electron transaction from ligand (LUMO) to metal is increased with an amount of pvp but further increase act as an impurity resulting in a suppress of the electron transaction.

Figure 4 shows the mechanoluminescence of thin film with 11 %w/w of pvp, which shows the similar emission to its photoluminescence. The intensity also depends on the amount of pvp in a same manner of that of photoluminescence.

The XRD measurements reveal that a crystallinity of the thin films is improved by an increase of the amount of pvp from 4 to 11 %w/w but it is deteriorated by further increase of pvp. The results suggest that the intensity of both PL and ML, namely electron transaction from the ligand to metal is strongly related to the crystallinity of thin films.

Fig. 1 TEM image of the synthesized EuD_4TEA

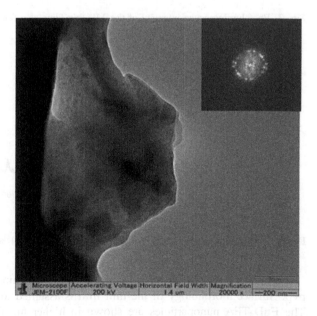

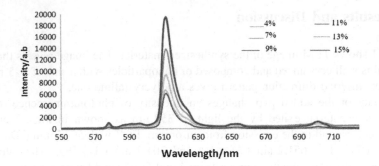

Fig. 2 Photoluminescence spectra of the ML thin films with different w/w% of pvp

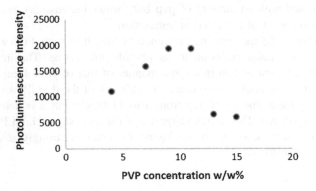

Fig. 3 Photoluminescence intensity of the ML thin films as a function of the pvp concentration

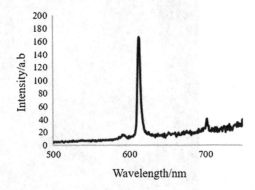

Fig. 4 Mechanoluminescence spectrum of the thin film with 11 %w/w of pvp

Figure 5 shows SEM micrographs of the ML thin films with 11 %w/w of pvp. Surface morphology of the thin film is assumed to reflect the pvp (Fig. 5a). The EuD$_4$TEA nanoparticles are shown in higher magnified SEM image (in the circle of Fig. 5b).

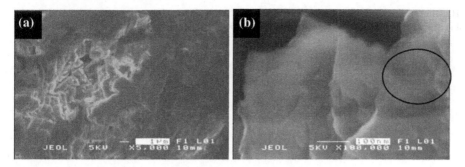

Fig. 5 SEM micrographs of the fabricated ML thin film with 11 % w/w of pvp

4 Conclusion

We are successfully formed the organic mechanoluminescent thin film without any binder or resin. Intensity of both the mechanoluminescence and the photoluminescence increases with the addition of polyvinylpirrolidone from 4 to 11 %w/w but further increase of pvp decreases the intensities due to the deteriorated crystallinity of material. Both the ML and the PL (excited by 357 nm) give emissions at 592.7, 612, 651.2 and 701.3 nm which correspond to $^5D_0$ to ^7F_n (n = 1, 2, 3, 4) electron transactions induced by the europium doping.

Acknowledgments This work was carried out under the Cooperative Research Project Program of Research Institute of Electronics, Shizuoka University.

References

1. Akiyama, M., et al.: Recovery phenomenon of mechanoluminescence from $Ca_2Al_2SiO_7$: Ce by irradiation with ultraviolet light. Appl. Phys. Lett. **75**(17), 2548–2550 (1999)
2. Chandra, B.P.: Mechanoluminescence. In: Vij, D.R. (ed.) Luminescence of Solids, pp. 361–389. Springer, New York (1998)
3. Fontenot, R., et al.: Luminescent properties of lanthanide dibenzoylmethide triethylammonium compounds. J. Theor. Appl. Phys. **7**(1), 1–10 (2013)
4. Chandra, V.K., Chandra, B.P.: Suitable materials for elastico mechanoluminescence-based stress sensors. Opt. Mater. **34**(1), 194–200 (2011)
5. Liu, Y., Xu, C.-N.: Influence of calcining temperature on photoluminescence and triboluminescence of europium-doped strontium aluminate particles prepared by sol–gel process. J. Phys. Chem. B **107**(17), 3991–3995 (2003)
6. Kaur, J., et al.: Optical properties of rare earth-doped barium aluminate synthesized by different methods—A review. Res. Chem. Intermed. **41**(4), 2317–2343 (2015)
7. Chandra, B.P., et al.: Mechanoluminescence glow curves of rare-earth doped strontium aluminate phosphors. Opt. Mater. **33**(3), 444–451 (2011)
8. Watanabe, T., Xu, C., Akiyama, M.: Triboluminescent inorganic material and a method for preparation thereof. Google Patents (2000)
9. Fontenot, R.S., et al.: J. Lumin. **132**, 1812–1818 (2012)

10. Hurt, C.R., et al.: Nature **212**, 179–180 (1966)
11. Takada, N., et al.: Mechanoluminescent properties of europium complexes. Synth. Met. **91** (1–3), 351–354 (1997)

Part II
Nanotechnology, Nanometrology, Nanoelectronics

Toward Room Temperature Operation of Dopant Atom Transistors

Michiharu Tabe, Arup Samanta and Daniel Moraru

Abstract As an extremely miniaturized Si transistor (MOSFET) close to the atomic scale, Dopant Atom Transistor is one of the promising candidates and the research field has been rapidly growing in the last decade. The dopant atom transistor consists of a dopant-induced quantum dot in the channel and its carrier transport is tunnelling of electrons or holes from the source to the drain through the single dopant atom. Until now, operation temperature has been mostly limited to 20 K or even below, since the ground state of the dopant is too shallow. In order to resolve this issue, we study potential deepening effect by application of a "cluster" or a "molecule" of dopant atoms, which is a number of dopant atoms closely gathering. As a result, it is shown that operation temperature approaches room temperature. In this work, a guiding principle for high temperature operation will be shown.

Keywords Single dopant · MOSFET · Dopant atom transistor · Single donor

1 Dopant Atom Transistors

In the last decade, several groups including our group have developed novel metal-oxide-semiconductor field-effect transistors (MOSFETs) dominated by a single dopant-atom [1–7], in which carrier transport mechanism is single-electron (or single-hole) tunneling via the single dopant-atom (Fig. 1). The single dopant-atom FET is promising for future electronics, because it could be miniaturized to an extreme limit close to atomic size, as well as because it offers the possibility of extremely low energy consumption due to a small number of carriers.

M. Tabe (✉) · A. Samanta · D. Moraru
Research Institute of Electronics, Shizuoka University,
3-5-1 Johoku, Naka-Ku, Hamamatsu 432-8011, Japan
e-mail: tabe.michiharu@shizuoka.ac.jp

© Springer International Publishing AG 2017
R. Jabłoński and R. Szewczyk (eds.), *Recent Global Research and Education:
Technological Challenges*, Advances in Intelligent Systems and Computing 519,
DOI 10.1007/978-3-319-46490-9_12

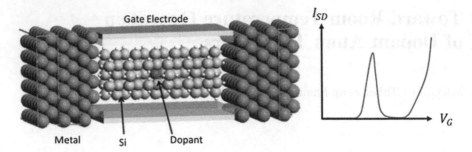

Fig. 1 Schematic view of a single dopant-atom FET and its characteristics

In most of the related papers, however, operation temperatures reported are limited to low temperatures, below ~20 K, since binding energy (BE) of the conventional single dopant is not large enough compared with the thermal energy $k_B T$ (≈25 meV) at 300 K and electrons (or holes) are easily thermally-excited over the tunnel barrier at room temperature. This is a key problem to be solved for practical application, and, therefore, we study to find a way to increase BE in order to elevate operation temperature toward room temperature. In this work, we present our recent research work with guiding principles for high-temperature tunneling operation.

2 Increase of the Dopant Binding Energy

2.1 A Single Donor Quantum Dot: Dielectric and Quantum Confinement Effect

When we focus on a phosphorous (P) donor atom, it is known that the ionization energy, or BE, with respect to the Si conduction band minimum, is ~45 meV. Therefore, single-electron tunneling P-atom devices can operate only at low temperatures. However, when a dopant is embedded in sufficiently small Si structures, BE is increased due to dielectric and quantum size confinement effects (Fig. 2) [8–10], leading to high-temperature operation. Recently, we have demonstrated 100 K operation in specifically-designed nano-stub-channels [11]. A good correlation is found between such high operation temperature and large BE (≅100 meV).

It should be noted, however, that this result is still far from our target, i.e., room temperature operation. In this work, we challenge to find another way for deepening of the dopant ground states, i.e., clustered dopants (or dopant "molecule") forming a quantum well with significantly deep ground states, as described below.

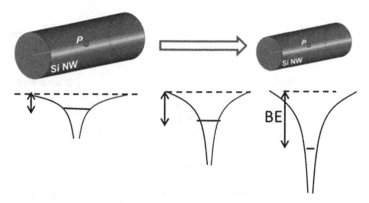

Fig. 2 Phosphorous donor potential and its ground state becomes deeper with decreasing radius of Si nanowire because of less screening of the Coulomb potential by a lack of surrounding Si

2.2 A Quantum Dot Formed by Clustered Donors

The concept of "clustered donors" is quite simple. If two or more P donors are close enough and their potential is simply overlapped, *BE* of the ground state electron becomes deeper and the barrier height for electron tunneling becomes higher. Experimentally, the channel of SOI-MOSFET (Fig. 3) was selectively doped within an area of ~ 30 nm in width by the conventional diffusion process. Even in such a classical doping process, the number of dopants in the doped area can be roughly controlled to be around 10 (Fig. 3). I–V characteristics measured for these selectively-doped FETs are consistent with the model [12], further supported by our analysis of dopant-induced potential modulation [13]. Most recently, we have examined high temperature characteristics and eventually observed preliminary results of tunneling characteristics at room temperature. The detailed results and their analysis will be reported elsewhere. In this paper, theoretical background of the deepening effect in clustered donors is presented.

If inter-donor spacing *d* is zero (a number of donors are located at the same positions) and it is hypothetically assumed that they maintain their "donor" nature, *BE* is expected to increase significantly with increasing number of donors, as

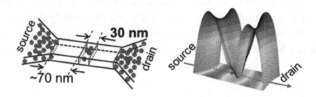

Fig. 3 Phosphorous (P) donors are selectively-doped near the center of the channel and its corresponding potential. When the inter-donor spacing is small enough in the doped area, larger BE will be obtained

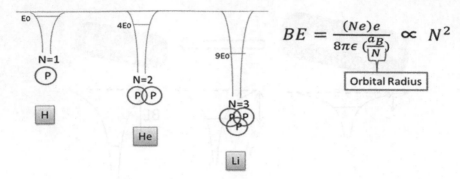

Fig. 4 *BE* of clustered donors are shown in an extreme limit of *d* = 0. In this case, the ground state can be expressed by an equation, which is similar to that for chemical elements in the periodic table

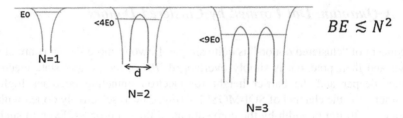

Fig. 5 *BE* of clustered donors in case of *d* ≠ 0 is shown for different *N*. Corresponding *BE* is smaller than those in Fig. 4

expressed by the equation in Fig. 4. This is, however, hypothetical extreme limit like the periodic table of "chemical elements". More realistic potential model (*d* ≠ 0) is shown in Fig. 5 and deepening effect of *BE* would be weaker than described in Fig. 4. Even in this case, however, it is certain that *BE* would be enhanced compared with the case of a single (isolated) donor.

N and d dependence of *BE* is plotted together in Fig. 6. It is found that small d and relatively large N is required to produce sufficiently large *BE* above 100 meV. If d is as small as 2–3 nm, i.e., Bohr radius for P-donor, $N = 3$–5 would be enough and more donors are not needed. In this technology, d and N cannot be deterministically controlled, but statistically controlled with some fluctuations. If volume concentration of P is around 5×10^{19} cm^{-3}, d is approximately 2–3 nm in average. Also, even if selective doping volume contains several tens of donors, a smaller number of donors below 10 will effectively form a deepest quantum well.

Fig. 6 *BE* of clustered donors in *d-N* plane. In the *purple-colored area*, large *BE* suitable for high temperature operation is expected

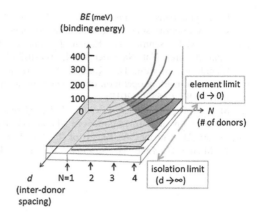

3 Conclusions

A unique approach to achieve high tunneling-transport temperature in dopant atom FETs is studied by using selectively-doped clustered donors as a quantum dot. In this method, it is critical to optimize two key parameters, i.e., inter-donor spacing d and the number of coupled donors N. According to our preliminary experiment, this method is promising for room temperature operation.

References

1. Moraru, D., Udhiarto, A., Anwar, M., Nowak, R., Jabłoński, R., Hamid, E., Tarido, J.C., Mizuno, T., Tabe, M.: Nanoscale Res. Lett. **6**, 479 (2011)
2. Sellier, H., Lansbergen, G.P., Caro, J., Rogge, S., Collaert, N., Ferain, I., Jurczak, M., Biesemans, S.: Transport spectroscopy of a single dopant in a gated silicon nanowire. Phys. Rev. Lett. **97**, 206805 (2006)
3. Lansbergen, G.P., Rahman, R., Wellard, C.J., Woo, I., Caro, J., Collaert, N., Biesemans, S., Klimeck, G., Hollenberg, L.C.L., Rogge, S.: Gate-induced quantum-confinement transition of a single dopant atom in a silicon FinFET. Nat. Phys. **4**, 656–661 (2008)
4. Ono, Y., Nishiguchi, K., Fujiwara, A., Yamaguchi, H., Inokawa, H., Takahashi, Y.: Conductance modulation by individual acceptors in Si nanoscale field-effect transistors. Appl. Phys. Lett. **90**, 102106 (2007)
5. Tabe, M., Moraru, D., Ligowski, M., Anwar, M., Jablonski, R., Ono, Y., Mizuno, T.: Single-electron transport through single dopants in a dopant-rich environment. Phys. Rev. Lett. **105**, 016803 (2010)
6. Pierre, M., Wacquez, R., Jehl, X., Sanquer, M., Vinet, M., Cueto, O.: Single-donor ionization energies in a nanoscale CMOS channel. Nat. Nanotechnol. **5**, 133–137 (2010)
7. Prati, E., Hori, M., Guagliardo, F., Ferrari, G., Shinada, T.: Anderson-Mott transition in arrays of a few dopant atoms in a silicon transistor. Nat. Nanotechnol. **7**, 443–447 (2012)
8. Diarra, M., Niquet, Y.M., Delerue, C., Allan, G.: Ionization energy of donor and acceptor impurities in semiconductor nanowires: importance of dielectric confinement. Phys Rev B. **75**, 045301 (2007)

9. Li, B., Partoens, B., Peeters, F.M., Magnus, W.: Dielectric mismatch effect on coupled shallow impurity states in a semiconductor nanowire. Phys. Rev. B. **79**, 085306 (2009)
10. Björk, M.T., Schmid, H., Knoch, J., Riel, H., Riess, W.: Donor deactivation in silicon nanostructures. Nat. Nanotechnol. **4**, 103–107 (2009)
11. Hamid, E., Moraru, D., Kuzuya, Y., Mizuno, T., Anh, L.T., Mizuta, H., Tabe, M.: Electron-tunneling operation of single-donor-atom transistors at elevated temperatures. Phys. Rev. B. **87**, 085420 (2013)
12. Moraru, D., Samanta, A., Anh, L.T., Mizuno, T., Mizuta, H., Tabe, M.: Transport spectroscopy of coupled donors in silicon nano-transistors. Sci. Rep. **4**, 6219 (2014)
13. Moraru, D., Samanta, A., Krzysztof, T., Anh, L.T., Muruganathan, M., Mizuno, T., Jabłoński, R., Mizuta, H., Tabe, M.: Nanoscale Res. Lett. **10**, 372 (2015)

EDMR on Recombination Process in Silicon MOSFETs at Room Temperature

Masahiro Hori and Yukinori Ono

Abstract Electrically detected magnetic resonance (EDMR) enables us to obtain magnetic properties of the localized states in devices. Here we developed the EDMR measurement system with a low noise, and applied it to the defect states in a silicon metal-oxide-semiconductor field-effect-transistor. The EDMR signal was observed for the spin dependent recombination process at room temperature. The gate voltage dependence of the signal intensity indicated that the signal originates from the silicon dangling bonds at the Si/SiO$_2$ interface.

Keywords EDMR · ESR · Spin · Silicon · MOSFET · Interface defects

1 Introduction

In recent years, localized states formed at the interface and at the dopant sites are intensively studied from viewpoints of manipulating electronic charges and spins in a silicon transistor [1–5]. Most studies, however, focus on the shallow energy states, which restrict the device operation to a unipolar type at cryogenic temperatures. The electron-hole recombination (i.e., the bipolar operation) at deep energy states, on the other hand, is immune to the thermal agitation, and permits the high-temperature operation.

In order to manipulate the electronic charges/spins at room temperature, the electrically detected magnetic resonance (EDMR) [6–18] is one of the key methods. EDMR monitors a change in the current through localized states due to the electron spin resonance (ESR). Here, we develop an EDMR measurement system with a high sensitivity, and apply it to the detection of electronic spins at the Si/SiO$_2$ interface in the recombination process.

M. Hori (✉) · Y. Ono
Research Institute of Electronics, Shizuoka University, Hamamatsu, Japan
e-mail: hori.masahiro@shizuoka.ac.jp

Y. Ono
e-mail: ono.yukinori@shizuoka.ac.jp

© Springer International Publishing AG 2017
R. Jabłoński and R. Szewczyk (eds.), *Recent Global Research and Education:
Technological Challenges*, Advances in Intelligent Systems and Computing 519,
DOI 10.1007/978-3-319-46490-9_13

2 Experimental

An n-type metal-oxide-semiconductor field-effect-transistor (MOSFET) fabricated on a silicon (100) substrate was used as a test device. The channel length, width and oxide thickness were respectively 500 μm, 500 μm, and 30 nm. The interface defects were induced by the Fowler-Nordheim stress at 9 MV/cm. The resultant interface defect density was 1×10^{11} cm^{-2}, which was evaluated by the charge pumping method [19–22].

We used the ESR system of the X-band spectrometer with a TE100 cylindrical microwave cavity [23], in which the MOSFET was mounted. EDMR detects a small change ΔI in the device current I (typically ΔI/I $\sim 10^{-5}$ to 10^{-4}) due to the ESR, and the signal is obtained using the lock-in detection technique. All measurements were done at room temperature with the magnetic field modulation of 80 Hz, and the microwave power of 200 mW.

For the EDMR measurements, the MOSFET was operated as a gated diode as shown in Fig. 1. To obtain the recombination current between the source/drain and the substrate, the PN junctions in the MOSFET were forward biased (VF = −0.24 V) with the source and drain shorted and the substrate grounded. We used the gate voltage Vg as a parameter to control the depletion layer formation, which governs the recombination activity at the MOS interface.

3 Results and Discussions

Figure 2a shows the EDMR signal for a fixed V_g of −3.8 V. The resultant I was 10 nA. The g-value was estimated to be 2.007, from which we ascribed the origin of the signal to one of the silicon dangling bonds at (100) interface, the P_{b0} center [8, 9, 12, 15].

In order to estimate the intrinsic noise level and signal to noise ratio, we reconstructed the original EDMR current by integrating the raw data (Fig. 2a) obtained by the field-modulation technique. The results are shown in Fig. 2b, c. The current modulation (the peak height in (b)) ΔI is about 0.5 pA, which means

Fig. 1 Measurement setup for the EDMR. The MOSFET is used as a gated diode. The source and drain terminals are shorted, and are negatively biased so that the PN junctions in the MOSFET are forward biased. The gate voltage is applied so as to form the depletion layer

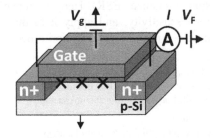

Fig. 2 Results of the EDMR measurement at $V_g = -3.8$ V. **a** Output current, **b** integrated signal, and **c** the noise floor. I_{p-p} and ΔI respectively denote the output current amplitude and the integrated signal intensity

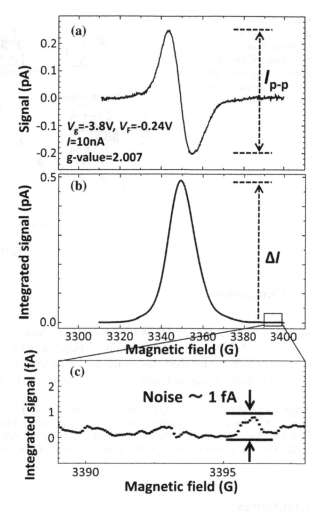

$\Delta I/I \approx 5 \times 10^{-5}$. The noise is on the order of 1 fA, which is lower than that of the preceding studies.

We next compare, in Fig. 3, the output current amplitude I_{p-p} (open circles, see Fig. 2a for the definition) to the device current I (closed circles) as a function of V_g. In the MOS gated diode, the forward current of the PN junction flows regardless of the interface condition, i.e., accumulation, depletion, or inversion (see the schematics in Fig. 3). An extra current flows only when the MOS interface is depleted due to the recombination at the interface defects (see the center schematic). Therefore, I has a peak at the depletion region. As one can see, I_{p-p} changes in a similar manner to I as a function of V_g. This demonstrates that the electron spins contributing to the EDMR signal are located at the MOS interface.

Fig. 3 Gate voltage V_g
dependence of I and I_{p-p}. The
schematics show the
conditions of the MOS
interface, accumulation (*left*),
depletion (*center*), and
inversion (*right*)

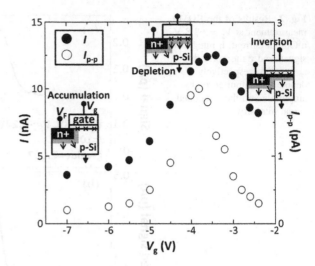

4 Conclusions

We have developed the EDMR measurement system with a low noise. The signal
was successfully obtained in the spin dependent recombination process at room
temperature. We attributed this signal to the Si/SiO_2 interface defects by investigating the gate voltage dependence of the signal intensity.

Acknowledgments The authors would like to thank Prof. T. Tsuchiya for providing test devices
and Dr. A. Fujiwara for his supports. This work was partially supported by the Grants-in-Aid for
Scientific Research Nos. 23226009, 25289098, 25600015, 25706003, 26289105, 15K13970,
16H02339 and 16H06087 from the Japan Society for the Promotion of Science.

References

1. Lansbergen, G., Ono, Y., Fujiwara, A.: Donor-based single electron pumps with tunable
   donor binding energy. Nano Lett. **12**, 763–768 (2012)
2. Prati, E., Hori, M., Guagliardo, F., Ferrari, G., Shinada, T.: Anderson-Mott transition in arrays
   of a few dopant atoms in a silicon transistor. Nat. Nanotech. **7**, 443–447 (2012)
3. Yamahata, G., Nishiguchi, K., Fujiwara, A.: Gigahertz single-trap electron pumps in silicon.
   Nat. Commun. **5**, 1–7, Article No. 5038 (2014)
4. Prati, E., Kumagai, K., Hori, M., Shinada, T.: Band transport across a chain of dopant sites in
   silicon over micron distances and high temperatures. Sci. Rep. **6**, 1–8, Article
   No. 19704 (2016)
5. Xiao, M., Martin, I., Yablonovitch, E., Jiang, H.W.: Electrical detection of the spin resonance
   of a single electron in a silicon field-effect transistor. Nature **430**, 435–439 (2004)
6. Lepine, D.J.: Spin-dependent recombination on silicon surface. Phys. Rev. B **6**, 436–441
   (1972)

7. Vranch, R.L., Henderson, B., Pepper, M.: Spin-dependent recombination in irradiated Si/SiO$_2$ device structures. Appl. Phys. Lett. **52**, 1161–1163 (1988)

8. Jupina, M.A., Lenahan, P.M.: A spin dependent recombination study of radiation induced defects at and near the Si/SiO$_2$ interface. IEEE Trans. Nucl. Sci. **36**, 1800–1807 (1989)

9. Henderson, B., Pepper, M., Vranch, R.L.: Spin-dependent and localisation effects at Si/SiO$_2$ device interfaces. Semicond. Sci. Technol. **4**, 1045–1060 (1989)

10. Jupina, M.A., Lenahan, P.M.: Spin dependent recombination: a ^{29}Si hyperfine study of radiation-induced P$_b$ centers at the Si/SiO$_2$ interface. IEEE Trans. Nucl. Sci. **37**, 24–31 (1990)

11. Krick, J.T., Lenahan, P.M., Dunn, G.J.: Direct observation of interfacial point defects generated by channel hot hole injection in N-channel metal oxide silicon field effect transistors. Appl. Phys. Lett. **59**, 3437–3439 (1991)

12. Stathis, J.H., DiMaria, D.J.: Identification of an interface defect generated by hot electrons in SiO$_2$. Appl. Phys. Lett. **61**, 2887–2889 (1992)

13. Stathis, J.H.: Microscopic mechanisms of interface state generation by electrical stress. Microelectron. Eng. **22**, 191–196 (1993)

14. Gabrys, J.W., Lenahan, P.M., Weber, W.: High resolution spin dependent recombination study of hot carrier damage in short channel MOSFETs: ^{29}Si hyperfine spectra. Microelectron. Eng. **22**, 273–276 (1993)

15. Vuillaume, D., Deresmes, D., Stievenard, D.: Temperature-dependent study of spin-dependent recombination at silicon dangling bonds. Appl. Phys. Lett. **64**, 1690–1692 (1994)

16. Lenahan, P.M.: Atomic scale defects involved in MOS reliability problems. Microelectron. Eng. **69**, 173–181 (2003)

17. Lenahan, P.M.: deep level defects involved in MOS device instabilities. Microelectron. Reliab. **47**, 890–898 (2007)

18. Yonamoto, Y.: Recovery behavior in negative bias temperature instability. Microelectron. Reliab. **54**, 520–528 (2014)

19. Brugler, J.S., Jespers, P.G.A.: Charge pumping in MOS devices. IEEE Trans. Electron Devices **16**, 297–302 (1969)

20. Groeseneken, G., Maes, H.E., Beltran, N., DeKeersmaecker, R.F.: A reliable approach to charge-pumping measurements in MOS transistors. IEEE Trans. Electron Devices **31**, 42–53 (1984)

21. Hori, M., Watanabe, T., Tsuchiya, T., Ono, Y.: Analysis of electron capture process in charge pumping sequence using time domain measurements. Appl. Phys. Lett. **105**, 1–4, Article No. 261602 (2014)

22. Hori, M., Watanabe, T., Tsuchiya, T., Ono, Y.: Direct observation of electron emission and recombination processes by time domain measurements of charge pumping current. Appl. Phys. Lett. **106**, 1–4, Article No. 041603 (2015)

23. Hori, M., Uematsu, M., Fujiwara, A., Ono, Y.: Electrical activation and electron spin resonance measurements of arsenic implanted in silicon. Appl. Phys. Lett. **106**, 1–4, Article No. 142105 (2015)

Inter-band Current Enhancement by Dopant-Atoms in Low-Dimensional *pn* Tunnel Diodes

Daniel Moraru, Manoharan Muruganathan, Le The Anh,
Ratno Nuryadi, Hiroshi Mizuta and Michiharu Tabe

Abstract Inter-band tunneling current is an attractive transport mechanism for future generations of electronics. However, this mechanism is limited by the momentum conservation law which requires phonon assistance in tunneling due to the indirect-bandgap nature of Si. Here, we show that in low-dimensional *pn* Esaki diodes, inter-band tunneling current can be enhanced by the resonance of discrete dopants with deepened energy levels. Current enhancement is comparable with the background direct inter-band tunneling and can be modulated by the applied biases, suggesting a pathway for controlling atomic-level resonances for practical purposes.

Keywords Inter-band tunneling · P-B pair · Tunnel diode · Atomic *pn* diode

1 Inter-band Tunneling in Si

As a result of continuous downscaling of semiconductor electronic devices, we enter a regime in which quantum tunneling becomes a practical mechanism in transport operation. Among some applications, Esaki tunnel diodes [1, 2] and tunnel field-effect transistors (TFETs) [3] use the basic *pn* junction design. In such

D. Moraru (✉) · M. Tabe
Research Institute of Electronics, Shizuoka University, 3-5-1 Johoku,
Naka-Ku, Hamamatsu 432-8011, Japan
e-mail: moraru.daniel@shizuoka.ac.jp

M. Muruganathan · L.T. Anh · H. Mizuta
Japan Advanced Institute of Science and Technology (JAIST),
1-1 Asahidai, Nomi 923-1292, Japan

R. Nuryadi
Agency for Assessment and Application of Technology,
South Tangerang, Indonesia

H. Mizuta
Faculty of Physical Sciences and Engineering, University of Southampton,
High Field, Southampton SO17 1BJ, UK

© Springer International Publishing AG 2017
R. Jabłoński and R. Szewczyk (eds.), *Recent Global Research and Education:
Technological Challenges*, Advances in Intelligent Systems and Computing 519,
DOI 10.1007/978-3-319-46490-9_14

95

Fig. 1 a Schematic band diagram for an Esaki (tunnel) diode, showing electron tunneling between conduction band and valence band at small forward bias. **b** *E-k* space diagram illustrating the effect of indirect bandgap nature on inter-band tunneling, involving absorption or emission of phonons

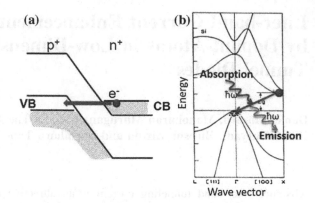

devices, inter-band tunneling (between conduction band and valence band) is seen as a possible candidate for improving the operation of next-generation of transistors. Since Si is the most suitable platform for electronics, it becomes essential to identify and demonstrate such applications for Si nanodevices. In that sense, Si has the disadvantage that it is an indirect-bandgap material and inter-band tunneling has to be assisted by phonons (emission or absorption) to satisfy the momentum conservation law, as illustrated in Fig. 1. This limits the current drivability of such devices. Hence, many researchers have been recently looking for various mechanisms to improve the inter-band tunneling current.

Among such current-enhancement mechanisms, most are relying on the engineering of unconventional impurities (such as isoelectronic traps) [4], defects or dislocations in heterostructures [5], all of which require the development of relatively new technologies that may not be compatible with the Si platform. It is desirable to identify "intrinsic" mechanisms for current enhancement that can be ascribed directly to the properties of Si and that can be, thus, naturally incorporated within the present Si technology.

In this work, we show the possibility of achieving enhancement of inter-band tunneling current naturally due to the effects of low-dimensionality, i.e., quantum and/or dielectric confinement. As a prototype device, we focus on 2D Esaki tunnel diodes, in which inter-band tunneling takes place via a narrow depletion region (Fig. 1). Most importantly, when a pair of discrete dopants with deeper energy levels are in resonance in the depletion region, a significant current enhancement can be observed. This can be further tuned by applying substrate voltage similarly to a gate in a TFET structure.

2 Fabrication and Structure of 2D Lateral Tunnel Diodes

In order to investigate inter-band tunneling, we designed and fabricated Si Esaki (tunnel) diodes with a lateral layout, as shown in Fig. 2a. The *pn* diode is formed in a thin silicon-on-insulator (SOI) layer, using specific doping masks in the

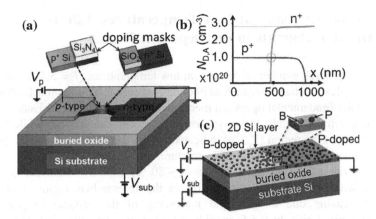

Fig. 2 **a** Device structure and measurement circuit near the nanoscale constriction region, with the specific masks used for p^+ and n^+ doping. **b** Doping profiles for p^+ and n^+ regions, with an overlap to control the position of the junction. **c** Schematic illustration of depletion layer area formed in a 2D Si layer. A P-B dopant pair is shown as a fundamental unit useful for inter-band tunneling

thermal-diffusion doping process to realize extremely high doping concentrations (on the order of 10^{20} cm^{-3}). During B-doping (p-type), a 15-nm-thick Si_3N_4 mask protects the nominally n-type region, while during P-doping (n-type) a $\sim$ 10-nm-thick SiO_2 layer protects the nominally p-type region (as shown in the insets). Doping profiles, as illustrated in Fig. 2b, are overlapped to ensure that the depletion region is formed within a channel patterned by an electron-beam lithography (EBL) technique. The length of this region is 1000 nm, while its width is changed as a parameter on the order of $\sim$ 100 nm. Finally, a thin SiO_2 layer ($\sim$ 2 nm) is formed for passivation.

For such high doping concentrations ($N_D \approx 2.7 \times 10^{20}$ cm^{-3}, $N_A \approx 0.9 \times 10^{20}$ cm^{-3}), we evaluate that the depletion region width (W_{depl}) is on the order of $\sim$ 10 nm. Therefore, inter-band tunneling is expected to be a main transport mechanism. A schematic picture of this region is shown in Fig. 2c. As an inset, we also illustrate the simplest unit that may work in such a narrow depletion region, i.e., a pair of dopants, most likely a P-donor near the n-type edge and a B-acceptor near the p-type edge. Because of the optimum positions of such dopants and, in addition, because of confinement effects deepening their energy levels, such a fundamental unit could provide a dramatic enhancement of the tunneling current. We have estimated that such a P-B deep-level dopant pair can be found with sufficiently high probability in our devices [6].

3 I–V Characteristics at Low Temperatures: Effects of Discrete Dopants in the Depletion Region

We measured a large number of devices, at low temperatures ($T = 5.5$ K), in order to avoid complex mechanisms, such as phonon absorption. This approach allows us to focus on the fundamental operation mechanism of these thin lateral Esaki diodes. In the measurement circuit, the n-type region is set to the ground, while bias is applied to the p-type region (V_p), as indicated in Fig. 2a. The substrate voltage can be used similarly to a gate in a TFET structure.

The majority of the measured devices (~ 80 %) exhibit a behavior basically consistent with regular Esaki diodes [6]. In the reverse-bias region, current is quickly increasing due to the rapid narrowing of the depletion region with increasing applied bias. In the forward-bias region, we typically observe the negative differential resistance (NDR) peak which is a signature of the Esaki tunneling mechanism. Due to the low-dimensionality, these devices also exhibit prominent 2D-quantization effects (discrete energy levels). A full analysis of these effects, in correlation with phonon assistance, was presented elsewhere [6].

Figure 3 shows I_p-V_p characteristics obtained for several devices (~ 20 % of the total number of measured devices), which exhibit an intriguing behavior. Superimposed on the NDR peak due to the direct inter-band tunneling, sharply-enhanced current peak(s) can be observed. This significant current enhancement can be attributed to a resonance between most likely two energy levels [7] located in the narrow depletion region.

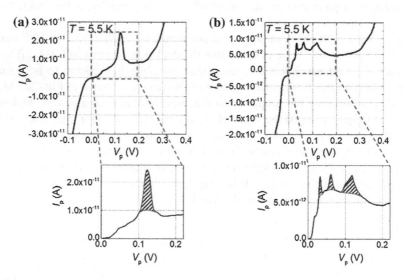

Fig. 3 **a** I_p-V_p characteristics at low temperature ($T = 5.5$ K) in forward/reverse bias regions, typical for ~ 20 % of measured devices. In forward-bias region, sharply-enhanced current peaks are observed: **a** single peak; **b** multiple peaks

According to our statistical evaluation, such levels can be ascribed to the resonance of two dopants' levels: a P-donor near the n-type edge and a B-acceptor near the p-type edge (Fig. 3a). For the case of several sharply-enhanced current peaks, several such P-B pairs may work at slightly different V_p values (Fig. 3b). Dopants with deeper ground-state energy levels (>45 meV) must be considered to match the experimental yield. This deepening of the dopant levels is expected due to dielectric and quantum-size confinement, when the dimensionality is reduced from 3D (bulk) to lower dimensionality [8]. This effect is schematically represented in Fig. 4, in correlation with inter-band tunneling.

4   Tunability of Current Enhancement by Applied Biases

We can confirm that the current enhancement comes from resonance between energy levels located in the depletion region by analyzing the effect of the substrate voltage, V_{sub} (used practically as a gate). The change is shown only by an example of two different V_{sub} values, for simplicity, as illustrated in Fig. 5a. In Fig. 5b, the relative increase of the current (I_p) compared with the background current (I_p^0) is indicated. Generally, a significant increase of the current, comparable with the background current, can be obtained at the position of the sharply-enhanced current peaks.

In Fig. 5c, we illustrate schematically the model for explaining this enhanced current peak. On the background of inter-band tunneling, which gives rise to the broad NDR peak, a resonance between a P-donor-level and a B-acceptor-level can explain the sharp enhancement of the tunnel current. As described earlier, it is even more likely that the dopants have deeper energy levels due to the confinement effects, as indicated in Fig. 4. The two levels have different couplings to the biases, i.e., V_p and V_{sub}, due to the different positions within the depletion region. Thus, we can expect that the levels can be adjusted differently by each bias, giving us the ability to tune the levels into resonance almost independently. Further analysis is required to fully understand the interplay between the resonant levels of the P-B

Fig. 4 Resonant-tunneling transport via a P-B pair in a narrow, thin depletion layer. The P-donor and B-acceptor levels may be deeper than in bulk (3D) due to quantum and/or dielectric confinement effects, specific for 2D structures (insets)

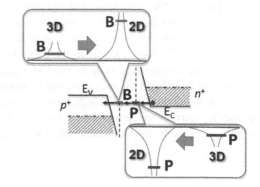

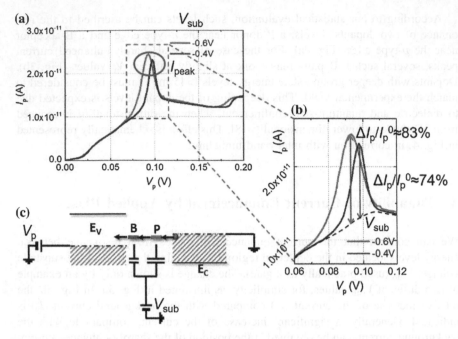

Fig. 5 **a** Sharply-enhanced current peaks for different values of V_{sub}. **b** Lower panel shows a zoom-in on the enhanced peak, indicating a significant relative current increase compared with the background current ($\Delta I_p/I_p^0$). **c** Mechanism of tuning of the resonance between the P- and B-levels using V_p and V_{sub}, with different coupling to different dopants

pair and the density-of-states (DOS) of the nearby leads. Such study could provide new information about different approaches for optimizing the current enhancement at atomic scale.

5 Conclusions

We reported on the fabrication and characterization of lateral two-dimensional (2D) *pn* Esaki (tunnel) diodes in Si. In such devices, specific phenomena emerge, such as 2D quantization of the *p*- and *n*-leads, as well as resonance between deepened-energy dopant levels in the depletion region. Here, we further show that the substrate voltage (V_{sub}) can be used as a gate together with the in-plane bias (V_p) to tune the inter-band tunneling current by controlling the P-B resonance within the transport window. These findings suggest a pathway for optimizing the inter-band tunneling current by engineering the atomic-level resonances in the depletion region, eventually in correlation with the nearby nanoscale leads.

References

1. Esaki, L.: New phenomenon in narrow Germanium p-n junctions. Phys. Rev. **109**, 603 (1958)
2. Chynoweth, A.G., Logan, R.A., Thomas, D.E.: Phonon-assisted tunneling in silicon and germanium Esaki junctions. Phys. Rev. **125**, 877 (1962)
3. Ionescu, A.M., Riel, H.: Tunnel field-effect transistors as energy-efficient electronic switches. Nature **479**, 329 (2011)
4. Mori, T., et al.: Study of tunneling transport in Si-based tunnel field-effect transistors with ON current enhancement utilizing isoelectronic trap. Appl. Phys. Lett. **106**, 083501 (2015)
5. Bessire, C.D., Björk, M.T., Schmid, H., Schenk, A., Reuter, K.B., Riel, H.: Trap-assisted tunneling in Si-InAs nanowire heterojunction tunnel diodes. Nano Lett. **11**, 4195 (2011)
6. Tabe, M., Tan, H.N., Mizuno, T., Muruganathan, M., Anh, L.T., Mizuta, H., Nuryadi, R., Moraru, D.: Atomistic nature in band-to-band tunneling in two-dimensional silicon pn tunnel diodes. Appl. Phys. Lett. **108**, 093502 (2016)
7. Savchenko, A.K., Kuznetsov, V.V., Woolfe, A., Mace, D.R., Pepper, M., Ritchie, D.A., Jones, G.A.C.: Resonant tunneling through two impurities in disordered barriers. Phys. Rev. B **52**, R17021(R) (1995)
8. Diarra, M., Niquet, Y.M., Delerue, C., Allan, G.: Ionization energy of donor and acceptor impurities in semiconductor nanowires: importance of dielectric confinement. Phys. Rev. B **75**, 045301 (2007)

Ferroelectric Properties of Nanostructured SBTN Sol-Gel Layers

V.V. Sidsky, A.V. Semchenko, S.A. Khakhomov, A.N. Morozovska,
N.V. Morozovsky, V.V. Kolos, A.S. Turtsevich, A.N. Pyatlitski,
Yu M. Pleskachevsky, S.V. Shil'ko and E.M. Petrokovets

Abstract One of the urgent problems of modern engineering is to obtain a ferroelectric capacitor for various structures. A promising material in the class of ferroelectrics with perovskite structure is strontium bismuth tantalate-niobate $Sr_xBi_yNb_zTa_{2-z}O_9$ (SBTN). The synthesized layers should be homogeneous on structure and thickness, to have good fatigue characteristics (i.e. to not change value of remanent polarization after repeated carrying out of cycles inclusion/deenergizing) and to possess small leakage current.

Keywords Sol-gel method · Surface structure · Crystallization · Ferroelectrics

1 Introduction

FRAM is the new type of non-volatile memory with the same low-voltage, high-speed random access characteristics as DRAM and SRAM, while maintaining the nonvolatile data characteristics of Flash Memory and EEPROM. At the same time, FRAM is a media that features a different storage method that allows it to consume less power during operation. This type of memory can be implemented in

V.V. Sidsky (✉) · A.V. Semchenko · S.A. Khakhomov
Francisk Scorina Gomel State University, Gomel, Belarus
e-mail: sidsky@gsu.by

A.N. Morozovska · N.V. Morozovsky
Institute of Physics, Kiev, Ukraine

V.V. Kolos · A.S. Turtsevich · A.N. Pyatlitski
JSC "INTEGRAL", Minsk, Belarus

Y.M. Pleskachevsky · S.V. Shil'ko · E.M. Petrokovets
V.A. Belyi Metal-Polimer Research Institute N.A., Gomel, Belarus

© Springer International Publishing AG 2017   103
R. Jabłoński and R. Szewczyk (eds.), *Recent Global Research and Education:
Technological Challenges*, Advances in Intelligent Systems and Computing 519,
DOI 10.1007/978-3-319-46490-9_15

particular on the base of $SrBi_2Ta_2O_9$ compound synthesized by sol-gel method. Test samples were prepared as both powders and films synthesized by sol-gel method [1, 2].

Sol-gel films with the general formula $(SrBi_2(Ta_xNb_{1-x})_2O_9)$ (SBTN-film) were synthesized by sol-gel method. Gelation took place under the effect of centrifugal force. The films were deposited on $Pt/TiO_2/BPSG/SiO_2/Si$ sublayers by spin-coating at different substrate speeds (500–1000 r/min) or by dip-coating. In order to achieve the desired thickness (200–300 nm) sol was applied 2–3 times followed by heat treatment of each layer at the temperature of 800–1000 °C. Annealing of SBTN-films in order to form perovskite structure was made at 800 °C for 40 min.

Diffraction peaks identification was performed using database JCPDS (with software Search Match). Diffractograms were processed using JANA 2006 software. X-ray diffraction profiles obtained for SBT and $SrBi_2Ta_{2-y}Nb_yO_9$ with the ratio Nb/Ta $y = 10$ wt% (SBTN10) and $y = 30$ wt% (SBTN30) are shown in Fig. 1.

Comparison of the intensity and peak positions of crystallographic planes (115), (200) and (0010) showed that the SBTN10 composition after annealing at temperature of 800 °C has the structure nearest to perovskite type. These qualitative estimations were confirmed by quantitative analysis of the phase composition performed using the JANA 2006 software (Table 1).

The results of scanning force microscopy (with SOLVER 47-PRO processed by the Gwiddion software) are presented in Fig. 2. For all the films the grain effect is visible. The increase of Nb content leads to change of the shape and size of the grains and their size dispersion. At that average grain size increases from 50 nm for SBT to 100 nm for SBTN10 with rounded off grains and to 300 × 700 nm for SBTN30 with preferable orientation of oblong grains (see Fig. 2a–c).

Results of scanning electron micrography of chips (cross section) and the surface morphology of annealed two-layer films of SBTN are presented in Fig. 3. Layered structure (Fig. 3a, c) corresponds to the sequence (SBT or SBTN)/$Pt/TiO_2/SiO_2$ at the thicknesses in the range (220–230) nm for SBT films and (260–270) nm for SBTN films. SBT films (Fig. 3a, b) consist of partially fused granules with sizes ranging from 70 to 200 nm separated by pores with sizes from 20 to 50 nm. SBTN films (Fig. 3c, d) consist substantially of fused grains with sizes from 300 to 500 nm.

Dynamic charge-voltage and current-voltage loops for unannealed films of SBT and SBTN with different contents of Nb are shown in Fig. 4.

Shape of charge-voltage loops (Fig. 4, top row) is characteristic of the nonlinear dielectric with losses [3]. Form of current-voltage loops (Fig. 4, bottom row) reflects the rate of change of the charge, maximal in the vicinity of maxima. The

Fig. 1 XRD patterns of
SBTN films annealed at
various temperatures

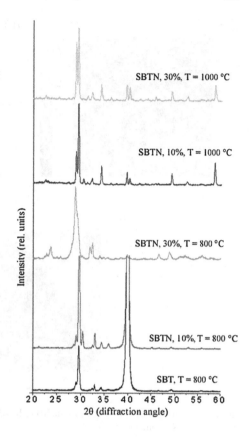

Table 1 Results of quantitative analysis of the phase composition of SBTN films with different contents of Nb, prepared at different annealing temperatures

Material	Content of Nb (%)	Annealing temperature (°C)	Content of the perovskite phase (%)
SBT	0	800	65.2
SBTN	10	800	85.2
SBTN	30	800	77.3
SBTN	10	1000	67.2
SBTN	30	1000	59.3

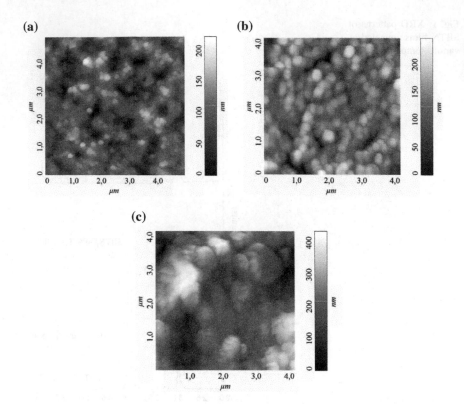

Fig. 2 AFM images of different films: **a** SBT, **b** SBTN with 10 % of Nb content and **c** SBTN with 30 % of Nb content

values of the coercive voltage for charge-voltage loops are in the range (12–16) V (600–800 kV/cm) at 50 Hz, 50 V sine wave. This well exceeds the known values of 1 V and 50 kV/cm for SBT films and SBN/BTN ceramics [4].

Estimation of SBT and SBTN relative permittivity from measurements of the capacitance (at 1 kHz) of non-annealed films gives a value of E 15, which is almost an order of magnitude less than the known value for SBT ceramics E 130 [5]. This allows considering the possible existence of local ferroelectric regions in non-annealed SBTN films. At that only a small fraction of the applied voltage drops on ferroelectrically active inclusions, which leads to the increase of apparent coercive voltage.

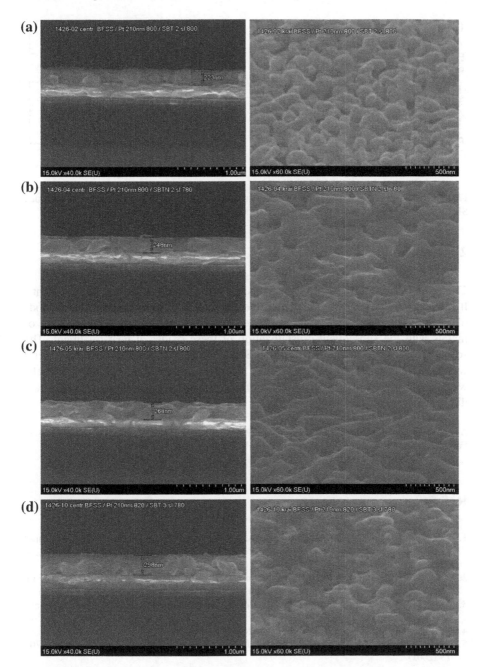

Fig. 3 SEM images of chips (*left side*) and surfaces (*right side*) of two-layer SBT (**a, d**) and SBTN (**b, c**) films annealed at 800 °C (**a, c**) and 780 °C (**b, d**). Bottom Pt electrodes were annealed at 800 °C (**a–c**) and 780 °C (**d**). The sequences (SBT or SBTN)/Pt/TiO$_2$/SiO$_2$ are well distinguished

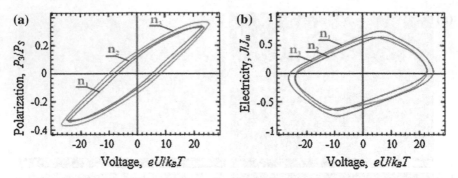

Fig. 4 Dynamic charge-voltage (*top row*) and current-voltage (*bottom row*) loops for no annealed SBT and SBTN films with different content of niobium ($n_1 = 0$ %, $n_2 = 10$ %, $n_3 = 20$ %)

2 Conclusion

The ferroelectric $Sr_xBi_yNb_zTa_{2-z}O_9$ satisfies to the majority of these requirements. It can be produced, in particular, by sol-gel method. It is one of perspective methods of crystal nanostructures synthesis, based on joint sedimentation hydroxides of metals with the subsequent crystallization of these oxides at heat treatment. The sol-gel layers with the general formula $Sr_xBi_yNb_zTa_{2-z}O_9$ (SBTN) (SBTN-layers) were annealed at 800 and 1000 °C. As starting compounds inorganic metal salts were used. Ferroelectric and nanostructured properties of synthesized layers are discussed.

Acknowledgments This research was partially conducted in the framework by the Belarusian Foundation for Basic Research, project №. T15UA/A-067

References

1. Sidsky, V.V., Semchenko, A.V., Tyulenkova, O.I., Soroka, S.A., Sudnik, L.V.: Nanostructured Sr(Bi$_x$Ta$_x$)O$_9$ materials synthesized by sol-gel method. Metallophysics New Technolo. **33**, 21–30 (2011)
2. Bhaskar, S., Majumder, S.B., Das, R.R., Dobal, P.S., Katiyar, R.S.: Electrical properties of La graded heterostructure of Pb$_{1-x}$La$_x$TiO$_3$ thin films. J. Mater. Sci. Eng. (B) **86**, 172 (2001)
3. Lines, M.E., Glass, A.M.: Principles and Applications of Ferroelectrics and Related Materials. Clarendon Press, Oxford (1977)
4. Moure, A., Pardo, L.: J. Electroceram. **15**, 243–250 (2005)
5. Wu, Y., Forbess, M.J., Seraji, S., Limmer, S.J., Chou, T.P., Cao, G.: Mater. Sci. Eng. B **86**, 70–78 (2001)

Scanning Nanopipette Probe Microscope for Nanofabrication Using Atmospheric Pressure Plasma Jet

Futoshi Iwata, Daisuke Morimatsu, Hiromitsu Sugimoto, Atsushi Nakamura, Akihisa Ogino and Masaaki Nagatsu

Abstract We developed a novel scanning probe microscope (SPM) technique for fine processing of material surface using a localized fine atmospheric pressure plasma jet (APPJ) generated from a nanopipette. The nanopipette is a tapered glass capillary with an aperture of sub-micrometer. Using the nanopipette as a nozzle, it was possible to localize irradiation area of the APPJ. The nanopipette could also be used as a probe for a scanning probe microscope operated with shear-force feedback control, which is capable of positioning the pipette edge in the vicinity of material surfaces for APPJ processing. By using the SPM system, sub-micrometer holes were successfully fabricated.

Keywords Atmospheric pressure plasma jets · Scanning probe microscope · Nanopipette · Nano fabrication

1 Introduction

Atmospheric pressure plasma jets (APPJs) are jet-like plasmas generated and emitted under atmospheric pressure [1, 2]. APPJs have been studied for many applications in the fields of engineering, medicine and biology. By irradiating APPJ to material surface, various surface treatments such as surface modification, sterilization [3] and mass spectrometry [4] have been achieved. Material processing such as removal [5] and deposition also have been performed. However, as for the

F. Iwata (✉) · D. Morimatsu · H. Sugimoto · A. Nakamura · A. Ogino
Graduate School of Integrated Science and Technology, Shizuoka University, Hamamatsu, Japan
e-mail: iwata.futoshi@shizuoka.ac.jp

F. Iwata · M. Nagatsu
Research Institute of Electronics, Shizuoka University, Hamamatsu, Japan

M. Nagatsu
Graduate School of Science and Technology, Shizuoka University, Hamamatsu, Japan

© Springer International Publishing AG 2017   109
R. Jabłoński and R. Szewczyk (eds.), *Recent Global Research and Education: Technological Challenges*, Advances in Intelligent Systems and Computing 519,
DOI 10.1007/978-3-319-46490-9_16

reduction of the processing size, these technologies have remained on the millimeter order.

In previous research, our group introduced an APPJ with a nanopipette as a nozzle. The nanopipette is a tapered glass capillary with an aperture of sub-micrometer or smaller in diameter. By reducing the aperture size of the nozzle to confine the irradiation range, APPJs can be applied to fine fabrications. We reported some local surface treatments using this technique such as material etching [6], modification of amino groups onto polymer [7], functionalization of carbon nanotube array [8] and a surface modification of biomolecule film [9]. However, this APPJ ignition technique showed a drawback in that the processing amount could not be controlled because it did not have distance control function between the pipette edge and the sample surface. Thus, to control APPJ processing and achieve smaller fabrication size, this technique should be equipped with a function of positioning the pipette edge precisely.

In this paper, we describe a novel APPJ processing system by coupling it with a scanning probe microscope (SPM). For positing the pipette edge in the vicinity of the surface, two methods were evaluated to detect the pipette oscillation under shear force feedback control. One is a method using an optical detection sensor, and another is a method using a piezoelectric quartz tuning forks. The robustness to detect the pipette oscillation while generating APPJ was investigated. Using the system, sub-micrometer holes were fabricated on a photoresist.

2   Experimental Method

2.1   Principles of APPJ Ignition and Processing

In this research, a nanopipette probe was used to localize the irradiation area of an APPJ. The nanopipette used in our experiment was prepared as follows. A capillary borosilicate glass tube with an inner diameter of 0.6 mm and outer diameter of 1.0 mm (Narishige) was thermally pulled using a commercial micropipette puller (Sutter Instrument, P-2000). Figure 1a shows an SEM image of the nanopipette edge. As shown in the figure, the aperture size of the nanopipette was approximately from 200 to 300 nm. As for an APPJ processing matter, a positive-type photoresist coated on a glass substrate was employed. Figure 1b shows an illustration of the APPJ fine processing of the photoresist film. To ignite the atmospheric pressure plasma, a high-voltage pulse wave was applied to a thin copper tape electrode that was wound around the outside of the nanopipette while helium gas was introduced into the nanopipette. When the APPJ is ignited, the excited radicals of the APPJ decompose the organic matter of the photoresist film to CO_2 and H_2O and release these gases into the atmosphere. As a result, concave structures such as hole or line patterns can be processed.

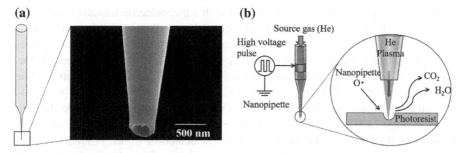

Fig. 1 Nanopipette to localize APPJ for fine processing. **a** SEM image of nanopipette aperture. **b** Principles of APPJ ignition and nano fabrication

2.2 Experimental Apparatus of an APPJ System Couples with an SPM

Figure 2a shows a schematic diagram of the experimental setup. The left hand of the schematic shows the setup for the SPM observation system and the right hand shows the setup for the APPJ ignition system. For the APPJ ignition system, helium gas was made to flow into a nanopipette at a fixed gas-flow rate using a mass flow controller. A pulse wave signal generated by a function generator was amplified to a high voltage (several kV) using a high-voltage amplifier, which was applied to a thin copper tape electrode wound around the nanopipette. In this system, the edge of the nanopipette was monitored during the APPJ processing using an optical microscope.

In the SPM system, we employed a shear-force feedback control technique for positioning the nanopipette and imaging the topographical surface. When the probe, oscillated laterally at its resonance frequency, was positioned in the vicinity of the surface, the resonance frequency of the probe vibration shifted due to the shear

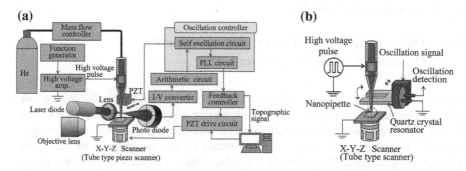

Fig. 2 Schematic of experimental set up for APPJ fine processing and SPM observation. **a** Detection method of the nanopipette oscillation using an optical sensor. **b** Detection method of the nanopipette oscillation using a quartz tuning forks sensor

force. The nanopipette probe was oscillated with a piezoelectric transducer (PZT) at its resonance frequency. In this study, to detect the oscillation signal, we tried two methods, i.e. one is an optical detection sensor, and another is a detection system using piezoelectric quartz tuning forks. The two methods were evaluated for the nanopipette positioning under the generation of APPJ inside the pipette. With regard to the optical detection system, the edge of the oscillated probe was illuminated using a laser beam with a 670 nm wavelength from a laser diode (LD) as shown in Fig. 2a. The shadow of the probe edge oscillation in the beam spot was projected onto a split photodetector (PD) for detection of the vibration. After passing through a home-made I–V convertor and an arithmetic circuit, the oscillation signal can be detected. With regard to the oscillation detection method using the piezoelectric quartz tuning forks, the simple setup is shown in Fig. 2b. The tuning forks are one of the best mechanical oscillator which have very high quality factors.

Piezoelectric quartz tuning forks were introduced into scanning probe microscopy. The piezoelectric effect of quartz allows to excite and detect the vibration of the nanopipette by contacting with the tuning forks. In both cases of the optical sensor and the tuning forks sensor, the detected signal of the vibration was input into a self-oscillated circuit. The excitation voltage of the PZT was generated by the self-oscillated circuit. At that time, the probe could always be oscillated due to this closed-loop system. To detect the frequency shift, a phase-locked loop circuit was employed. The signal of the frequency shift was input into the control circuit. Then the output signal was amplified by a PZT driver that drives the z-axis of a three-axis tube type piezoscanner. By maintaining the frequency shift at a constant predefined level, the probe–surface distance could be controlled with nanometer scale accuracy. In this study, the nanopipette probe was positioned in the vicinity of the surface by setting the frequency shift in the shear-force feedback control at 100 Hz. Under this feedback condition, the topography of the material surface after APPJ irradiation could be imaged by scanning the nanopipette probe on the surface.

3 Experimental Results

3.1 Influence of APPJ Generation on Pipette Oscillation Detection Sensors

Figure 3a shows the signals while generating APPJ, detected using the tuning fork sensor. The oscillation signals were recorded using a digital oscilloscope. As shown in the figure, the oscillation signal detected with the tuning forks was very disturbed. The tuning forks sensor was contacted on the nanopipette in which the plasma was generated. Thus, the plasma induced large electrical noise of the quartz tuning forks sensor placed just near the plasma source. Furthermore, the pipette edge was often damaged by the plasma enhanced at the metal electrodes of the

(a)   (b)

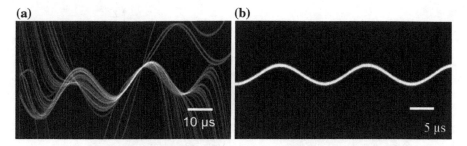

Fig. 3 Oscillation signal of the nanopipette under shear force feedback while generating APPJ. **a** The oscillation signal detected with the tuning fork sensor. **b** The oscillation signal detected with the optical sensor

tuning forks contacted on the pipette. On the other hand, Fig. 3b shows the signal while generating APPJ, detected using the optical sensor. As shown in the figure, even while generating APPJ, the oscillation signal could be detected without disturbing the signal.

In the setup using the optical detection system, the photodetector could be placed at significantly far position from the plasma source inside the nanopipatte, which might reduce the electrical noise of the photodetector. Therefore, we employed the optical sensing system to detect the pipette oscillation for pipette-surface distance control under shear force feedback.

3.2 Hole Pattern Fabrication

Using the system, holes were fabricated. Figure 4a, b show a topographical 2D and 3D images of the processed holes on the photoresist film, respectively. The pipette edge was positioned in the vicinity of the surface. The amplitude and frequency of applied pulse voltage were 6.5 kV_{p-p} and 5 kHz, respectively. APPJ irradiation time was 0.5 s. After the 9 holes were fabricated, the processed surface was imaged by scanning with the same nanopipette used for the fabrication. As shown in the figures, 9 holes were successfully fabricated. The size of the holes were almost the same. Figure 4c shows a cross-sectional profile of the hole. The width at half maximum of the hole was 250 nm. In the vicinity of the pipette aperture, the APPJ is well localized and the density of APPJ is higher. Thus, the processing efficiency might increase as the pipette-surface distance decreases. In this system, it was possible to position the pipette aperture in the vicinity of the photoresist surface using the SPM function of the accurate pipette-surface distance control. Therefore, using the system, sub-micrometer holes could be fabricated with high reproducibility.

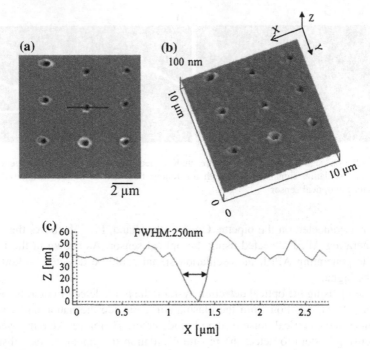

Fig. 4 Holes processed using the SPM system. **a** 2D Topographical image. **b** 3D topographical image. **c** Cross-sectional profile of *line* in image (**a**)

4 Conclusion

In this study, we developed a novel fine-processing system combining an APPJ ignition device with an SPM. A photoresist film was employed for the APPJ processing matter. For positioning the nanopipette probe in the vicinity of the surface using the shear-force feedback of the SPM technique, two methods using an optical sensor and a quartz tuning forks sensor were respectively evaluated to detect the pipette oscillation for shear force feedback under generating APPJ. The oscillation signal detected using the tuning forks sensor was disturbed while generating APPJ. On the other hand, the oscillation signal was robustly detected using the optical sensor even while generating APPJ. Using the fabrication system equipped with the optical sensor, the photoresist surface was fabricated by localized APPJ irradiated from the nanopipette aperture positioned in the vicinity of the surface. Sub-micrometer holes were successfully fabricated with reproducibility. This technique would be applicable not only to fabrications of micro and nano devices but also to various applications of local plasma treatments in the fields of engineering, medicine and biology.

References

1. Schütze, A., Jeong, J.Y., Babayan, S.E., Park, J., Selwyn, G.S., Hicks, R.F.: The atmospheric pressure plasma jet: a review and comparison to other plasma sources. IEEE Trans. Plasma Sci. **26**(6), 1685–1694 (1998)
2. Kitano, K., Taniguchi, K., Takagi, K., Namihira, Hattori T.K.: Let's obtain an atmospheric pressure plasma. J. Plasma Fusion Res. **84**(1), 19–28 (2008)
3. Akishev, Y., Grushin, M., Karalnik, V., Truchkin, N., Kholodenko, V., Chugunov, V., Kobzev, E., Zhirkova, N., Irkhina, I., Kireev, G.: Atmospheric-pressure, nonthermal plasma sterilization of microorganisms in liquids and on surfaces. Pure Appl. Chem. **80**(9), 1953–1969 (2008)
4. Lee, H.J., Oh, J.-S., Heo, S.W., Moon, J.H., Kim, J.-H., Park, S.G., Park, B.C., Kweon, G.R., Yim, Y.-H.: Peltier heating-assisted low temperature plasma ionization for ambient mass spectrometry. Mass Spectrum. Lett. **6**(3), 71–74 (2015)
5. Ideno, T., Ichiki, T.: Maskless etching of microstructures using a scanning microplasma etcher. Thin Solid Films **506–507**, 235–238 (2007)
6. Kakei, R., Ogino, A., Iwata, F., Nagatsu, M.: Production of ultrafine atmospheric pressure plasma jet with nano-capillary. Thin Solid Films **518**, 3457–3460 (2010)
7. Motrescu, I., Ogino, K., Nagatsu, M.: Micro-patterning of functional groups onto polymer surface using capillary atmospheric pressure plasma jet. J. Photopolym. Sci. Technol. **25**, 529–534 (2012)
8. Abuzairi, T., Okada, M., Mochizuki, Y., Poespawati, N.R., Wigajatri, R., Nagatsu, M.: Maskless functionalization of a carbon nanotube dot array biosensor using an ultrafine atmospheric pressure plasma jet. Carbon **89**, 208–216 (2015)
9. Topala, I., Nagatsu, M.: Capillary plasma jet: a low volume plasma source for life science applications. Appl. Phys. Let. **106**, 054105 (2015)

Fabrication of 2D TiO$_2$ Nanopatterns by Plasma Colloidal Lithography

Alexandra Demeter, Alexandra Besleaga,
Vasile Tiron and Lucel Sirghi

Abstract The present study investigates the capability of plasma colloidal lithography technique consisting on three fabrication steps, colloidal mask deposition, thin film deposition with colloidal masks and colloidal mask lift off, to fabricate 2D patterns of TiO$_2$ on silicon substrate. For the plasma assisted thin film deposition step, we used reactive high power impulse magnetron sputtering, a technique known to yield highly compact thin films with good control of composition and crystallinity. The fabricated 2D nanopatterns were investigated by atomic force microscopy and scanning force microscopy. Highly ordered 2D crystal array of TiO$_2$ triangular islands with the maximum height of about 10 nm were obtained independently of the TiO$_2$ thin film deposition time (thickness). The relatively low value of TiO$_2$ pattern height is explained by the shadow effects of the mask, a large number of particles from the gas phase contributing on the film deposited on the mask, until complete filling of the void spaces between nanoparticles of the mask.

Keywords Nanopatterns · Plasma colloidal lithography

1 Introduction

Colloidal lithography (CL) is a powerful and cost-effective method of obtaining 2D nano-patterns with controlled shape, size, and spacing over relatively large surface area. Therefore, CL is a popular technique used for fabrication of a large number of 2D patterns on flat substrates. In this study we use colloidal particles of polystyrene in a hexagonally packed array as a colloidal mask (CM) and plasma thin film deposition for fabrication of TiO$_2$ patterns on a silicon substrate [1, 2]. Usually, the CM is fabricated by self-assembly depositions of size monodispersed colloidal spheres to form a 2D colloidal crystal mask on a substrate. The self-assembly

A. Demeter (✉) · A. Besleaga · V. Tiron · L. Sirghi
Iasi Plasma Advanced Research Center (IPARC), Faculty of Physics,
Alexandru Ioan Cuza University of Iasi, 700506 Iasi, Romania
e-mail: alexandra_demeter@yahoo.com

© Springer International Publishing AG 2017
R. Jabłoński and R. Szewczyk (eds.), *Recent Global Research and Education:
Technological Challenges*, Advances in Intelligent Systems and Computing 519,
DOI 10.1007/978-3-319-46490-9_17

techniques of colloidal mask fabrication are diverse comprising dip-coating, floating on an interface, electrophoresis deposition, physical and chemical template-guided self-assembly, and spin-coating [3], which was the method chosen for our study. Spin-coating provides advantages for both scaling up and mass production since the process is rapid and compatible with wafer processing. Surface-patterning methods employing templates prepared by self-assembly processes are highly efficient in fabricating 2D patterned nanostructures on large surface areas since they are basically time-saving approaches with low equipment cost, compared with the conventional lithography methods [4].

Colloidal lithography technique based on these colloidal masks have attracted great attention because of their potential applications in technological fields such as biosensors, data storage and optoelectronic devices [5]. The advancement and development of many such 'nano' applications strongly rely on the design and controlled fabrication of nanoparticle arrays and patterned films. For example, metal films deposited over colloidal masks constitute a low-cost periodic structure as an interesting photonic material with plasmonic properties [6].

In this paper, we investigate the capability of plasma colloidal lithography (PCL) technique to fabricate 2D patterns of TiO_2 on silicon substrate. The PCL used in this study consists of three fabrication steps: (1) colloidal mask deposition; (2) thin film deposition with colloidal masks, and (3) ultra sound lift off of the colloidal mask with deposited TiO_2 film. For the plasma assisted thin film deposition step, we used reactive high power impulse magnetron sputtering (HiPIMS), a technique known to yield highly compact thin films with good control of composition and crystallinity.

2 Experimental Process. Materials and Methods

The TiO_2 2D patterns fabrication consisted in three steps illustrated in Fig. 1. Firstly, the silicon substrate has been cleaned and hydroxylated by exposure for 10 min to low-pressure plasma of water vapour and air mixture gases [7]. Then, a drop (90 μl) of aqueous colloidal solution (concentration of 0.609×10^{14} particle/mL) was deposited and spread instantaneously on the whole surface of the substrate. The nanoparticles of polystyrene (polystyrene Latex Beads = LB-5) with a diameter of 500 nm were purchased from Sigma Aldrich as a 15 mL vial colloidal solution with a stock concentration of 1828×10^{14} particles/mL. The substrate with the colloidal solution of PS beads has been loaded on a spin coater machine (LAURELL TECHNOLOGIES WS-650) and spinned in four steps with speeds of 200, 400, 800 and 3800 rotations/minute, respectively. The third step consists in deposition of TiO_2 thin film (with thickness 100 nm) by High Power Impulse Magnetron Sputtering (HiPIMS) working in

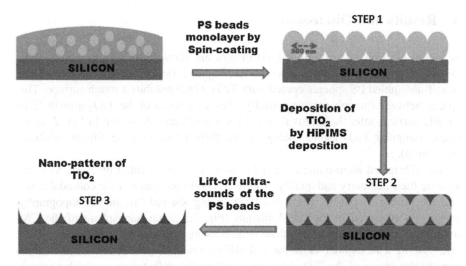

Fig. 1 Schematic diagram of TiO$_2$ nano-pattern fabrication by plasma colloidal lithography. *Step 1* spin coat a hexagonally close packed monolayer of polystyrene beads on substrates. *Step 2* deposition of metals oxides by HiPIMS method. *Step 3* lift-off by ultra-sound of the colloidal mask

multipulse mode (m-HiPIMS). A pure titanium target (99.95 % Ti purity, diameter of 50 mm and thickness of 5 mm) has been sputtered in reactive working gas (Ar and O$_2$ with mass flow rates of 50 and 0.1 sccm, respectively, at the total gas pressure of 30 mTorr) in order to deposit TiO$_2$ thin films. Sequences of three high voltage (-950 V) pulses with 3 μs in width and 50 μs in delay time were applied to the cathode with a repetition frequency ranged between 800 Hz. The distance between the target and the substrate was 10 cm and the substrates were at room temperature. More details on HiPIMS deposition of TiO$_2$ thin films are found in [8]. In the final fabrication step, the colloidal mask with the deposited film has been lift-off by sonication in deionized water. As results, the 2D ordered nanoaptterns of TiO$_2$ remained on the silicon substrate.

The morphology of the deposited TiO$_2$ patterns was investigated by atomic force microscopy (AFM) operated in tapping mode using a silicon AFM probe (HQ: NSC35/NoAl from μmasch) with tip curvature radius of 10 nm (nominal value) and cantilever frequency of 388 kHz. The topography images were levelled by in house software in order to correctly distinguish the substrate from the TiO$_2$ structures. Scanning electron microscopy (SEM) analysis was carried out to investigate the surface morphology of the deposited TiO$_2$ patterns, by a Hitachi S-3400N scanning electronic microscope working with a monoenergetic electron beam with the energy of 20 keV.

3   Results and Discussion

Scanning electron microscope (SEM) images are shown in Fig. 2 to demonstrate the result of steps 2 and 3 described above. Figure 2 (left side) shows a monolayer of self-assembled PS spheres coated with TiO_2 which exhibits a rough surface. The spaces between the spheres are partially filled as a result of the TiO_2 growth. The sample surface, after the removal of the PS nanospheres, is shown in Fig. 2 (right side), triangular TiO_2 islands being visible (bright filed) on the silicon substrate (dark field).

The fabricated nano-patterns of TiO_2 were also investigated by AFM to characterize their geometry and quality. Figure 3a shows an image of the colloidal mask and the profile of polystyrene nano beads along the red line on the topography image. As observed in the AFM images (Fig. 3a), each nanosphere of the CM monolayer is in contact with six close neighbouring nanospheres, which means they arranged in a hexagonal close-packed (HCP) order. Figure 3b shows the AFM topography image of the TiO_2 film deposited on the polystyrene nanosphere mask (before lift-off step). The film surface shows a high roughness indicating a columnar structure of the deposited film.

In the HiPIMS deposition process, the TiO_2 patterns were formed on the silicon substrate with positions and boundaries defined by the open spaces between nanospheres of the colloidal mask. The shape and thickness of the fabricated TiO_2 nanopatterns depended on the void spaces between the beads of the masks and the shadow effects of the mask with the deposited film. Thus, the pattern on the substrate has grown until the void spaces between beads were filled up by the deposited film. The AFM topography images of the fabricated patterns have shown that, for the deposited films with thickness larger than 45 nm, the pattern maximum height remained constant (approximately, 11 nm) because at this film thickness the

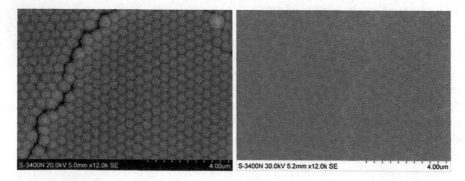

Fig. 2 (*left*) SEM image of TiO_2 thin film (thickness of 60 nm) deposited on the colloidal mask of PS beads with diameter of 500 nm (a crack in the film is visible along a line of defects in the colloidal mask) and (*right*) SEM image of TiO_2 nano-pattern (*bright area*) remained on silicon substrate (*dark area*) after lift-off of the colloidal mask

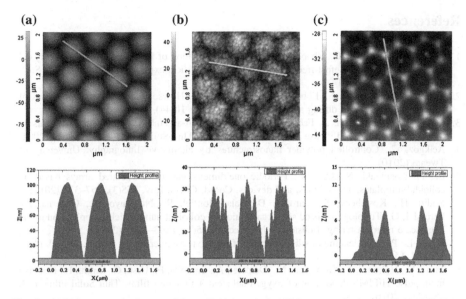

Fig. 3 **a** AFM topography image of the colloidal mask. **b** AFM topography image of the colloidal mask covered by a titanium dioxide thin film. **c** AFM topography image of TiO₂ nano-pattern fabricated by m-HiPIMS deposition with a colloidal mask with beads with diameter of 500 nm. The height profiles (*along red line*) for each topography image were also represented

opening spaces between the nanobeads were closed by the deposited film and there was no further deposition on patterns after this happened.

4   Conclusion

In this study, we have investigated HiPIMS deposition with colloidal masks as a fabrication technique of patterns consisting an ordered 2D hexagonal lattice of approximately triangular nano-islands of TiO₂ on silicon substrate. The fabricated patterns were investigated by atomic force microscopy and scanning electron microscopy. As an important result of this study, we have found that for colloidal mask of PS beads with diameter of 500 nm, the maximum height of TiO₂ nanopatterns cannot exceed 15 nm. This is happening because for films thicker than 45 nm the void spaces between the beads of the mask were completely filled by the deposited film and no further deposition on the pattern could take place.

Acknowledgments This work was supported by JOINT RESEARCH PROJECTS PN-II-ID-JRP-2012-RO-FR-0161.

References

1. Cheng, Y., Jönsson, P.G., Zhao, Z.: Controllable fabrication of large-area 2D colloidal crystal masks with large size defect-free domains based on statistical experimental design. Appl. Surf. Sci. **313**, 144–151 (2014)
2. Toolan, D.T.W., Syuji, F., Ebbens, S.J., Nakamura, Y., Howse, J.R.: On the mechanisms of colloidal self-assembly during spin-coating. Soft Matter **10**(44), 8804–8812 (2014)
3. Ko, H., Lee, H., Moon, J.: Fabrication of colloidal self-assembled monolayer (SAM) using monodisperse silica and its use as a lithographic mask. Thin Solid Films **447**, 638–644 (2004)
4. Acikgoz, C.: Controlled polymer nanostructures by alternative lithography. University of Twente (2010)
5. Li, Y., Koshizaki, N., Cai, W.: Periodic one-dimensional nanostructured arrays based on colloidal templates, applications, and devices. Coord. Chem. Rev. **255**(3), 357–373 (2011)
6. Kadiri, H., Kostcheev, S., Turover, D., Salas-Montiel, R., Nomenyo, K., Gokarna, A., Lerondel, G.: Topology assisted self-organization of colloidal nanoparticles: application to 2D large-scale nanomastering. Beilstein J. Nanotechnol. **5**(1), 1203–1209 (2014)
7. Sirghi, L.: Plasma cleaning of silicon surface of atomic force microscopy probes. Rom. J. Phys. **56**, 144–148 (2011)
8. Tiron, V., Velicu, I.-L., Dobromir, M., Demeter, A., Samoila, F., Ursu, C., Sirghi, L.: Reactive multi-pulse HiPIMS deposition of oxygen-deficient TiO x thin films. Thin Solid Films **603**, 255–261 (2016)

Pulse-Driven, Photon-Coupled, Protein-Based Logic Circuits

Balázs Rakos

Abstract The present study discusses the potential feasibility of protein-based computing circuits consisting of photon-coupled, photoswitchable protein molecules, driven by appropriate photon pulses. According to our considerations, the proposed architectures may be promising possible building blocks in nanoscale, terahertz-frequency computing processors of the future with low energy consumption and dissipation. The operational principle takes advantage of the experimentally demonstrated switching mechanism of photoswitchable proteins.

Keywords Molecular electronics · Organic electronics · Protein · Computing architectures · Logic circuits · Nanoelectronics · Photon coupling

1 Introduction

The possible application of photon pulse-driven molecular architectures for computing and digital signal processing has been examined by several different works [1–4]. Due to their advantageous properties, proteins are promising building elements for such purposes [5–7]. Coulomb coupling is a possible method for the integration of individual molecules into logic circuits, and in a previous paper [5] we proposed dipole-dipole coupled, photoswitchable protein computing arrangements. However, we suggest that photon coupling is another, more promising way of integration in such circuits with many advantages, since it is stronger than dipole-dipole interaction, which decays rapidly with distance and extremely sensitive to the separation between neighboring molecules, and permits more stable, robust operation.

B. Rakos (✉)
Department of Automation and Applied Informatics, Budapest University
of Technology and Economics, Budapest, Hungary
e-mail: balazs.rakos@gmail.com

B. Rakos
MTA-BME Control Engineering Research Group, Budapest, Hungary

© Springer International Publishing AG 2017
R. Jabłoński and R. Szewczyk (eds.), *Recent Global Research and Education:
Technological Challenges*, Advances in Intelligent Systems and Computing 519,
DOI 10.1007/978-3-319-46490-9_18

2 Operational Principle

Photoswitchable proteins, like Dronpa [8], an artificially engineered protein, can be switched between two forms by electromagnetic radiation with well-defined frequencies (radiation with a certain frequency switches the molecule to form2, and radiation with a different frequency switches it back to form1). Each form can emit radiation (in the case of Dronpa only one of the forms is fluorescent, the other one is "dark").

Let us assume now that we have got a hypothetical photoswitchable protein (the evolution of protein engineering might permit the realization of such proteins in the future) with two different forms (form1, and form2), it can be switched to form2 by radiation with f12, and to form1 with f21 frequency. We also assume that both forms are fluorescent, form1 can be excited by f1e, and form2 can be excited by f2e. The frequencies of the emitted fluorescent radiations in the case of form1 and form2 are f1g and f2g, respectively. The schematic diagram of photoswitching in the case of such hypothetical protein is displayed in Fig. 1.

Digital signal propagation in a molecular chain is possible with the application of three different proteins (see Fig. 2 for illustration of the process). Protein A is the input, and protein C is the output molecule. The rest of the chain consists of an arbitrary number of protein B. Form1 corresponds to logic '0', form2 to logic '1'. The sequence of the operation is the following: radiation with fA12 switches the input molecule to form2, the subsequent fA2e excites form2 of protein A resulting in a radiation with fA2g. This radiation matches the one needed for switching protein B to form2, thus the neighbor of the input molecule is switched to form2. The rest of the process is straightforward by looking at Fig. 2, by taking into account that both proteins B and C are designed to switch to their form2 due to irradiation by fB2g (this also means that fA2g = fB2g).

The two-input OR logic gate can be realized with the aid of three proteins with well-defined switching frequencies. The proteins are placed next to each other (see Fig. 3), molecules A and B are the inputs, and protein C in the middle is the output. In the case of each molecule, form1 corresponds to logic '0', and form2 corresponds to logic '1'. Input1 can be set to '1' by irradiating the entire structure with fA12, input2 can be switched to '1' by irradiating the arrangement with fB12. If, after setting the inputs to their desired logic states, the gate is irradiated with fA2e

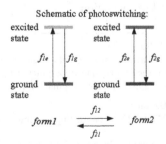

Fig. 1 Schematic of photoswitching in the case of a hypothetical photoswitchable protein

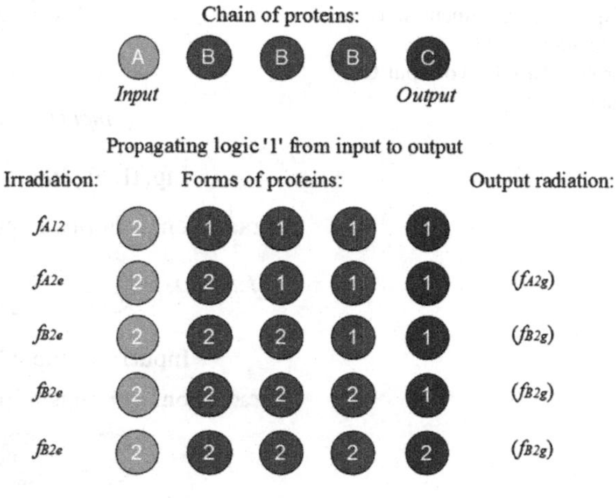

Fig. 2 Propagating a bit
through a one-dimensional
protein chain

Chain of proteins:

Propagating logic '1' from input to output

Irradiation: Forms of proteins: Output radiation:

and fB2e simultaneously, and both proteins A and B are designed to emit a
fluorescent radiation with a frequency of fC12 (fA2g = fB2g = fC12), protein C
will switch to form2 only if input1 or input2 or both input molecules are set to '1'.
The output can be read out by irradiating the gate with fC1e and fC2e and detecting
the frequency of the resulting fluorescent radiation (fC1g corresponds to logic '0'
and fC2g to logic '1'). The process is displayed in Fig. 3.

Figure 4 displays the arrangement and operation of the inverting logic gate
consisting of two different proteins (proteins A and B). If fA1g = fB12 and
fA2g = fB21 the architecture is operational.

3 Discussion

Experiments already showed that switching between two different forms is possible
in photoswitchable proteins with the application of irradiation with well-defined
frequencies [8]. Since protein engineering enables one to design molecules with the
desired properties, we can assume that realization of proteins with properties
described in the previous section may be possible in the future. Since, as we showed
in this paper, digital signal propagation, the OR and inverting logic gates (in this
way universal binary logic operations are possible) are potentially realizable with
the aid of photon-coupled, photoswitchable proteins, they are promising potential
elements in computing architectures of the future. The average switching speed of
Dronpa, an existing photoswitchable protein is on the order of milliseconds, which
is not too impressive considering our purposes, however, with well-designed
molecules speed on the order of picoseconds may be achieved (this would enable
THz-frequency operations), since structural rearrangements of proteins are on that
time-scale [7, 9].

Fig. 3 Arrangement and operation of the protein-based, two-input OR gate

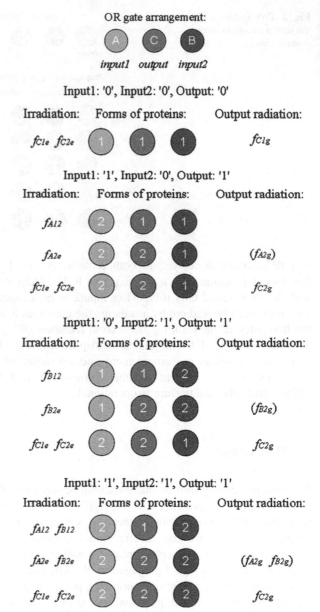

OR gate arrangement:

A   C   B

input1   output   input2

Input1: '0', Input2: '0', Output: '0'

Irradiation:	Forms of proteins:	Output radiation:
$f_{C1e}\ f_{C2e}$	1 1 1	f_{C1g}

Input1: '1', Input2: '0', Output: '1'

Irradiation:	Forms of proteins:	Output radiation:
f_{A12}	2 1 1	
f_{A2e}	2 2 1	(f_{A2g})
$f_{C1e}\ f_{C2e}$	2 2 1	f_{C2g}

Input1: '0', Input2: '1', Output: '1'

Irradiation:	Forms of proteins:	Output radiation:
f_{B12}	1 1 2	
f_{B2e}	1 2 2	(f_{B2g})
$f_{C1e}\ f_{C2e}$	2 2 1	f_{C2g}

Input1: '1', Input2: '1', Output: '1'

Irradiation:	Forms of proteins:	Output radiation:
$f_{A12}\ f_{B12}$	2 1 2	
$f_{A2e}\ f_{B2e}$	2 2 2	$(f_{A2g}\ f_{B2g})$
$f_{C1e}\ f_{C2e}$	2 2 2	f_{C2g}

Fig. 4 Protein-based
inverting gate arrangement
and operation

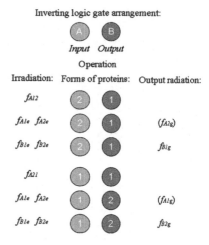

Inverting logic gate arrangement:

A   B

Input   Output

Operation

Irradiation:	Forms of proteins:	Output radiation:
f_{A12}	2 1	
f_{A1e} f_{A2e}	2 1	(f_{A2g})
f_{B1e} f_{B2e}	2 1	f_{B1g}
f_{A21}	1 1	
f_{A1e} f_{A2e}	1 2	(f_{A1g})
f_{B1e} f_{B2e}	1 2	f_{B2g}

4 Conclusion

We proposed a method by which photon-coupled, photon pulse-driven proteins can be utilized for computing purposes. We showed that such proteins with well-designed properties can potentially realize universal binary logic gates, and enable digital signal propagation, as well. The structures may operate at terahertz frequencies if the molecules are designed accordingly.

References

1. Körner, H., Mahler, G.: Optically driven quantum networks: applications in molecular electronics. Phys. Rev. B **48**(4), 2335–2346 (1993). doi:10.1103/PhysRevB.48.2335
2. Lloyd, S.: Programming pulse driven quantum computers, 17 Dec 1999. arXiv:quant-ph=9912086
3. Csurgay, Á.I., Porod, W., Rakos, B.: Signal processing by pulse-driven molecular arrays. Int. J. Circuit Theory Appl. **31**, 55–66 (2003). doi:10.1002/cta.225
4. Rakos, B., Porod, W., Csurgay, Á.I.: Computing by pulse-driven nanodevice arrays. Semicond. Sci. Technol. **19**, 472–474 (2004). doi:10.1088/0268-1242/19/4/155
5. Rakos, B.: Simulation of Coulomb-coupled, protein-based logic. J. Autom. Mobile Robot. Intell. Syst. **3**(4), 46–48 (2009)
6. Rakos, B.: Coulomb-coupled, protein-based computing arrays. Adv. Mater. Res. **222**, 181–184 (2011). doi:10.4028/www.scientific.net/AMR.222.181
7. Rakos, B.: Modeling of dipole-dipole-coupled, electric field-driven, protein-based computing architectures. Int. J. Circuit Theory Appl. **43**, 60–72 (2015). doi:10.1002/cta.1924
8. Habuchi, S., Ando, R., Dedecker, P., Verheijen, W., Mizuno, H., Miyawaki, A., Hofkens, J.: Reversible single-molecule photoswitching in the GFP-like fluorescent protein Dronpa. Proc. Natl. Acad. Sci. **102**(27), 9511–9516 (2005). doi:10.1073/pnas.0500489102
9. Xu, D., Phillips, J.C., Schulten, K.: Protein response to external electric fields: relaxation, hysteresis, and echo. J. Phys. Chem. **100**, 12108–12121 (1996). doi:10.1021/jp960076a

Nanosilica Suspensions for Monocrystalline Silicon Wafers CMP Surface for Micro- and Nanoelectronics

Yanina Kasianok, Vladimir Gaishun, Olga Tyulenkova
and Sergey Khakhomov

Abstract Method of chemical-mechanical polishing (CMP) approach to production submicron and deep submicron levels. For monocrystalline silicon wafer by CMP, we have developed a suspension, containing solid phase, amorphous, spherical particles of nanosized aerosil. SiO_2 particles are homogenously distribute in an alkaline medium. Removing material in CMP is the result of combined effects of chemically less active (etching) environment and mechanical particles of suspension and pad. In the study of the surface of silicon wafers after the final stage of the chemical- mechanical polishing using AFM the presence of surface defects and destruction of the surface layer is not established. Roughness of surface of monocrystalline silicon wafer (R_a) after polishing by silica suspension is 0.2–0.3 nm.

Keywords Chemical-mechanical polishing · Nanosilica suspensions · Monocrystalline silicon wafer · Aerosil

1 Introduction

The growth production of semiconductor devices and integrated circuits requires improving the quality of the various specific materials for electronic equipment, as well as production of materials that meet higher technology requirements. Now are semiconductor substrates with atomic dimensions microrelief. Therefore, there are constantly searching a new treatment methods and polishing materials for processing of silicon wafers.

Chemical mechanical polishing (CMP) is a technology, widely applied for the production of elements of devices of micro- and nanoelectronics [1]. CMP of semiconductor wafers is a most critical process stage for producing integrated circuits. The unique combination of chemical and mechanical effect of the working

Y. Kasianok (✉) · V. Gaishun · O. Tyulenkova · S. Khakhomov
Francisk Scorina Gomel State University, Gomel, Belarus
e-mail: ykosenok@gsu.by

fluid, solid particles and a relatively soft pad with a polished surface leads to an effective polishing and high purity obtained surfaces [2]. One type of CMP materials is based on the use of suspensions containing a solid amorphous spherical nanoparticles of silica dioxide, fumed obtained either by high temperature synthesis (aerosil), or are in the liquid phase sols. The main attention is on the crystalline perfection of surface and subsurface layers, physicochemical purity of the surface, which characterized by mechanical and chemical impurities and the surface roughness. Today is requires processing of semiconductor wafers with average height of micro relief near atomic sizes, so continuously searching of new processing methods and materials for their implementation. The investigation of new compositions for CMP suspensions of semiconductor and other materials and study the topography of the surface of monocrystalline silicon wafers after polishing suspensions are actually.

2 Experimental

We developed new polishing stabilized suspension base on fumed silica dioxide type OX-50 (Degussa, Germany) and investigated the surface of silicon wafers after CMP using atomic force microscopy (AFM). The aerosil OX-50 (SBET ≈ 50 m^2/g) can form stable suspensions in water with dispersed phase from individual primary particles ($d \approx 40$ nm) (Fig. 1). Methods of preparation of water polishing compositions includes three stages: mixing of the initial components, ultrasonic dispersing and purification of the composition from process impurities by centrifugation. Ethylene glycol is use as a surfactant [3]. With the introduction of surfactant wettability of the surface of wafers during polishing increases, and the rate of material removal increases. The ethylenediamine used as stabilizers for suspensions used for polishing stage I, and sodium hydroxide for suspensions used for polishing stage II (Table 1).

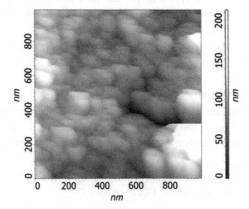

Fig. 1 AFM image of dry suspension residue

Table 1 Characteristics of polishing suspensions for CMP of semiconductor silicon wafers	Characteristic	Type of suspensions	
		SPS-55M	SPS-81M
	Stabilizer	NaOH	Ethylenediamine
	SiO$_2$ (mass.%)	25.0	23.0
	Particle diameter (nm)	80–100	60–80
	Density (g/cm^3)	1.168–1.172	1.138–1.142
	pH at 20 °C	10.4–10.8	12.4–12.8
	Viscosity (mPa s)	3.5	3.1
	Appearance	Milky white liquid	
	Expiry date	At least 6 months	

The tests performed of the polishing ability suspensions are produce on semi-automatic YU1MZ.105.004. The silicon wafers type KDB 12 used for test polishing. The suspension is flowing at a rate of 100–150 drops/min. The surface of the monocrystalline silicon wafers was study by high-resolution atomic force microscope SOLVER P 47—PRO (NT-MDT, Russia).

3 Results and Discussion

At CMP by polishing suspensions must take into account the connection of colloid-chemical properties of the polishing composition with the properties of the treated surface. Thus, the input factors are the dispersion medium, the size and shape of the cha-particles of the solid phase, the viscosity of the composition, as well as the ambient temperature. Figure 2 shows AFM image of the initial grinded silicon wafer (Table 2).

The basis of the chemical-mechanical polishing process are chemical reactions between the components of the liquid medium and polished material. At the initial stage of the substrate surface is etched, rough scratches and pits formed by defects depending on the properties of the dislocations. Using CMP alkaline conditions results in the formation of oxides on the surface of semiconductors, which are well

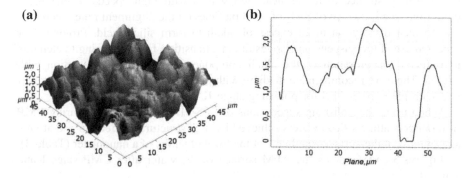

Fig. 2 The initial Si wafer surface after grinding: **a** AFM image; **b** surface profile

Table 2 CMP parameters of polishing suspensions

Characteristic	Type of suspensions	
	SPS-55M	SPS-55M
Stage	I	II
Recommended dilution	1:10	1:10
pH after dilution	11.2–11.6	10.4–10.6
Density after dilution (g/cm^3)	1.005	1.01
Pressure (kgf/cm^2)	0.4	1.0
Suspension flow (ml/min)	50	200
Polishing temperature (°C)	51–55	48–50
Processing time (min)	50	8
The amount of material removal (μm)	30	1
Recommended type pad	Politan	Segal

soluble in alkalis or friable low strength hydroxides that can be easily removed mechanically. For alkaline environments were used alkali—ethylenediamine $(CH_2)_2(NH_2)_2$ (for the polishing stage I) and NaOH (for polishing stage II). The chemical reaction interaction can be written as follows:

$$Si + 2OH^- + H_2O \rightarrow SiO_3^{2-} + 2H_2 \uparrow$$

$$2Na^+ + SiO_3^{2-} \rightarrow Na_2SiO_3$$

$$NH_2 - (CH_2)_2 - NH_2 + 2H_2O \rightarrow HONH_3 - (CH_2)_2$$
$$- NH_3OH \rightarrow NH_3^+ - (CH_2)_2 - NH_3^+ + 2OH$$

$$2nNH_3^+ - (CH_2)_2 - NH_3^+ + 2nSiO_3^{2-}$$
$$\rightarrow -[NH_3 - (CH_2)_2 - NH_3 - SiO_3 - NH_3 - (CH_2)_2 - NH_3 - SiO_3]_n$$

The pH in this case plays an important role. At pH < 10, a chemical reaction is unstable formation of alkaline compounds and solid phase particles can contact the surface. Conversely, high concentration of alkali causes a gradual transition from the polishing surface to a chemical etch wafers with high speeds. Under these conditions, the influence of SiO_2 of solid particles on the alignment relief decreases due to their dissolution in an excess of alkali to form silicic acid. Prolonged or repeated use of the suspension also results in a transition from polishing to etch due to reduced concentration of the alkali component by chemical interaction with silicon. There are recommended medium with pH = 11.0–13.0 for polishing stage I and with pH = 10.5–11.0 for polishing stage II.

When using the polishing suspensions of SPS-81M and SPS-55M during CMP of monocrystalline silicon wafers achieved by high structural perfection and atomic smoothness of the surface roughness at the level of tenths of a nanometer (Table 3).

Figures 3 and 4 shows the AFM surface of the wafer after CMP stage I and stage II.

Table 3 The parameters of semiconductor wafers after polishing by suspensions SPS-81 and SPS-55M

Wedge (µm)		6–12
Micro-scratches (%)		10
Restoration (%)		10
The surface roughness after CMP stage I,	R_a (nm)	0.55–1.07
	R_z (nm)	1.21–1.35
	R_{max} (nm)	3.67
The surface roughness after CMP stage II	R_a (nm)	0.195–0.302
	R_z (nm)	0.90
	R_{max} (nm)	0.54–0.61
Defectiveness of the surface layer		Complete absence of the damaged layer

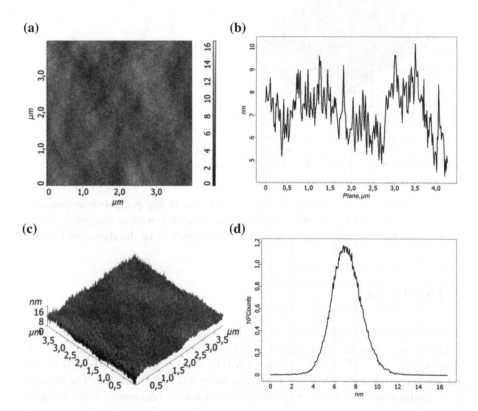

Fig. 3 The Si wafer surface after polishing stage I: **a** 2d AFM image; **b** surface profile; **c** 3d AFM image and **d** histograms of the distribution of surface roughness

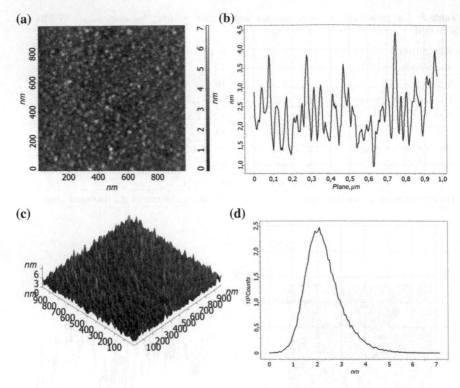

Fig. 4 The Si wafer surface after polishing stage II: **a** 2d AFM image; **b** surface profile; **c** 3d AFM image and **d** histograms of the distribution of surface roughness

The optimum ratio of component concentrations in the polishing suspensions gives effective polishing monocrystalline silicon wafers with a roughness at the level of tenths of a nanometer, and the complete absence of the damaged layer.

4   Conclusion

The developed suspensions are apply at the stage of finishing and superfinishing polishing of monocrystalline silicon wafers. Roughness of surface of monocrystalline silicon wafer (Ra) after finishing CMP by silica suspension is 0.55–1.07 nm and after superfinishing CMP by silica suspension is 0.195–0.302 nm. The increase of the surface quality of monocrystalline silicon wafers (decrease of the damaged layer) after polishing by suspensions is provided to use of aerosil with an average particle size of 40 nm, as well as the original manufacturing technology of the suspensions.

References

1. Zantye, P.B., Kumar, A., Sikder, A.K.: Chemical mechanical planarisation for microelectronics applications. Mater. Sci. Eng. R. **45**, 89–220 (2004)
2. Gaishun, V.E., et al.: Chemistry, Physics and Technology of Surface, vol. 14, p. 423 (2008)
3. Kosenok, Y.A., Gaishun, V.E., Tyulenkova, O.I.: The composition for polishing semiconductor materials. Belarus Patent Application № 20130711, 06 June 2013

Manipulation of Single Charges Using Dopant Atoms in Silicon—Interplay with Intervalley Phonon Emission

Yukinori Ono, Masahiro Hori, Gabriel P. Lansbergen and Akira Fujiwara

Abstract We make a brief review about our research on single electron manipulation using individual donors in silicon. The device we developed essentially consists of a silicon nanowire MOS field-effect transistor with local arsenic implantation between a set of fine gates. We demonstrate that, by tuning the gate voltages, electrons are exclusively transferred via individual donor atoms and the number of transferred electrons is tunable. Control of single phonon emission using the present device is also discussed.

Keywords Dopant · Silicon · Transistor · Phonon

1 Introduction

Manipulation of single charges using dopant atoms in silicon has recently attracted much attention from the viewpoint of quantum information technology and the so-called beyond CMOS technology [1, 2]. Electron transport via single and minute number of dopants have thus been intensively investigated by several groups [3–10]. Furthermore, fabrication techniques to create devices where the position of each dopant atom in the active device area is controlled are being developed taking both top-down and bottom-up approaches [11, 12].

In this article, we make a brief introduction about our device concept based on single dopant functionality from the viewpoint of single charge and phonon control. We demonstrate multiple-donor based single electron pumps [13]; the device can pump a quantized number of electrons from the source to drain via a tunable number of individual donors. We also discuss the possibility of applying the device

Y. Ono (✉) · M. Hori
Research Institute of Electronics, Shizuoka University, Hamamatsu, Japan
e-mail: ono.yukinori@shizuoka.ac.jp

G.P. Lansbergen · A. Fujiwara
NTT Basic Research Laboratory, Kanagawa, Japan

© Springer International Publishing AG 2017
R. Jabłoński and R. Szewczyk (eds.), *Recent Global Research and Education: Technological Challenges*, Advances in Intelligent Systems and Computing 519, DOI 10.1007/978-3-319-46490-9_20

to the emitter of single intervalley phonons, the control of which is critical for heat management in silicon transistors [14].

2 Device Concept and Fabrication

Figure 1 shows the concept of the charge transfer scheme by dopant atoms. A small number of dopants are introduced in a silicon channel and the dopant charge states are controlled by the two fine gates placed between the doped regions. The devices consisting of Si-wire MOSFETs in series were fabricated on a silicon-on-insulator (SOI) wafer (Fig. 2). A stacked gate layer structure is employed; the lower layer consist of three fine gates (LG, MG, and RG) defined by electron beam lithography and the top layer consist of a large single upper gate (UG). The fine gates are used to induce and control local electron barriers in the silicon nanowire. Donor atoms (arsenic) are introduced by ion implantation in the region between MG and RG before fine gate fabrication. The donors were implanted through a 60 nm wide aperture (with its width along the nanowire direction) in a predesigned e-beam mask. The upper gate layer is used to control the electron density in the silicon;

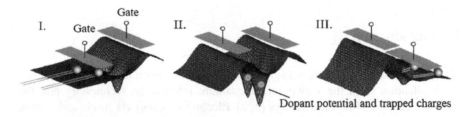

Fig. 1 Charge transfer using dopant atoms

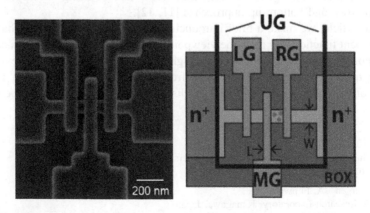

Fig. 2 Top view of the device. SEM image (*left*) and the schematic view (*right*)

application of a positive voltage leads to electron inversion in the undoped SOI layer beneath. The wide UG layer furthermore serves as a mask during an ion implantation step which forms the n+-type contact areas.

3 Single Electron Pump Operation

Figure 3 shows the charge transfer characteristics I_{SD} versus V_{UG} for increasingly negative right fine gate voltage (V_{RG}). We observe a decrease in I_{SD} for more negative V_{RG} combined with clear integer steps developing as a function of V_{UG}. When a negative potential is applied to V_{RG} (at fixed V_{UG}) it induces a depletion region extending well into the island region, making less donors available for charge transfer (see the band diagrams in Fig. 3). When we subsequently increase V_{UG} to higher positive values, in some cases donor are neutralized again, shifting the potential where inversion occurs to higher positive values. As Fig. 3 shows, the right fine gate voltages can thus be employed to tune the number of donors participating in the transport. Due to the different capacitive coupling between the donors in the depletion layer and the Si/SiO$_2$ interface where inversion takes place, the average slope of $I_{SD} - V_{UG}$ for $V_{UG} < V_{UG,is}$ is different from the slope at $V_{UG} > V_{UG,is}$, where $V_{UG,is}$ is the threshold voltage for inversion at the island region. These results indicate that the amount of donors participating in the charge transfer can be tuned by means of the fine gate voltages.

4 Intervalley Phonon Emission

We here discuss the electron capture process to a donor atom (step I–II in Fig. 1). The electron capture to a donor atom (more generally, to a Coulomb attractive center) has been intensively studied since mid 20th. In this process, the electron

Fig. 3 Band diagrams (*left*) and the pumping current with V_{RG} as a parameter (*right*)

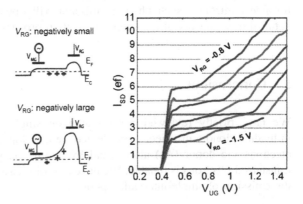

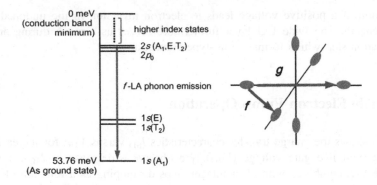

Fig. 4 *Left* Energy states for As in Si. The *vertical arrow* indicates the transition due to the *f*-LA phonon emission. *Right* Six degenerated valleys of Si conduction band. Two possible intervalley scattering are denoted by *f* and *g*

loses its energy by emitting either photon(s) or phonon(s). It was originally suggested the cascade-type relaxation, where multiple acoustic phonon emission plays a major role for the energy relaxation [15]. However, the energy relaxation process is governed by the energy/momentum conservation laws and symmetry of the electron wavefunction [16, 17], which restrict possible channels for transitions in particular between low lying energy states with a large gap. In fact, the relaxation mechanism strongly depends on the donor species. For example, THz photon emission governs the transition from $2p_0$ for P and Sb, while such an optical transition is not observed for As and Bi. This is because As and Bi have a strong coupling (resonant donor-phonon interaction [18]) with intervalley phonons (*f*-LA for As [19] and *f*-TO and *g*-LO for Bi [20]) with their transition time much shorter than that for the optical transitions. The energy levels and the possible transition with the phonon emission are depicted in Fig. 4 for the case of As, together with the conduction-band valley structure of Si.

We anticipate that the elaborate control of the electron capture process in the present device permits us to emit single intervalley phonons. Since the intervalley phonons are major source of the heat generation in Si nMOS transistors, the method for fine control of such phonon emission will open up a way to elucidate the detailed mechanisms of the energy relaxation process in the transistor.

5 Conclusions

We have briefly reviewed our research on single electron manipulation using individual donors in silicon. The device we developed allows us to convey single electrons via donor (As) atoms, and by tuning the gate voltage, the number of transferred electrons is tunable. We have also discussed possibility of the control of the emission of single intervalley phonons.

Acknowledgments This work was partially supported by the Grants-in-Aid for Scientific Research Nos. 23226009, 25289098, 25600015, 25706003, 26289105, 15K13970, 16H02339 and 16H06087 from the Japan Society for the Promotion of Science.

References

1. Zwanenburg, F.A., Dzurak, A.S., Morello, A., Simmons, M.Y., Hollenberg, L.C.L., Klimeck, G., Rogge, S., Coppersmith, S.N., Eriksson, M.A.: Silicon quantum electronics. Rev. Mod. Phys. **85**, 961–1019 (2013)
2. Jehll, X., Niquet, Y.-M., Sanquer, M.: Single donor electronics and quantum functionalities with advanced CMOS technology. J. Phys.: Condens. Matter **28**, 1–18, Article No. 103001 (2016)
3. Sellier, H., Lansbergen, G.P., Caro, J., Rogge, S., Collaert, N., Ferain, I., Jurczak, M., Biesemans, S.: Transport spectroscopy of a single dopant in a gated silicon nanowire. Phys. Rev. Lett. **97**, 1–4, Article No. 206805 (2006)
4. Calvet, L., Wheeler, R., Reed, M.: Observation of the linear stark effect in a single acceptor in Si. Phys. Rev. Lett. **98**, 1–4, Article No. 96805 (2007)
5. Ono, Y., Nishiguchi, K., Fujiwara, A., Yamaguchi, H., Takahashi, Y.: Conductance modulation by individual acceptors in Si nanoscale field-effect transistors. Appl. Phys. Lett. **90**, 1–3, Article No. 102106 (2007)
6. Khalafalla, M.A.H., Ono, Y., Nishiguchi, K., Fujiwara, A.: Identification of single and coupled acceptors in silicon nano-field-effect transistors. Appl. Phys. Lett. **94**, 1–3, Article No. 223501 (2007)
7. Pierre, M., Wacquez, R., Jehl, X., Sanquer, M., Vinet, M., Cueto, O.: Single-donor ionization energies in a nanoscale CMOS channel. Nat. Nanotechnol. **5**, 133–137 (2009)
8. Fuechsle, M., Miwa, J.A., Mahapatra, S., Ryu, H., Lee, S., Warschkow, O., Hollenberg, L.C.L., Klimeck, G., Simmons, M.Y.: A single-atom transistor. Nat. Nanotechnol. **10**, 1–5, Article No. 1038 (2012)
9. Tabe, M., Moraru, D., Ligowski, M., Anwar, M., Jablonski, R., Ono, Y., Mizuno, T.: Single-electron transport through single dopants in a dopant-rich environment. Phys. Rev. Lett. **105**, 1–4, Article No. 016803 (2010)
10. Moraru, D., Samanta, A., Anh, L.T., Mizuno, T., Mizuta, H., Tabe, M.: Transport spectroscopy of coupled donors in silicon nano-transistors. Sci. Rep. **4**, 1–6, Article No. 6219 (2014)
11. Shinada, T., Okamoto, S., Kobayashi, T., Ohdomari, I.: Enhancing semiconductor device performance using ordered dopant arrays. Nature **437**, 1128–1131 (2005)
12. Ho, J.C., Yerushalmi, R., Jacobson, Z.A., Fan, Z., Alley, R.L., Javey, A.: Controlled nanoscale doping of semiconductors via molecular monolayers. Nat. Mater. **7**, 62–67 (2008)
13. Lansbergen, G., Ono, Y., Fujiwara, A.: Donor-based single electron pumps with tunable donor binding energy. Nano Lett. **12**, 763–768 (2012)
14. Pop, E.: Energy dissipation and transport in nanoscale devices. Nano Res. **3**, 147–169 (2010)
15. Lax, M.: Cascade capture of electrons in solids. Phys. Rev. **119**, 1502–1523 (1960)
16. Hasegawa, H.: Spin-lattice relaxation of shallow donor states in Ge and Si through a direct phonon process. Phys. Rev. **118**, 1523–1534 (1960)
17. Griffin, A., Carruthers, P.: Thermal conductivity of solids IV: resonance fluorescence scattering of phonons by donor electrons in germanium. Phys. Rev. **131**, 1976–1995 (1963)
18. Pavlov, S.G., Zhukavin, R.K., Shastin, V.N., Hubers, H.-W.: The physical principles of terahertz silicon lasers based on intracenter transitions. Phys. Status Solidi B **250**, 9–36 (2013)
19. Hubers, H.-W., Pavlov, S.G., Zhukavin, R.K., Riemann, H., Abrosimov, N.V., Shastin, V.N.: Stimulated terahertz emission from arsenic donors in silicon. Appl. Phys. Lett. **84**, 3600–3602 (2004)
20. Onton, A., Fisher, P., Ramdas, A.K.: Anomalous width of some photoexcitation lines of impurities in silicon. Phys. Rev. Lett. **19**, 781–783 (1967)

Doped Two-Dimensional Silicon Nanostructures as a Platform for Next-Generation Sensors

Roland Nowak, Krzysztof Tyszka and Ryszard Jablonski

Abstract In this work, we provide results of studies devoted to nano-sensors based on doped two-dimensional silicon nanotransistors and p-n nanojunctions. Based on obtained results and analysis, we demonstrate that, under certain conditions, both these structures can be ultra-sensitive: nanotransistors can resolve single photons; whereas p-n nanojunctions are able to detect single charge. Detection mechanism, involving appearance of individual dopants and aspects related to the reduced dimensionality, has been proposed. Elaborated model can be successfully applied to nanotransistor- and nanojunction-based sensors.

Keywords Silicon · Dopant · Nanosensor · Nanojunction · Nanotransistor · Kelvin probe force microscope · Random telegraph signal

1 Introduction

Rapid progress of nanotechnology observed in the past two decades would not be possible without high precision measurements. Ground-breaking developments related to such measurements consequently gave birth to nanometrology, which nowadays is a principal part of nanotechnology-related disciplines, and is also strongly connected with ongoing works on new SI system [1].

Nanometrology, in order to be a well-founded discipline, requires intense efforts of scientists and engineers world-wide. As of today, it suffers from severe issues related to fundamental aspects, such as: standards, calibration and validation procedures, reliability evaluation etc. [2]. Nevertheless, research work is progressing and scientific community hopes to overcome mentioned difficulties within coming years.

R. Nowak (✉) · K. Tyszka · R. Jablonski
Division of Sensors and Measuring Systems, Warsaw University
of Technology, Sw. A. Boboli 8, 02-525 Warsaw, Poland
e-mail: nowak@mchtr.pw.edu.pl

© Springer International Publishing AG 2017
R. Jabłoński and R. Szewczyk (eds.), *Recent Global Research and Education:
Technological Challenges*, Advances in Intelligent Systems and Computing 519,
DOI 10.1007/978-3-319-46490-9_21

From the practical application point of view, nanometrology is mainly focused on the development of the state-of-the-art metrological equipment—nanosensors [3]. Such sensors are ultimate measuring tools, able to detect and transduce information about object (e.g., nanoparticle) or phenomena (e.g., quantum tunneling) at nanometer scale. Family of nanosensors is growing rapidly, constantly finding new potential applications in various fields.

Considering mentioned roadblocks, it is strongly desired to produce nanosensors with (1) the highest possible degree of controllability in terms manufacturing process and with (2) well-understood governing physics. Taking into account present status of semiconductor technology, silicon seems to be the best candidate as a host material for novel nanosensors. On one hand, this ubiquitous material has been receiving steady improvements in terms of CMOS processing technology. As a result, it is possible to produce structures with the accuracy of several nanometers. On the other, number of research works devoted to silicon-based structures is tremendous; hence knowledge about their properties is well-established.

In this paper, we demonstrate examples of novel silicon-based nanosensors, able to detect single photons or charges—main objects of interest in case of nanophotonics and nanoelectronics. Prototype devices, nanotransistors and p-n nanojunctions, utilize two major factors related to the nanoscale: (a) active role of single dopant atoms and (b) two dimensional properties of host silicon.

2 Properties of Ultrathin Silicon

Materials with strongly scaled-down dimensions (i.e., with reduced dimensionality) may exhibit properties that are not observable in case of bulk structures. Hence, in order to characterize low-dimensional structure, it is of critical importance to correctly recognize emergence of particular phenomena.

The same is applicable to silicon—when brought to a nanometer scale, number of effects may occur:

- dielectric confinement (or dielectric mismatch) [4]
- quantum confinement (size effect) [5]
- prominent surface-to-volume ratio [6]
- appearance of single dopants [7]
- significant effect of lattice orientation on carrier transport [8]
- ballistic or quasi-ballistic (phonon) transport [9]
- energy bandgap widening [10]

These effects may have serious consequences: dopant ionization energy can be altered (important for transistors) [11]; electric field can penetrate outwards of the depletion layer (in case of p-n junctions) [12]; transport mechanism, under some conditions, may be totally governed by quantum tunneling etc. As it turns out, appearance of these new aspects can be efficiently utilized in Si-based sensors, as it will be described in the next section.

3 Platforms for New-Generation of Sensors— Nanotransistors and p-n Nanojunctions

Nanotransistor structure is schematically presented in Fig. 1a. Phosphorous (P) concentration is estimated to be $N_D \approx 1 \times 10^{18}$ cm^{-3}. Low-temperature (≈ 15 K) I–V characteristics, as shown in Fig. 1b, was taken under visible light illumination (l = 550 nm). We then investigated first observed current peak, which exhibited irregularity (indicated by an arrow). It is well-known that such isolated peak can be ascribed to a single-electron tunneling through ground-energy level of a donor atom [13]. Next, with V fixed at first peak, we measured in dark I-t characteristic (current-time dependence). As a result, no current changes were observed —Fig. 1c, lower part. Thus, with no light irradiation, background charges do not exhibit significant fluctuations. On the other hand, due to light illumination, the I-t characteristic is remarkably different: two-level Random Telegraph Signal (RTS) [14] can be observed, as seen in upper part of Fig. 1c. This demonstrates that photon-induced carriers can be trapped sufficiently long, so that the detection is possible. Since RTS is mainly of two-levels, most likely single trap is responsible for the observed current switching. Hence, we concluded that single donor works as a trap—it changes charge state from ionized to neutral. This is due to single electron capturing, as depicted in Fig. 1d: photon with sufficiently high energy generates an electron, which is then trapped by a donor. Therefore, single photon detection can be achieved, utilizing donors working as quantum dots [15].

Observation described above was confirmed by performing surface potential measurements by Kelvin prove force microscopy (KFM) [16]. Figure 2 shows

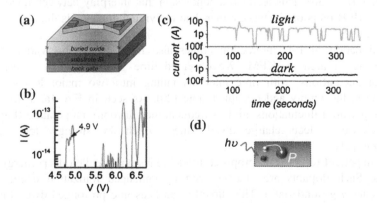

Fig. 1 **a** Nanotransistor structure. **b** Low-temperature I–V curve. **c** RTS induced by a change of donor charge state. **d** Schematic of photon-generated electron trapping [15]

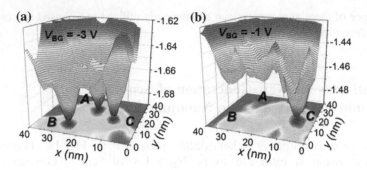

Fig. 2 **a** Three potential wells due to ionized phosphorus donors. **b** Disappearance of two potential wells due to neutralization of two donors (electron trapping) [17]

example of measurement performed directly over three isolated donors. Initially, we were able to clearly see three dopant-induced potential wells (Fig. 2a). Then, by increasing gate voltage (and consequently by injection of electrons), we observed that two potential wells disappeared (Fig. 2b)—donors changed charge state from ionized to neutral by capture of electrons [17].

Figure 2a shows the p-n nanojunction structure. Phosphorous (P) and boron (B) concentrations are estimated to be $N_D \approx 1 \times 10^{18}$ cm^{-3} and $N_A \approx 1.5 \times 10^{18}$ cm^{-3}, respectively. I–V characteristic, as shown in Fig. 2b, was taken at low temperature (≈ 15 K), for reverse and forward bias conditions. As observed, device shows typical diode-like behavior. Next, we measured I-t characteristic, for few different values of voltage. As a result, again mostly two-level RTS was observed (Fig. 2c). In p-n junctions, however, two types of dopants are present—donors and acceptors, with different doping concentrations (i.e., higher acceptor concentration). Therefore, in the active part of p-n junction—depletion region—it is probable to observe dopant clustering or formation of P-B pairs [18]. Hence, it is likely that one donor atom can be embedded within a cluster of acceptors, resulting in a mutual interaction (Fig. 3d). This leads to a conclusion that interplay between donors and acceptors (P-B pairs or clusters), rather than individual dopants, is the origin of the observed RTS.

Confirmation of P-B interaction comes also from direct observation of p-n junction active area by KFM. We observed time- and temperature dependent potential fluctuations, with time intervals falling into two major levels, corresponding to the empty and occupied state [20]. As seen in Fig. 4a, at low temperature potential fluctuations are less frequent than at room temperature (Fig. 4b). This behavior reflects charge dwell times, which are strongly temperature-dependent [21].

It is important to note that proposed model relies on so-called deep energy level dopants. Such dopants are characterized by enhanced ionization energy (i.e., deeper-energy ground state). This directly translates into prolonged dwell times—otherwise detection would not be possible. Existence of dopants with deeper-energy

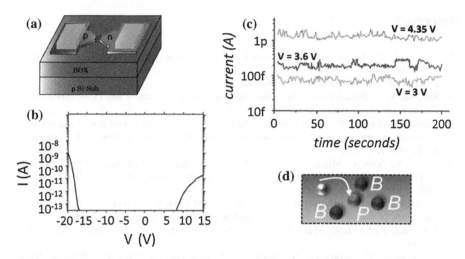

Fig. 3 **a** p-n nanojunction structure. **b** Low-temperature I–V curve. **c** RTS for three voltage values. **d** Electron trapping by a donor embedded within a cluster of acceptors [19]

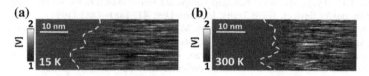

Fig. 4 KFM maps taken at the depletion region boundary at 15 K (**a**) and 300 K (**b**). Potential fluctuations have been ascribed to the continuous trapping and detrapping events [19, 20]

levels is mainly attributed to the quantum and dielectric confinements, both due to reduced dimensionality (thickness) of host silicon.

4 Conclusions

In summary, we have demonstrated examples of silicon-based photon and charge nanosensors, fabricated by means of standard CMOS processes. Devices are based on transistor and p-n junction nanostructures. Detection mechanism has primarily been attributed to the emergence of individual dopants and reduced dimensionality of host silicon. We believe that examples presented here can serve as universal platforms for other types of nanosensors—important family of novel sensing devices.

References

1. Nawrocki, W.: The quantum SI—Towards the new system of units. Metrol. Meas. Syst. **2**, 139–150 (2010)
2. Eight Nanoforum Report. www.nanoforum.org
3. Khanna, V.K.: Nanosensors: Physical, Chemical, and Biological. CRC Press, Boca Raton (2011)
4. Diarra, M., Niquet, Y.-M., Delerue, C., Allan, G.: Ionization energy of donor and acceptor impurities in semiconductor nanowires: importance of dielectric confinement. Phys. Rev. B **75**, 045301 (2007)
5. Mol, J.A., Salfi, J., Miwa, J.A., Simmons, M.Y., Rogge, S.: Interplay between quantum confinement and dielectric mismatch for ultra-shallow dopants. Phys. Rev. B **87**, 245417 (2013)
6. Niklas, E., Juhasz, R., Sychugov, I., Engfeldt, T., Karlström, A.E., Linnros, J.: Surface charge sensitivity of silicon nanowires: size dependence. Nano Lett. **7**, 2608–2612 (2007)
7. Koenraad, P.M., Flatté, M.E.: Single dopants in semiconductors. Nat. Mater. **10**, 91–100 (2011)
8. Karamitaheri, H., Neophytou, N., Kosina, H.: Ballistic phonon transport in ultra-thin silicon layers: effects of confinement and orientation. J. Appl. Phys. **113**, 204305 (2013)
9. Markussen, T., Rurali, R., Jauho, A.-P., Brandbyge, M.: Transport in silicon nanowires: role of radial dopant profile. J. Comput. Electron. **7**, 324–327 (2008)
10. Chow, T.P., Tyagi, R.: Wide bandgap compound semiconductors for superior high-voltage unipolar power devices. IEEE Trans. Electron Dev. **41**, 1481–1483 (1994)
11. Pierre, M., Wacquez, R., Jehl, X., Sanquer, M., Vinet, M., Cueto, O.: Single-donor ionization energies in a nanoscale CMOS channel. Nat. Nanotechnol. **5**, 133 (2009)
12. Achoyan, ASh, Yesayan, A.É., Kazaryan, É.M., Petrosyan, S.G.: Two-dimensional p-n junction under equilibrium conditions. Semiconductors **36**, 903–907 (2002)
13. Tabe, M., Moraru, D., Ligowski, M., Anwar, M., Jablonski, R., Ono, Y., Mizuno, T.: Single-electron transport through single dopants in a dopant-rich environment. Phys. Rev. Lett. **105**, 016803 (2010)
14. Simoen, E., Dierickx, B., Claeys, C.L., Declerck, G.J.: Explaining the amplitude of RTS noise in submicrometer MOSFETs. IEEE Trans. Electron Devices **39**, 422 (1992)
15. Udhiarto, A., Moraru, D., Mizuno, T., Tabe, M.: Trapping of a photoexcited electron by a donor in nanometer-scale phosphorus-doped silicon-on-insulator field-effect transistors. Appl. Phys. Lett. **99**, 113108 (2011)
16. Nonnenmacher, M., O'Boyle, M.P., Wickramasinghe, H.K.: Kelvin probe force microscopy. Appl. Phys. Lett. **58**, 2921 (1991)
17. Nowak, R., Anwar, M., Moraru, D., Mizuno, T., Jablonski, R., Tabe, M.: Electron filling in phosphorus donors embedded in silicon nanostructures observed by KFM technique. J. Appl. Res. Phys. **3**, 021202 (2012)
18. Udhiarto, A., Moraru, D., Purwiyanti, S., Mizuno, T., Tabe, M.: Photon-induced random telegraph signal due to potential fluctuation of a single donor-acceptor pair in nanoscale Si p–n junctions. Appl. Phys. Express **5**, 112201 (2012)
19. Nowak, R., Jablonski, R.: Nanoscale lateral p-i-n junction as a dopant-based charge sensor. Submitted (2016)
20. Nowak, R., Moraru, D., Mizuno, T., Jablonski, R., Tabe, M.: Effects of deep-level dopants on the electronic potential of thin Si pn junctions observed by Kelvin probe force microscope. Appl. Phys. Lett. **102**, 083109 (2013)
21. Foty, D.: Impurity ionization in MOSFETs at very low temperature. Cryogenics **30**, 1056 (1990)

Part III
Biotechnology, Bioengineering, Environmental Engineering

Numerical Investigation of the Effect of Fluid Flow on Biofilm Formation in a Channel with Varying Cross-Section

Y. Okano, Y. Takagi, T. Ohata, Z.K. Sanchez and K. Kimbara

Abstract In order to shed light on the effect of fluid flow on biofilm formation observed in micro channel devices, a fluid flow in a millimeter-size channel with varying cross-section was numerically simulated. Simulation results show that the rate of film formation in the narrower section is higher, compared with that in the larger section, due to enhanced mass transport by the faster flow in this section. It appears that faster flow gives rise to the transport of more nutrients to the seed leading to more growth. It was also found that the inlet nutrient concentration strongly affects the grown biofilm structure in terms of size and density.

Keywords Biofilm · Fluid flow · Numerical simulation · Phase field method

1 Introduction

Biofilms are ubiquitous in nature and may develop on variety of materials and bodies such as metals, plastics, medical implants, plants, body tissues, etc., wherever moisture and nutrients exist. Biofilm formation may have both positive and negative effects. For instance, biofilm formation in engineering systems may lead to energy losses, product contamination, and structural damage. It may also cause infection in the human body by growing on medical implants and devices. However, the biofilm formation may also be utilized to our advantage in many areas of applications such as tolerance towards antibiotic treatment, host immune responses, biocide treatment, biofilm dispersal as well as waste water treatment and

Y. Okano (✉) · Y. Takagi · T. Ohata
Department of Materials Engineering Science, Osaka University,
Toyonaka, Japan
e-mail: okano@cheng.es.osaka-u.ac.jp

Z.K. Sanchez · K. Kimbara
Department of Applied Chemistry and Biochemical Engineering,
Shizuoka University, Hamamatsu, Japan

© Springer International Publishing AG 2017
R. Jabłoński and R. Szewczyk (eds.), *Recent Global Research and Education:
Technological Challenges*, Advances in Intelligent Systems and Computing 519,
DOI 10.1007/978-3-319-46490-9_22

biofiltering. To have a better understanding for the effects of biofilm formation, fundamental research must continue.

In order to investigate biofilm formation and its behavior in a flow channel, we have developed a new multichannel device with varying crosssection in which biofilm formation under channel flow can easily be monitored [1]. The underlying fundamentals of such experimental observations can be verified and be better understood by numerical modelling and simulations. Particularly, such simulations may not only provide a broader and deeper understanding for the complex process of biofilm formation but also qualitative and quantitative predictions that may serve as guidelines for device and system designs. In this direction, we have carried a numerical simulation modelling study for the effect of fluid flow on biofilm formation in a channel with varying crosssection.

Recent studies have incorporated computational fluid dynamics (CFD) into biofilm modelling. They mainly focussed on the effect of fluid flow on biofilm formation. It was found that the bulk fluid flow plays an important role in biofilm formation (film structure) due to enhanced nutrient transport by convection [2, 3]. Zhang et al. conducted a 2-D analysis using the phase field method through a full continuum approach, and investigated the behavior of biofilms in a cavity flow [4, 5]. The phase field method allowed the interface conditions to be dramatically simplified since the interface was not separated from the rest of the system. They predicted a viscous limit through a general viscoelastic model and represented deformation and detachment of biofilms under fluid flow.

In spite of the availability of the models mentioned above, a model with relatively low computation cost, such as a continuum model with the phase field method without the explicit treatment of phase interface, would be very useful to analyze the formation of complex structures such as dendrites and biofilms. Thus, such an objective in mind, we have carried out the present numerical simulation study using the phase field method to examine biofilm formation in a channel with three different cross-sections. First, the behavior of biofilm formation in each section was examined for a fixed inlet concentration. Then, simulations were performed at two concentration levels to evaluate the effect of the inlet concentration on growth rate and biofilm structure.

2 Numerical Method

Figure 1 shows schematically the computational model domain. A 2-D analysis was considered sufficient since the Reynolds number in the system was small. Channel geometry and dimensions were selected to correspond closely to those of the actual experimental set up for applicability in biofilm studies using flow cells of [1]: channel depth $L_y = 2 \times 10^{-3}$ m at the inlet and channel length $L_x = 28 \times 10^{-3}$ m. The channel was divided into three sections of different depths with proper transition slopes, as seen in Fig. 1. In the discretization of the computational domain, the slope geometry was handled by rectilinear-cut cells.

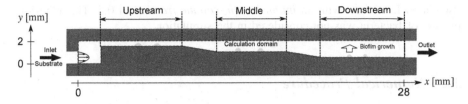

Fig. 1 Numerical model domain, and its dimensions

2.1 Model Assumptions and Governing Equations

We assume a laminar and incompressible Newtonian flow in the computational domain shown in Fig. 1. Biofilm seeds were initially randomly placed on the bottom of the channel. The fluid flow in the domain (solvent) was governed by the well-known equations of continuity, momentum balance, and the species mass balance for the nutrient. The Cahn-Hilliard equation, considering the endogenous metabolism, was adopted for the volume fraction of biomaterials [6, 7]. The computational domain consists of two phases: the liquid phase (solvent) and the solid phase (biomass or biofilm). The term representing friction between these two phases (solvent and biofilm) is added to the momentum balance to consider the influence of biofilm on the flow field. The biofilm is assumed to consist of single-species aerobic bacteria, and oxygen is considered as one of the nutrients. Biofilm is assumed to be stationary on the substratum with no deformation and detachment because of slow fluid flow, and biofilm growth obeys the substrate-limitation Monod equation. For substrate consumption, the sum of the Monod rate and cell maintenance rate was used. The phase field method used in the study treats the phase interface as a diffusive surface.

In the present numerical model, we have made a number of assumptions and selected appropriate boundary conditions for the computational domain. These are summarized below. At the inlet ($x = 0$), we considered a fully developed flow with a parabolic velocity profile (known as the so-called Hagen-Poiseuille flow) with a Dirichelet boundary condition on the flow velocity. At the outlet, the flow velocity gradient was set to zero. At the channel wall surface, no-slip boundary condition was assumed for the flow velocity (i.e., zero tangential flow velocity component). The pressure gradient normal to all boundaries, except the outlet boundary, was set to zero. Pressure was assumed to be zero at the outlet boundary. On the inlet boundary, we assumed that the fluid (solvent) contained nutrient at two concentration values: 1×10^{-4} and 3×10^{-5} kgm^{-3} (will be called *inlet concentration* onward). We assumed that the channel wall boundaries are impermeable to mass transport (i.e. no-diffusion condition for nutrient substrate). Similarly, no-flux boundary conditions for the volume fraction of biofilm and the chemical potential of the biofilm were applied at all the boundaries. The fully developed velocity profile at the inlet was assumed prescribed. The *inlet concentration* was taken as uniform. A total of six biofilm seeds were placed on the channel bottom surface (two in each section) with an *initial biofilm volume fraction* of 0.5. The adhesion process of bacteria was not considered in the simulation.

2.2 Numerical Procedure

The steady-state flow velocity was determined when the error of computation was less than a threshold of the maximum error (10^{-3}–10^{-5}). In solving the governing equations, the second order central difference method and fourth order Runge-Kutta method were used for spatial and time discretization, respectively. The Navier-Stokes equations were solved by the SMAC method. The SOR method was used to solve the Poisson equation for the pressure field. The uniform structured grids with grid width of 1/64 mm were used. The Reynolds number which is defined as UD/ν (where U is the average velocity of Hagen-Poiseuille type flow, D is the channel depth and ν is the kinetic viscosity of the solvent) was set at the value of 3.0 in the simulations, based on actual experimental conditions of [1].

3 Results and Discussion

3.1 Biofilm Formation Along the Channel

The computed (a) volume fraction of biofilm and (b) nutrient concentration distribution in the liquid are presented in Fig. 2. Computations were carried out at the dimensionless inlet concentration of $c = 0.10$ and results were presented at time $t = 1200$ s. Nutrient starvation occurs in biofilm formation due to the consumption of nutrient by biomaterials. The biofilm seeds located in the upstream section have developed into larger and denser biofilms compared with those in the middle and downstream sections. This is due to the effect of fluid flow velocity distribution. The faster fluid flow in this narrower section (upstream) enhanced the biofilm growth on the seeds located in this section since the mass transport boundary layer around the seed was thinner compared with that of the seed in the downstream section due to slower fluid flow. Consequently, the growth is less on the seeds in the down stream section.

The flow velocity drastically decreases near the biofilms since the solvent hardly penetrates into biofilms due to the friction force between the solvent and biofilm.

Biofilm thickness increases in the upstream section because of the higher nutrient availability due to the faster flow in this section. The flow velocity profiles almost remain parabolic while the velocity in the region near the biofilm is negligibly small. We observe a steeper flow velocity gradient near the liquid-film interface due to the higher flow velocity in the upstream section. We also observe that the flow velocity increases along the channel since the channel becomes

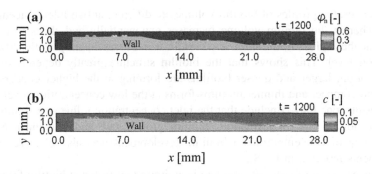

Fig. 2 **a** Volume fraction of biofilm and **b** nutrient concentration distribution computed at the inlet dimensionless concentration of 0.1 and time $t = 1\ 200$ s

narrower as biofilms grow. These computed velocity profiles in the channel qualitatively agree with the measured velocity profile in the experiments [8].

3.2 Effect of Inlet Concentration on Biofilm Formation

Biofilm development is determined by the balance between nutrient availability and endogenous metabolism as expressed in the Cahn-Hilliard equation. The effect of inlet concentration (at two values: $c = 0.10$ and 0.03) on biofilm formation is examined. The solid line in Fig. 3 represents the value at $c = 0.03$. Results show that the higher inlet concentration results in a higher growth rate while a lower growth rate at the low concentration level. This implies that by choosing a low inlet concentration we may better control the undesired effects of biofilm formation in a channel flow since the difference in overall biofilm volume in each section at the low inlet concentration level is negligible in comparison with that at the high concentration level.

Fig. 3 Time evolution of overall biofilm volume in each section of the channel

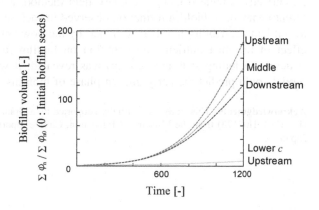

Moreover, the profiles of biofilm volume are different at two inlet concentration levels wherein the biofilm volume increases exponentially and shifts to a linear profile at the high concentration level while it increases linearly at the low concentration level. This shows that the biofilm structure greatly depends on inlet concentration; larger and denser biofilms are forming at the higher concentration level, and compact and thinner biofilms forms at the low concentration level. From these results, one may conclude that the inlet concentration influences significantly the behavior of biofilm formation (biofilm growth rate and structure). Such a dependency of concentration on biofilm development was also observed in the experiments reported in [1, 8].

The penetration of nutrient (species) is an important factor in biofilm formation. The penetration of nutrient into biofilms drastically decreases because of the consumption of nutrient by biomaterials and the friction force along the biofilm which prevents the penetration of nutrient into the biofilm. Therefore, the diffusion of nutrient and the reaction of nutrient and biofilm are the dominant phenomena in biofilm formation. At the higher inlet concentration level, the region of nutrient starvation arises near the substratum as biofilms grow and increase in size. This means biomaterial growth occurs only within a thin outer layer of the biofilm (the active layer or penetration depth). At the low inlet concentration level, however, even most of the biofilm formation is under less nutrient condition from an early stage, more penetration of nutrient into biofilms occurs because of a smaller size and lower density of biofilm which denote a low consumption of nutrient. Thus, results show that the inlet concentration level controls biofilm size and density that subsequently influence the penetration depth of nutrient and area of nutrient starvation; indicating that the biofilm formation is strongly affected by the flow velocity in the channel and the inlet nutrient concentration.

4 Conclusions

Biofilm formation under a fluid flow in a channel with varying crosssection was numerically simulated using the phase field method. Results showed that the different behavior of biofilm formation observed in each section of the channel is due to nutrient availability that is determined by the flow velocity in each section. The effect of the inlet nutrient concentration on biofilm formation (growth rate and structure including size and density) was revealed. It was found that the present model could predict the early growth phase of biofilms.

Acknowledgments This study was partially supported by a Grant-in-Aid for Scientific Research (B) (no 15H04173) from the Ministry of Education, Culture, Sports, Science and Technology of Japan.

References

1. Sanchez, Z., Tani, A., Suzuki, N., Kariyama, R., Kumon, H., Kimbara, K.: Assessment of change in biofilm architecture by nutrient concentration using a multichannel microdevice flow system. J. Biosci. Bioeng. **115**, 326–331 (2013)
2. Picioreanu, C., Loosdrecht, M.C., Heijnen, J.J.: Discrete-differential modelling of biofilm structure. Wat. Sci. Tech. **39**, 115–122 (1999)
3. Picioreanu, C., Loosdrecht, M.C., Heijnen, J.J.: Two-dimensional model of biofilm detachment caused by internal stress from liquid flow. Biotech. Bioeng. **72**, 205–218 (2001)
4. Zhang, T., Cogan, N., Wang, Q.: Phase field models for biofilms. I. Theory and one-dimensional simulations, SIAM. J. Appl. Math. **69**, 641–669 (2008)
5. Zhang, T., Cogan, N., Wang, Q.: Phase-field models for biofilms II. 2-D numerical simulations of biofilm-flow interaction. Commun. Comput. Phys. **4**, 72–101 (2008)
6. Fagerlind, M.G., Webb, J.S., Barraud, N., McDougald, D., Jansson, A., Nilsson, P., Harlen, M., Kjelleberg, S., Rice, S.A.: Dynamic modelling of cell death during biofilm development. J. Theor. Biol. **295**, 23–36 (2012)
7. Cahn, J.W., Hilliard, J.E.: Free energy of a nonuniform system. I: Interfacial free energy. J. Chem. Phys. **28**, 258–267 (1958)
8. Song, J.L., Au, K.H., Huynh, K.T., Packman, A.I.: Biofilm response to smooth flow fields and chemical gradient in novel microfluidic flow cells. Biotech. Bioeng. **111**, 597–607 (2014)

Decision Based Algorithm for Gene Markers Detection in the ISH Images

Tomasz Les, Tomasz Markiewicz, Marzena Jesiotr,
Wojciech Kozlowski and Urszula Brzoskowska

Abstract The article presents a decision based algorithm of an automatic identification of markers in microscope images—ISH (In Situ Hybridization). The ISH test allows a quick and inexpensive initial diagnosis of the breast cancer. The evaluation of a degree of the HER2 gene's amplification and the selection of the appropriate treatment require locating and counting markers in cell nuclei. This article presents a new heterogeneous algorithm based on decision making. The main idea is to analyze a portion of an image and decide which algorithm should be used for processing the given fragment. The different parts of the image can be analyzed by different types of algorithms. Tests and results of the experiment are presented and discussed in this article. They confirm higher efficiency of markers recognition by the homogeneous system, using many different algorithms, rather than in the case of using the method based on a single algorithm.

Keywords Signal and image processing · Object recognition · ISH images

1 Introduction

The paper presents a method of identifying biomarkers in ISH microscopic images. This complex method is based on an analysis of a portion of an image which leads to choose the best algorithm to identify biomarkers. ISH is a staining technique for marking HER2 and CEN17 gens in tissues microscope images. Identification of the HER2 and CEN17 genes allows to count gens in the cell nuclei and then allows to detect an amplification of this oncogene. Basing on this diagnosis, it is possible to choose an appropriate treatment of breast cancer. ISH is an inexpensive technique and is often used in research centers, hospitals and pathology institutions. This

T. Les (✉) · T. Markiewicz
Warsaw University of Technology, Warsaw, Poland
e-mail: lest@ee.pw.edu.pl

T. Markiewicz · M. Jesiotr · W. Kozlowski · U. Brzoskowska
Military Institute of Medicine, Warsaw, Poland

© Springer International Publishing AG 2017
R. Jabłoński and R. Szewczyk (eds.), *Recent Global Research and Education:
Technological Challenges*, Advances in Intelligent Systems and Computing 519,
DOI 10.1007/978-3-319-46490-9_23

technique uses non-fluorescent staining as opposed to alternative method—FISH (Fluorescent in situ hybridization) imaging. An ordinary optical microscope can be used in the analysis. Unfortunately ISH imaging is a method less precise than FISH. FISH technique, by contrast, is a method much more expensive and more time-consuming but at the same time the field of view obtained with this technique is more clear. Common analysis consists of the initial ISH diagnosis and, depending on the result, an appropriate treatment is applied or more detailed analysis based on of FISH technique is employed if needed.

2 Problem Statement

The main task in the diagnosis of breast cancer is the assessment of HER2 gene amplification. In the case of gene amplification Trastuzumab therapy should be applied. Trastuzumab reduces the risk of recurrence among patients with breast cancer. Because of its side effects—particularly negative effect on the heart—it should be used in limited cases. Therefore, a proper assessment of HER2 gene amplification is very important. In order to assess HER2 gene amplification, diagnostics count the number of occurrences of the HER2 and CEN17 genes in each cell nucleus. When the ratio is greater than 2.2—treatment with trastuzumab should be applied. When the ratio is lower than 1.8—there is no indications to trastuzumab treatment. When the value is between 1.8 and 2.2—new analysis is required with another field of view or using different technique. Two sample fields of view obtained with ISH technique are shown in Fig. 1.

In the above figure HER2 genes can be seen as black spots while CEN17 appear as red spots. There are many techniques of an automatic biomarkers detection in ISH and FISH images. Often those techniques, after minor modifications, can be used in both types of images (ISH and FISH). Identification of markers technique is presented in [1–5]. A solution which is very often used is to apply the "Top-hat"

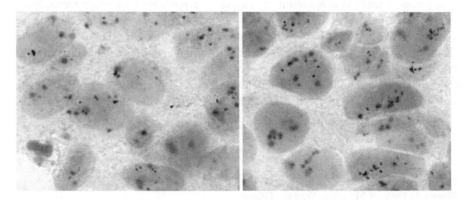

Fig. 1 Two example fields of view obtained with ISH technique

transform [1, 5] or a modified watershed method [2]. In [3, 4] genes localization is based on various classifiers, such as naive Bayes classifier or neural networks models. Interesting results can also be obtained using the color analysis [6] and 3D shape analysis of the RGB channels [7]. Each method has its weaknesses, and it is very difficult to find an universal method for the identification of biomarkers which works correctly for all types of images. Very good results of biomarkers localization in the FISH images were obtained after application of fusion classifiers based on weighted majority algorithm [8].

This article presents a new approach of biomarkers localization in ISH images. Our idea was to read a small portion of an image and to analyze its channel intensity and spots shape distribution in order to choose the best method for the analysis of these areas. For the identification of markers three very different methods were used: color analysis [6], 3D shape analysis [7] and the distance function. First two techniques allow to obtain satisfactory results in identification markers in the FISH images. ISH technique generates a less clear view with biomarkers which often differentiate significantly in shape, saturation and size. Application of a single technique for ISH analysis does not always give satisfactory results. This makes it necessary to apply a complex method, combining the advantages of different solutions.

3 Methods of Biomarkers Identification

An automatic identification of the HER2 and CEN17 genes is performed independently. Computer analysis can be reduced to detection of the black (HER2) and red (CEN17) spots. CEN17 generates local increase in the luminosity function of the red channel. HER2 genes generates local increase of all channels (R, G and B). This means that the identification of CEN17 genes can be applied directly by the same methods as in FISH images. For the detection of HER2 a new method must be employed. In ISH staining HER2 appears as a compact spot. CEN17 can appear both as compact and diffuse spots. Our proposal is to classify each region into one of three classes: HER2, CEN17 compact and CEN17 diffuse. Basing on this decision, one method is applied: color analysis, 3D shape analysis or distance function for final recognition of the spot.

3.1 Color Analysis

Localization of biomarkers in this method uses maps of colors. The map is created with horizontal and vertical lines representing different intensity values of the selected colors (in our case: red). The intensity value of each of three colors is related to the appropriate place on the map. Then, using K-NN technique (k-nearest neighbor), pixels are grouped into clusters in the shape of a circle with a fixed radius (in our experiments: 7 pixels). A cluster is classified as a spot when all pixels in the cluster are found in the map of colors.

3.1.1 3D Shape Analysis

In this method, the image is scanned to find the most similar area to the predetermined pattern. For each point a pattern (in our case: cone shape) is laid. Around the pattern a surrounding area in the form of a circle is also defined. To detect a spot the sum of internal pixels (within the pattern) and external pixels (within the surrounding area) needs to be counted. The internal sum is expressed as a sum of the differences between the intensity of an image pixel value and the value of the pattern function for all pixels which intensity is smaller than the intensity value of the pattern. The external sum is expressed as the sum of differences between the image pixel intensity value and the surrounding area for all pixels which intensity is higher than the intensity value of surrounding area. The point is classified as a spot when the internal and external sum fulfill tolerance conditions: ε_1 i ε_2 (in our case $\varepsilon_1 = 170$ and $\varepsilon_2 = 1000$).

3.2 Distance Function

The localization of HER2 gene is performed using distance function. This function measure a distance (in RGB color space) between the intensities of all three channels. The function is expressed as:

$$I(p) = abs\big(I_r(p) - I_g(p)\big) + abs\big(I_r(p) - I_b(p)\big) \\ + abs\big(I_g(p) - I_b(p)\big) \tag{1}$$

where $p = (x, y)$, and I_r, I_g, I_b represents succeeding channels of RGB. The Cluster is classified as a spot when at least 70 % of pixels, within the area of a circle of the specified radius of 7 pixels meet condition: $I(p) < \varepsilon$. In the experiments a tolerance $\varepsilon = 120$ was assumed.

3.3 Decision Based Method

In order to choose the best algorithm for spots identification we applied the morphological structuring element in the shape of a circle with a radius of 10 pixels. The size of the radius was selected on the basis of the greatest spots. For each point of an image, in the area of structural element, a difference between the maximum value of the red channel and the minimum pixel intensity of the red channel is calculated. If the difference exceeds the threshold value (in the experiments, equal to 70), then a more detailed, further analysis is required to identify final spot— HER2 or CEN17. Otherwise, another area should be analyzed.

In case of the further analyze a new structural element in the shape of circle with radius of 6 pixels is then applied in the center of the current point. If at least 70 % of pixels of red channel are greater than a predefined threshold (the value of 150 in the experiments), the red spot (CEN17) detection algorithm should be applied. Otherwise HER2 genes detection method should be applied. If HER2 algorithm was chosen, the distance function should be applied. If CEN17 algorithm was chosen, then a spot should be classified as diffuse or compact. For this purpose two areas with fixed radiuses r_1 and r_2 are examined (in experiments $r_1 = 10$ and $r_2 = 6$ pixels). The area A (external) and area B (internal) are situated in the way that the point which is being analyzed is placed in the center of these areas (Fig. 2)

If the 70 % of pixels with the lowest intensity of red channel values, located in the area B, are higher by ρ than 70 % of the highest intensity of the red channel values, located in the area A, it indicates that a compact spot can occur and the 3D shape algorithm should be applied. Otherwise, it indicates the possibility of the presence of a diffuse spot and the color analysis algorithm should be applied. In the experiments a value of $\rho = 15$ has been chosen. A full decision-making diagram of biomarkers detection algorithm is shown in Fig. 3.

4   Tests and Results

To assess the full system we performed the tests on the basis of 10 microscopic images ISH. Each image contains more than 150 different spots. The images were obtained from the Department of Pathomorphology in Military Institute in Warsaw. The ISH specimens were prepared with ROCHE/VENTANA INFORM HER2 Dual ISH DNA Probe Cocktail, UltraView Red ISH DIG and DNP SISH Detection Kits. For each image an expert marked manually the locations of the HER2 and CEN17 genes. Then the system automatically identified markers of both types in the same images, using three techniques: 3D shape analysis, color analysis and a complex system based on decision making. The results are shown in the Table 1.

Fig. 2 The area A (external) and area B with fixed radiuses r_1 and r_2

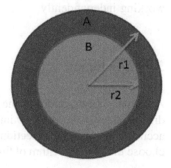

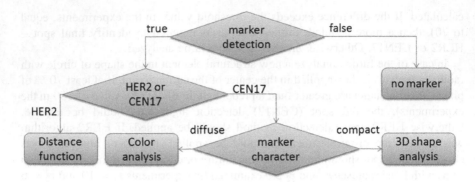

Fig. 3 Full decision-making diagram

Table 1 The results of the spots detection of individual systems

	3D shape analysis		Color analysis		Complex system	
Case	HER2	CEN17	HER2	CEN17	HER2	CEN17
1	10.22	17.53	**8.45**	10.54	**8.45**	**9.56**
2	12.54	14.86	**7.64**	9.34	10.56	**8.23**
3	**6.45**	12.67	20.45	20.43	13.75	**10.45**
4	**10.45**	13.34	16.45	**8.84**	12.45	10.95
5	16.45	**8.56**	18.04	11.84	**11.47**	10.56
6	14.37	13.65	12.55	18.34	**10.65**	**12.34**
7	**5.34**	8.45	14.45	17.73	9.64	9.68
8	**7.55**	14.45	17.04	16.45	11.23	**7.45**
9	10.96	15.65	16.86	14.94	**8.54**	**8.56**
10	16.44	17.45	**12.45**	22.45	12.56	**14.56**
Avg	11.08	13.66	14.44	15.09	**10.93**	**10.23**

The numerical results show an error expressed in the form: $1 - (s_{pos}/ex)$, where s_{pos} is a number of correctly identified markers by the system and ex is a total number of markers in the image. The results show that the accuracy of the system based on decision-making algorithm is greater than the accuracy of systems working independently.

5 Conclusion

Automatic localization of the HER2 and CEN17 genes in ISH images is more difficult than in the FISH staining. The article presents a method which increase the accuracy of markers detection by the application of decision process in order to choose the best algorithm of final spots recognition. The numerical results show that the accuracy of biomarkers detection in the ISH images by a complex system is better than the accuracy of using individual algorithms working independently.

Acknowledgment This study was supported by the National Centre for Research and Development, Poland (grant PBS2/A9/21/2013).

References

1. Solorzano, C.O., Santos, A., Vallcorba, I., Garcia-Sagredo, J.-M., Pozo, F.: Automated fish spot counting in interphase nuclei: statistical validation and data correction. Cytometry **31**(2), 93–99 (1998)
2. Pratt, W.K.: Digital image processing. Wiley, New York (1991)
3. Lerner, B., Clocksin, W.F., Dhanjal, S., Hulten, M.A., Bishop, C.M.: Feature representation and signal classification in fluorescence in-situ hybridization image analysis. IEEE Trans. Syst. Man Cybern. **31**, 655–665 (2001)
4. Clocksin, W.F., Lerner, B.: Automatic analysis of fluorescence in situ hybridization images. In: Proceedings of the 11th British Machine Vision Conference, pp. 666–674, Bristol, England, Sept 2000
5. Kasampalidis, J.N., Pitas, I., Karayannopolou, G.: FISH image analysis using a modified radial basis function network. Signals Circ. Syst. ISSCS **2**, 1–4 (2007)
6. Les, T., Markiewicz, T., Osowski, S., Cichowicz, M., Kozlowski, W.: Automatic evaluation system of fish images in breast cancer. In: Image and Signal Processing Lecture Notes in Computer Science, vol. 8509, pp. 332–339 (2014)
7. Les, T., Markiewicz, T., Osowski, S., Jesiotr, M., Kozlowski, W.: Localization of spots in FISH images of breast cancer using 3-D shape analysis. J. Microsc. **262**(3) (2016) http://dx.doi.org/10.1111/jmi.12360
8. Les, T., Markiewicz, T., Osowski, S., Kozlowski, W., Jesiotr, M.: Fusion of FISH image analysis methods of HER2 status determination in breast cancer. Expert Syst. Appl. (2016). doi:10.1016/j.eswa.2016.05.020

A Study of the Influence of Plasma Particles for Transdermal Drug Delivery

Jaroslav Kristof, An Nhat Tran, Marius Gabriel Blajan and Kazuo Shimizu

Abstract We studied the influences of argon plasma, nitrogen plasma, and argon with water vapors on the stratum corneum layer. We also investigated the effects of repeated exposure of the stratum corneum to Ar plasma. Changes caused by plasma in stratum corneum layer were observed by Attenuated Total reflectance-Fourier transform infrared (ATR-FTIR) spectroscopy. Disruption of lipid bilayer was observed by a wavenumber shift of vibrational stretch of CH_2. Changes in keratin structure were observed by amide I and amide II bands of ATR-FTIR spectra.

Keywords Plasma jet · Stratum corneum · Atmospheric argon discharge

1 Introduction

Transdermal delivery is a very attractive method for the application of drugs, but skin has a very low permeability for foreign molecules mainly because of the structure of layer called stratum corneum, a lipid-rich matrix with embedded corneocyte cells. This low permeability is especially problematic for relatively large drugs, which represent a large majority of active agents for therapeutic applications,

J. Kristof (✉) · K. Shimizu (✉)
Graduate School of Science and Technology, Department of Optoelectronics
and Nanostructure Science, Shizuoka University, Johoku, Hamamatsu, Shizuoka
432 8561, Japan
e-mail: jaroslav.kristof@gmail.com

K. Shimizu
e-mail: shimizu@cjr.shizuoka.ac.jp

A.N. Tran · K. Shimizu
Graduate School of Engineering, Shizuoka University, Johoku, Hamamatsu,
Shizuoka 432 8561, Japan

M.G. Blajan · K. Shimizu
Organization for Innovation and Social Collaboration, Shizuoka University,
Johoku, Hamamatsu, Shizuoka 432 8561, Japan

© Springer International Publishing AG 2017
R. Jabłoński and R. Szewczyk (eds.), *Recent Global Research and Education:
Technological Challenges*, Advances in Intelligent Systems and Computing 519,
DOI 10.1007/978-3-319-46490-9_24

167

and also for hydrophilic drugs. The enhancement of skin permeability was studied using chemical solutions, ablation, iontophoresis, electroporation, etc. [1]. The use of plasma is a relatively new technique in this field, and only several studies have thus far been conducted [2–4], though they show promising results. The elementary processes leading to the enhancement of skin permeability, and testing of the safety of the treatment(s), have yet to be investigated. A common method for studying skin permeability is Attenuated Total Reflectance-Fourier Transform Infrared spectrometry (ATR-FTIR) [5]. In this paper, we investigated the influence of used gas in plasma jet on stratum corneum of pig skin. Shifts in the maxima of asymmetric stretching bands of CH_2 and amide I and II were compared after treatment in Ar, N_2, or Ar/H_2O plasma. In addition, the delay of changes and the influence of repeated treatment was observed using an Ar plasma jet.

2 Principle and Experimental Set-up

2.1 Experimental Set-up

A diagram of the experimental set-up is shown in Fig. 1. The source of plasma (plasma jet) consists of a Pyrex tube (outer diameter 6 mm, length 100 mm), and a central tungsten (0.8 mm in diameter) high voltage (HV) electrode covered by a glass layer except for a 10-mm long region close to outlet of Pyrex tube. The grounded electrode is an aluminum ring (8 mm wide) located on outer surface of the Pyrex tube, 12 mm from the end of the plasma jet. The electrodes have been coupled by using a Neon transformer (ALPHA Neon M-5) with an AC frequency of 16 kHz. The voltage and current were monitored with a Tektronics P60015A high-voltage probe and a Tektronics P6021 current probe. The distance between the skin sample and the outlet of the plasma jet was set to 2 mm via a grounded micrometric sample holder. The sample was isolated from the holder by a 30-mm thick PVC isolator. The treatment time of the sample was set to 1 min, and the sample was in contact with plasma during this time. The gas is introduced into the plasma jet through a Yamato flow meter. Argon (purity 4.9) and nitrogen (purity

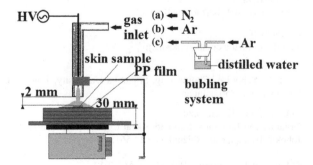

Fig. 1 Experimental setup with realisation of (*a*), (*b*) or (*c*) line

5.5) were used. A mixture of argon and water vapors was delivered to the discharge tube through the bubbling system (Fig. 1).

We investigated influence of used gas or gas mixture by several settings:

1. Argon as the working gas, electrode coupled by 4 kV, gas flow rate of 3 l/min.
2. Argon as the working gas, electrode coupled by 6 kV, gas flow rate of 10 l/min.
3. Nitrogen as the working gas, electrode coupled by 6 kV, gas flow rate of 10 l/min.
4. A mixture of argon gas and water vapors as the working gas, electrode coupled by 7 kV, argon gas flow rate of 3 l/min.

Attenuated total reflectance-Fourier transform infrared (ATR-FTIR) spectrometer (Jasco FT/IR 6300 with ATR PRO610P-S) with a diamond prism was used to observe the upper layer of stratum corneum of pig skin. Spectra were recorded with a resolution of 8 cm^{-1} and by accumulating 150 scans.

2.2 Sample Preparation

Pig skin of Yucatan micropig from Charles River Japan, Inc. (Yokohama, Japan) was used for investigating of the influence of the plasma treatment. The pig skins were stored at -20 °C in a freezer before the experiment. At first, the fat layer of the skin was removed by using a knife, then it was cut and soaked at 4 °C in phosphate buffered saline (PBS) for 3 h, and then after a bath in 60 °C PBS for 1 min, the epidermal layer of a thickness of 200 µm was peeled by using tweezers. Thereafter, it was dried for 30 min at room temperature and flattened between polypropylene films for another 30 min. Finally, the skin sample was cut to 3 × 3 mm pieces and attached to polypropylene film by using double-sided tape.

3 Results and Discussion

3.1 Effect of Used Gas

The stratum corneum layer of samples of pig skin was measured by ATR-FTIR spectrometer before and after irradiation by the plasma jet; thus, the spectra were compared. A typical example of the spectra of a sample of stratum corneum is shown in Fig. 2. In this study, we examined the asymmetric stretching band of CH$_2$ (2917–2923 cm^{-1}) because a shift in the maximum wavenumber of this band correlates with the degree of disorder of lipid acyl chains. In addition, we used the wavenumber shifts of amide I (~ 1646 cm^{-1}) and amide II (~ 1550 cm^{-1}) absorbance, which correlate with the level of hydrogen bonding in stratum corneum and the bands comes from keratin cells.

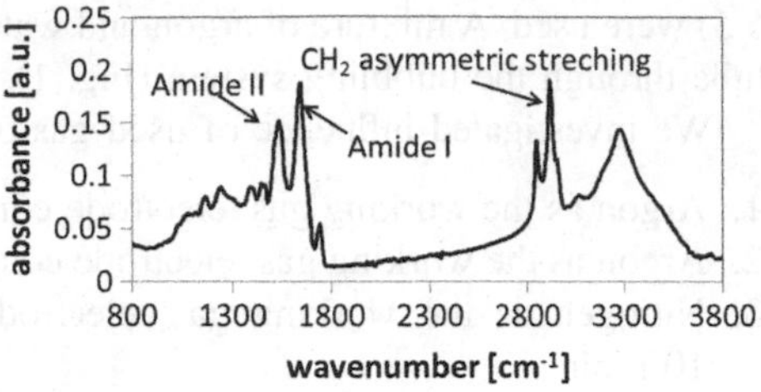

Fig. 2 Typical FTIR-ATR
spectra of stratum corneum

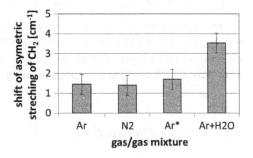

Fig. 3 Shift of asymmetric
stretching band of CH$_2$ to
higher wavenumbers for the
gases used in plasma jet
discharge

The positions of the maxima were determined by a second derivation of ATR-FTIR spectra following cubic spline interpolation to achieve more dense data in the spectra. The shift of wavenumbers of the asymmetric stretching band of CH$_2$ is shown in Fig. 3. In the case of argon, the observed shift of maxima was 1.45 cm^{-1} (conditions: 4 kV, 3 l/min). Increasing the voltage to 6 kV and the gas flow to 10 l/min did not yield a significant change (Fig. 3—Ar with asterisk). The measured shift was 1.7 cm^{-1}, which is not higher than error of measurement. Using nitrogen, we achieved a comparable result equal to 1.4 cm^{-1}. A significantly higher value of shift was observed when argon with water vapors was used. The asymmetric stretching band was shifted 3.5 cm^{-1} (Fig. 3). This result indicates that the stratum corneum became the most permeable after treatment of Ar plasma with water vapors. On the other side, we observed low wavenumber shifts of the maxima of amide I and amide II (Fig. 4). The plasma treatment also caused etching of the skin surface, decreasing the absorbance of amide and decreasing the absorbance of asymmetric stretching band. Argon plasma produces electrons, argon ions, metastable and excited states. These particles can react with the skin directly. But, if argon or nitrogen plasma is working in atmospheric air, it makes situation more complex because molecules of air (O$_2$, N$_2$, CO$_2$ and H$_2$O) can enter the volume of the plasma jet [6, 7]. In this case, argon can react with skin also indirectly through the excitation or dissociation or ionization process with air molecules. The reaction results in the creation of a number of species. It is difficult to find the molecule causing changes in skin. A comparison of argon and nitrogen plasma jets has shown that argon can play a role in reactions with molecules of stratum corneum, but a similar effect can be achieved by nitrogen plasma itself. Thus, further research is necessary to identify the process that disrupted lipid layer.

Fig. 4 Shift of maxima of
amide I and amide II

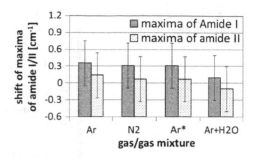

When argon flew through the water reservoir, the argon ensured a higher concentration of water vapors in the discharge. The high shift of the asymmetric stretching band of CH_2 indicates that H_2O and the created OH molecules can play important roles in increasing the shift of the asymmetric stretching band of CH_2 in stratum corneum. OH could be created mainly by two channels [8]:

$$Ar_m + H_2O \rightarrow OH + H + Ar \tag{1}$$

$$H_2O + e \rightarrow OH + H + e \tag{2}$$

Paal et al. [9] performed a simulation of the interaction of OH radicals with α-linolenic acid as a representative of fatty acid; their calculations predicted that OH radicals most typically abstract an H atom from the fatty acids, which can lead to the creation of a double bond and also to the incorporation of alcohols or aldehyde groups, increasing hydrophilic properties of fatty acids and changing the lipid composition of the skin, causing an increase of skin permeability.

3.2 Band Shift Evolution and Effect of Repeated Application of Plasma

After irradiation of the sample by Ar plasma, we observed an evolution of the wavenumber shift of the asymmetric stretching band of CH_2 over time. All samples were held at room temperature between experiments. All positions were compared with the sample's position before irradiation of the skin. As shown in Fig. 5, skin irradiated one time for 1 min returned to its initial state after 48 h (empty diamonds). However, when the sample was irradiated every day, the wavenumber shift was maintained for longer time. After 48 h, the shift was the same as that observed after the first plasma treatment (black diamond). After 120 h, the wavenumber shift was in its initial state for both forms of irradiation (every day and 1 time). Black circles show the evolution of the control condition: A non-treated sample where a slight increasing of shift to lower wavenumbers was observed. If the shift of the asymmetric stretching band of CH_2 is to a higher wavenumber (blue shift), it

Fig. 5 Evolution of shift of asymmetric band of CH_2 in time

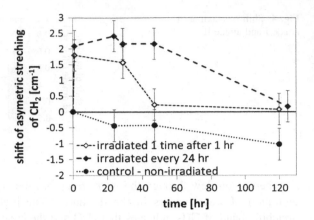

indicates SC membrane (lipid bilayer) fluidization and the barrier properties are disrupted, which cause higher permeability. If the shift is to a lower wavenumber, it means that the lipid groups are oriented again and barrier properties are strengthened [10].

4 Conclusion

Disruption of lipid bilayer was observed by using an atmospheric plasma jet. We found that Ar or N_2 plasma can cause this disruption but the most effective was a mixture of Ar with water vapors, probably because of OH radicals that can change structure and composition of lipids but also influence of heating of skin has to be evaluated in future research. We did not observe any changes in keratin caused by shifts of maxima of amide I or II. As demonstrated in case of Ar plasma, changes of the disruption of the lipid layer are temporary and are largely undone after 48 h. These changes can be prolonged by daily irradiation.

References

1. Rai, V., Ghosh, I., Bose, S., Silva, S.M.C., Chandra, P., Michniak-Kohn, B.: A transdermal review on permeation of drug formulations, modifier compounds and delivery methods. J. Drug Del. Sci. Tech. **20**(2), 75–87 (2010)
2. Lademann, J., Patzelt, A., Richter, H., Lademann, O., Baier, G., Breucker, L., Landfester, K.: Nanocapsules for drug delivery through the skin barrier by tissue-tolerable plasma. Laser Phys. Lett. **10**, 083001 (2013)
3. Shimizu, K., Hayashida, K., Blajan, M.: Novel method to improve transdermal drug delivery by atmospheric microplasma irradiation. Biointerphases **10**(2), 029517 (2015)
4. Kalghatgi, S., Tsai, C., Gray, R., Pappas, D.: Transdermal drug delivery using cold plasmas. In: 22nd International Symposium on Plasma Chemistry, 5–10 July 2015

5. Bommannan, D., Potts, R.O., Guy, R.H.: Examination of stratum corneum barrier function in vivo by infrared spectroscopy. J. Invest. Dermatol. **95**, 403–408 (1990)
6. Tsai, I.-H., Hsu, Ch-Ch.: Numerical simulation of downstream kinetics of an atmospheric-pressure nitrogen plasma jet. IEEE Trans. Plasma Sci. **38**(12), 3387–3392 (2010)
7. Van Gaens, W., Bogaerts, A.: Kinetic modelling for an atmospheric pressure argon plasma jet in humid air. J. Phys. D Appl. Phys. **46**, 275201 (2013)
8. Bruggeman, P., Schram, D.C.: On OH production in water containing atmospheric pressure plasmas. Plasma Sources Sci. Technol. **19**, 045025 (2010)
9. Van der Paal, J., Aernouts, S., van Duin, A.C.T., Neyts, E.C., Bogaerts, A.: Interaction of O and OH radicals with a simple model system for lipids in the skin barrier: a reactive molecular dynamics investigation for plasma medicine. J. Phys. D Appl. Phys. **46**, 395201 (2013)
10. Salimi, A., Hedayatipour, N., Moghimipour, E.: The effect of various vehicles on the naproxen permeability through rat skin: a mechanistic study by DSC and FT-IR techniques. Adv. Pharm. Bull. **6**(1), 9–16 (2016)

Automatic Method for Vessel Detection in Virtual Slide Images of Placental Villi

Żaneta Swiderska-Chadaj, Tomasz Markiewicz, Robert Koktysz and Wojciech Kozlowski

Abstract The purpose of this work is to design an algorithm for automatic vessel recognition and counting on placental histological images in order to support the pathomorphological diagnostic. The studied images of placental villi come from spontaneous miscarriages and they are stained with Hematoxylin and Eosin. The proposed algorithm is based on colour component analysis, mathematical morphology operations, and decision tree classification. The major problems are variability of vessels and presence of collagen which can surrounds a villi. Based on the proposed method, automatic identification and counting of vessels is realized. The presented method can be applied as a support to traditional examinations.

Keywords Vessel · Whole slide image · Placenta villi · Image processing

1 Introduction

Image processing methods and digital pathology is a rapidly developing area. Microscope specimens can be present in digital form, called whole slide images (WSI). Histopathological staining is a method used for visualization of biological structure in the specimen. The contextual analysis of digital histopathological images is a popular method of tissue examination [1, 2].

The placenta is an organ, which is formed in a woman ovary during pregnancy, and it is necessary for fetus to develops and to provide vital functions [3]. Anomalies in its structure and functionality can be associated with various severe fetal lesions and cause spontaneous miscarriage [4]. Histological study of the aborted placenta can explain reasons of spontaneous miscarriage and provides valuable information relating to the placenta and fetal pathologies [5]. Among the

Ż. Swiderska-Chadaj (✉) · T. Markiewicz
Warsaw University of Technology, Warsaw, Poland
e-mail: zaneta.swiderska@gmail.com

T. Markiewicz · R. Koktysz · W. Kozlowski
Military Institute of Medicine, Warsaw, Poland

© Springer International Publishing AG 2017
R. Jabłoński and R. Szewczyk (eds.), *Recent Global Research and Education: Technological Challenges*, Advances in Intelligent Systems and Computing 519, DOI 10.1007/978-3-319-46490-9_25

placenta pathologies we can distinguish: placental infarction (25 % of cases), chorioamnionitis (12.5 % of cases), molar changes (9 % of cases), hipervasculation, avasculation etc. [6]. One of the important step in the placenta examination is vessels counting in the separate villi, which commonly has be done manually through a costly and time-consuming process. The avasculartion can be associated with the problem with angiogenesis of the fetus, whereas a hipervasculation can be an effect of dysfunction in the osmotic transfer between the fetus and the mother.

Computerized villi structures and vessels identification is a complicated problem due to the large variability in the shape, colour, texture and staining intensity of a specimen. In a case of vessels detection, there is a problem with different ways of vessels visibility on the image. In the literature, there are only few studies which proposed the methods of recognition of an object on the placenta. Paper [7] presents the automatic method for the classification of placental tissues into four grades based on ultrasound images. Such images are significantly different from pathomorphological ones and an application of the same solutions is impossible. The paper [8] presents a method to automatically detection and extraction of blood vessels from histological image of the whole placenta. Its authors proposed the solution based on image processing techniques and neural networks. As far as we know, in the literature there is not method for automatic placenta villi and vessels detection on the pathomorphological image of placenta villi.

In this work, we propose an algorithm for automatic vessel recognition and counting in order to support the pathomorphological diagnostic procedure.

2　Materials

The images used to develop algorithm present placenta villi from spontaneous miscarriage subject to Hematoxylin and Eosin (H and E). They were obtained from the archives of the Department of Pathomorphology from the Military Institute of Medicine in Warsaw, Poland. The acquisition of WSI was performed on the 3DHistech Pannoramic 250 Flash II scanner. The RGB images were acquired under magnification $200\times$ with effective resolution 0.389 μm per pixel. To developing presented method the selected 1548×2070 pixels size field of view from WSI were used.

3　Methods

In this paper, we propose the method for the automatic vessel detection on the placenta villi images, based on mathematical morphology [9], colour analysis, and decision tree classification [10]. The vessel detection is a complicated case, because vessels and villi can have diverse colour, shape and size (Fig. 1).

On the placenta images other structures, as collagen (Fig. 1b), free mother blood, and Hofbauer cells [11] can be visible and stained on similar colour to the blood

(a)   (b)

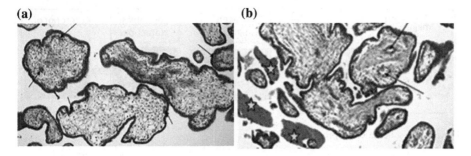

Fig. 1 Examples of placental villi with vessels marked by *arrows*. **a** Vessels with blood cells.
b Vessels empty inside and collagen structures marked by *stars*

cells or vessel walls. These structures could be wrongly classified as vessel areas.
Furthermore, vessels can be visible in two ways: as stained vessel walls when its
interior is empty, or as stained blood cells inside the vessels (vessel walls are only
partly visible). Moreover, blood cells can be present on the image not only inside
vessels, but also visible outside placenta villi, as free cells coming from the mother
blood circulation. In Fig. 1 there are presenting types of vessels which can be
visible on the placenta images.

In order to detect vessels the preprocessing step including villi detection and
collagen elimination is necessary. Then, in two steps the vessels are detected and
finally counted, according with algorithm presented in Fig. 2.

3.1 Pre-processing

The preprocessing step as a detection of vessels inside of villi is necessary. Figure 3
presents the details schema of proposed solution of this task.

It includes four steps: the map creation, the collagen detection and elimination,
the trophoblast detection, and interior of placental villi detection. The structure map
is created on the basis of image in grayscale after thresholding by the mathematical
morphology operations. Next, the detection and elimination of collagen from the
image is a crucial step because these areas are stained with a colour similar to the
blood cells and trophoblast. This can lead to inappropriate classification during

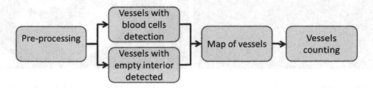

Fig. 2 The general schema of vessels detection

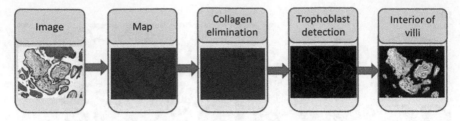

Fig. 3 The schema of preprocessing

the vessels detection. By applied mathematical morphology and decision tree on the *M* component image (from CMYK colour space) the algorithm allows for collagen detection. The one before last step is the detection of trophoblast, which creates villi walls. This operation is based on texture analysis, colour component analysis and region growing operation. What is important is that trophoblasts surrounding villi can have different thicknesses, and that the collagen might surround them. Also, trophoblast is stained with similar colour to the vessel walls, and due to this, it should be detected and eliminated from vessel counting area. The last step includes separation of interiors of villi, because vessels can exist only in these areas. Moreover, application of the map of villi interior allows for the reduction of time-calculation.

3.2 Vessels Detection

The study has shown that two type of image are useful for the vessel detection: image component *M* from CMYK colour space and image in gray scale (Fig. 4).

The detection of vessels with blood cells is performed on the *M* component image (Fig. 4a). The algorithm schema is presented on Fig. 5. In the first step, image is thresholded to detect potential areas with a blood. Next, the mathematical morphology operations are applied in order to combine neighboring areas. Finally,

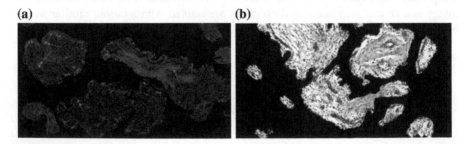

Fig. 4 Example of images used for vessel detection. **a** Image M from CMYK colour space with vessels marked by *arrows*. **b** Image in *grayscale* with vessels marked by *arrows*

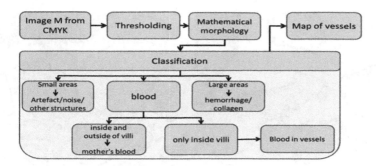

Fig. 5 The schema of algorithm for vessels with blood detection

the decision tree is used to classify detected areas. If an area of potential vessels is too small, then it is omitted. Vessels have to have minimal size, which includes at least few cells (the vessels are built from walls and lumen of vessel with blood cells) [4]. In the case of a large area size, it can be a hemorrhage or collagen structure. If the blood is located inside and outside of the villi interior, then it is not blood in vessels. Probably, it comes from the mother blood circulation or hemorrhage, and these areas should be eliminated. As a result, we achieve a map of detected vessels with blood cells.

The vessels with empty interior are detected separately. This process is conducted on gray scale images (Fig. 4b), and is based on vessel walls detection. After thresholding and mathematical operation, we achieve a map of potential vessel walls. Unfortunately, the intensity of vessel walls, Hofbauer cells, fibrosis and trophoblast are similar. These structures have different features:

- fibrosis create bigger areas with linear shape,
- Hofbauer cells are small, disc elements,
- vessels are medium structures with empty interior (vessels with blood are detected in previously step).

So, after detection of these structures, classification process is necessary. As a result, we achieved a map of vessels with empty interior. Combination of the map of vessels with empty interior with the map of vessels with blood, created a coherent map of various types of vessels. The center of each vessels is detected by mathematical morphology operations. Connection of the map of vessels with the map of villi allows for make calculation on the number of vessels in each of placenta villi.

4   Results

The presented method was applied for vessels detection for cases with low and high number of vessels. The examples of results are presented in Fig. 6.

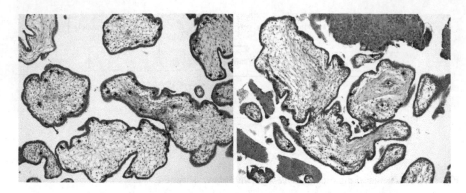

Fig. 6 Example of vessel detection, vessels are marked with *stars*

As can be visible, all detected objects are located inside of placenta villi. The number of vessels are variable. The most of them are localized closely to the trophoblast and only few highest ones (main vessels) are in the center of villi. In some villi no vessels exist. The number of vessels in each villi is calculated after combination of the map of vessel with the map of villi. As a result, we achieved index, which describes quantity of vessels in each of villi. The quantitative results were consulted with the medical expert, in order to confirm the correct detection.

5 Conclusion

The presented method, based on colour analysis, mathematical morphology and detection tree classification allows for correct vessel detection in the placenta images. The results of the algorithm performance were consulted with medical expert, who confirmed the correct solution. It should be noted that the vessels can have diverse of shape, size and staining, and number of vessels in each of villi can be different. Features of presented approach are short calculation time and repeatability of results. This algorithm can be use as support for medical diagnosis or in medical research. The future research include detection and estimation of index used in medical diagnose of pathomorphological images of placenta villi in whole slide images.

Acknowledgment This work has been supported by the National Science Centre (Poland) by the grant 2012/07/B/ST7/01203 in the years 2013–2016.

References

1. Kothari, S., Phan, J.H., Stokes, T.H., Wang, M.D.: Pathology imaging informatics for quantitative analysis of whole-slide images. J. Am. Med. Inform. Assoc. **20**, 1099–1108 (2013)
2. Yinhai, W., Danny, C., Jim, D., Peter, H., Richard, T.: Segmentation of squamous epithelium from ultra-large cervical histological virtual slildes. In: Proc. 29th IEEE Engineering in Medi-cine and Biology System (EMBS) Conference, pp. 775–778 (2007)
3. Wang, Y., Zhao, S.: Vascular Biology of the Placenta. Morgan & Claypool Life Sciences, San Rafael, CA (2010)
4. Stallmach, T., Hebisch, G.: Placental pathology: its impact on explaining prenataland perinatal death. Virchows Arch. **445**(1), 9–16 (2004)
5. Ernst, L.M., D, M., Gawron, L., Fritsch, M.K., Ernst, L.M.: Pathologic examination of fetal and placental tissue obtained by dilation and evacuation. Arch. Pathol. Lab. Med. **137**, 326–337 (2013)
6. Hassan, T.M.M., Hegazy, A.M.S., Mosaed, M.M.: Anatomical and histopathologic analysis of placenta in dilation and evacuation specimens. Forensic Med. Anathomy Res. **02**(02), 17–27 (2014)
7. Ayache, M., Khalil, M., Tranquart, F.: DWT to classify automatically the placental tissues development: neural network approach. J. Comput. Sci. **6**(6), 634–640 (2010)
8. Almoussa, N., Dutra, B., Lampe, B., Getreuer, P., Wittmanm, T., Salaa, C., Vese, L.: Automated vasculature extraction from placenta images. In: Proc. SPIE 7962, Medical Imaging (2011). doi:10.1117/12.878343
9. Soille, P.: Morphological Image Analysis, Principles and Applications, Springer, Berlin (2003)
10. Srivastava, A., Han, E.H.S., Singh, V., Kumar, V.: Parallel formulations of decision-tree classification algorithms. In: Proceedings International Conference on Parallel Processing, pp. 237–244 (1998)
11. Ingman, K., et al.: Characterisation of Hofbauer cells in first and second trimester placenta: incidence, phenotype, survival in vitro and motility. Placenta **31**, 535–544 (2010)

A Novel Particle Classification Technique Arising from Acoustic-Cavitation-Oriented Bubbles (ACOBs) Under kHz-Band Ultrasonic Irradiation in Water

Sayuri Yanai and Takayuki Saito

Abstract We have developed a novel particle classification technique using kHz-band ultrasound in water. By irradiating kHz-band ultrasound in a particle-liquid mixture, we found out that swarm-like flocculation of particles is formed in the liquid. This particle flocculation is caused by Acoustic-Cavitation-Oriented Bubbles (ACOBs) generated by the ultrasound. This novel technique can be applied to a wide range of particle sizes from several hundred μm to several mm. In the present study, we carefully manipulate the flocculation and intend to apply this to a new technique of particle classification.

Keywords Particle classification · kHz-band ultrasound · Particle flocculation · Acoustic-Cavitation-Oriented Bubbles (ACOBs)

1 Introduction

Many solid–liquid separation techniques have been proposed, and some of them are still used in industrial fields. For instance, sedimentation [1] is a very classical separation technique using the density difference between dispersed particles and mother liquid under the gravitational force or a centrifugal force. However, when the density difference between particles and liquid is small, it is very difficult to separate the particles from the mixture. Filtration [2] needs a multi-step process on separation using some filter materials; inevitably, this technique needs large

S. Yanai (✉)
Graduate School of Integrated Science and Technology,
Shizuoka University, Shizuoka, Japan
e-mail: yanai.sayuri.15@shizuoka.ac.jp

T. Saito
Research Institute of Green Science and Technology,
Shizuoka University, Shizuoka, Japan
e-mail: saito.takayuki@shizuoka.ac.jp

© Springer International Publishing AG 2017
R. Jabłoński and R. Szewczyk (eds.), *Recent Global Research and Education:
Technological Challenges*, Advances in Intelligent Systems and Computing 519,
DOI 10.1007/978-3-319-46490-9_26

apparatus, and erosion of the particles is unavoidable. In order to overcome these difficulties in conventional separation techniques, we grasped a breakthrough in separation technique, by irradiating kHz-band ultrasound to a particle-liquid mixture [3]. In the present study, we propose a promising unit process using this technique, which might replace a conventional particle classification process.

2 Mechanism of the Particle Flocculation

Figure 1 shows typical time-series snapshots of a moving particle forming a spherical particle flocculation in water. An Acoustic-Cavitation-Oriented bubble (ACOB) adhering to the particle surface can be observed. The acoustic radiation force toward the Large-Pressure-Amplitude-Region (LPAR) acts on the bubble; as a result, the particle moves toward LPAR and forms the spherical swarm-like flocculation. At this stationary position, the particle swarm is suspended by the equilibrium of the gravitational force, buoyancy force, acoustic radiation force and acoustic streaming.

3 Experimental Apparatus for the Particle Flocculation

A sine wave voltage signal generated by a function generator (SG4115, IWATSU, frequency: 20 kHz, output voltage: 1.00 V) was supplied into a bolt-clamped Langevin type transducer (HEC-45254M, Honda Electronics) through a power amplifier (2100L, E&I). The ultrasonic was irradiated upwardly-vertically from the bottom of an acrylic water vessel (inner dimension: 54 mm × 54 mm, height: 100 mm). Purified water was put in the vessel up to a depth of 60 mm. Polystyrene particles (density: 1.06 g/cm3; diameter: 1000 μm, terminal velocity: 15 mm/s, Reynolds number of the particle: 15) were dispersed in water; immediately, 20-kHz-ultrasonic was irradiated. A high-speed video camera (FASTCAM SA-X2,

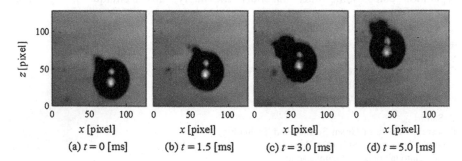

(a) $t = 0$ [ms]   (b) $t = 1.5$ [ms]   (c) $t = 3.0$ [ms]   (d) $t = 5.0$ [ms]

Fig. 1 Time series snapshots of the particle motion and ACOB motion

Photron) was used to visualize the particles' flocculation. The camera and an LED illumination were mounted on optical stages as facing each other. The spatial resolution, image size and frame rate were 53.1 µm, 1024 × 1024 pixels and 1000 fps, respectively. In addition, the ACOBs oscillation (expansion and constriction) was visualized as shadowgraph method using CW laser (wavelength: 532 nm, power: 1.2 W). The acoustic pressure profile was measured through a hydrophone probe (HPM1/1, Precision Acoustics LTD., diameter of the sensing tip: 1 mm).

4 Visualization Result of the Particle Flocculation

A spherical particle swarm was formed at z ≈ 19 mm under 20 kHz ultrasonic irradiation. The stationary position of spherical particle swarm corresponded with ACOBs' swarm position and Large-Pressure-Amplitude-Region (LPAR) in the acoustic field of 20 kHz ultrasound. Its time-average flocculation diameter d was 8.8 mm. This flocculation resulted from ACOBs generation at the particles' surface (Fig. 2).

5 Manipulation of the Particle Flocculation

A stick (manipulation stick) can manipulate the flocculated particles. The swarm moved in association with the stick motion (material: POM, density: 2.20 g/cm^3, size: $\phi 7 \times 270$ mm). As plotted in Fig. 3, we found out a relationship between a travel distance of the particle swarm and the total mass of the manipulated particles. There was a strong relationship among the travel distance of flocculation, ACOBs-contribution and limitation of flocculation. Once the rod was inserted in the liquid, the ACOBs emerge from the surface of the rod and attracted particles. However, when M > 0.3 g, the flocculating particle population (i.e. total mass of

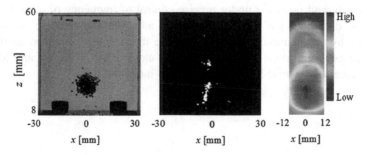

Fig. 2 a Particle flocculation. b ACOBs' swarm. c Acoustic pressure profile

Fig. 3 Travel distance of
particle swarm along the
vertical axis

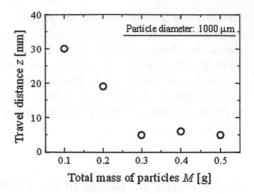

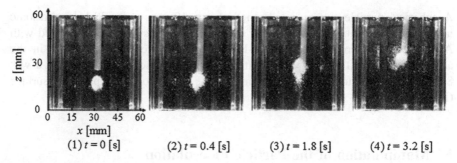

(1) $t = 0$ [s]　　(2) $t = 0.4$ [s]　　(3) $t = 1.8$ [s]　　(4) $t = 3.2$ [s]

Fig. 4 The particle swarm motion with the POM rod in the water

the particles) was already limited; hence, the travel distance steeply decreased after $M = 0.3$ g (Fig. 4).

6　Conclusion

We have developed a novel particle classification technique by the use of ACOBs irradiating 20 kHz ultrasound in water. This technique could separate particles from a liquid easily and precisely under precise control of frequency of ultrasound and volume of dissolved gas in the liquid. In addition, this particle swarm can manipulate by the manipulation stick because the ACOBs generated around the stick's surface would draw the particle swarm close to the stick. We aim to apply this technique to the solid-liquid separation in industrial field.

References

1. Richardson, J., Zaki, W.: Sedimentation and fluidization: Part I. Trans. Inst. Chem. Eng. **32**, 35–53 (1954)
2. Huq, A., Xu, B., Chowdhury, M.A.R., Islam, M., Montilla, R., Colwell, R.: A simple filtration method to remove plankton-associated *Vibrio cholerae* in raw water supplies in developing countries. Appl. Environ. Microbiol. **62**, 2505–2512 (1996)
3. Mizushima, Y., Nagami, Y., Nakamura, Y., Saito, T.: Interaction between acoustic cavitation bubbles and dispersed particles in a kHz-order-ultrasound-irradiated water. Chem. Eng. Sci. **93**, 395–400 (2013)

Numerical Investigation of Drag Reduction by Hydrogel with Trapped Water Layer

Petya V. Stoyanova, Youhei Takagi and Yasunori Okano

Abstract Hydrogels are novel materials that exhibit drag reduction properties in marine applications. It is suggested that they can reduce drag by trapping water between the dimples present on a rough coating but the precise mechanism hardly understood. In the present study, a series of direct numerical simulations (DNS) of turbulent channel flow were performed. The water trap effect of the hydrogel was modelled as Navier slip boundary condition on a flat surface with periodically varying slip length in the streamwise and spanwise directions. The results showed that drag reduction effect increases with larger slip lengths and that wavelengths comparable with the size of the vortex structures are effective for drag reduction.

Keywords Drag reduction · Hydrogel coating · Local slip velocity · Turbulent channel flow

1 Introduction

In the shipping industry losses due to frictional drag can amount to more than half of the total resistance [1]. Additionally, fouling of the ship's hull can increase drag considerably. To combat fouling build up, self-polishing coatings containing toxic compounds were used, however it was found that they are harmful to organisms other than the targeted and are currently strictly regulated. As a result, recent research effort is focused on developing environmentally-friendly paints for fouling control.

P.V. Stoyanova · Y. Takagi (✉) · Y. Okano
Department of Materials Engineering Science, Osaka University, Osaka, Japan
e-mail: st.petya@cheng.es.osaka-u.ac.jp

Y. Takagi
e-mail: takagi@cheng.es.osaka-u.ac.jp

Y. Okano
e-mail: okano@cheng.es.osaka-u.ac.jp

© Springer International Publishing AG 2017
R. Jabłoński and R. Szewczyk (eds.), *Recent Global Research and Education: Technological Challenges*, Advances in Intelligent Systems and Computing 519,
DOI 10.1007/978-3-319-46490-9_27

Hydrogels are novel materials inspired by living organisms like dolphins, porpoises, sea squirts, which can resist biofouling. Moreover, it was shown that hydrogel paintings can reduce drag [2]. However the mechanism is hardly understood. In real paint application the production of smooth coating is very difficult and often the surfaces are characterized with measurable roughness, which leads to increased skin friction. However it was found that hydrogel coatings reduce drag even though they possess hydrodynamic roughness. Other surfaces which also reduce drag despite being rough are superhydrophobic surfaces and surfaces with riblets, both of which have been studied in the literature [3–10]. However it is yet unclear whether hydrogels reduce drag by the same mechanism as riblets. It has been suggested [2] that hydrogels reduce friction by trapping water in the valleys of the rough coating. In the present study, a model of the water trapping effect is used to simulate the influence of hydrogel roughness on drag.

2 Numerical Method

Fully-developed turbulent channel flow between two infinite parallel plates was considered in order to investigate the effect of hydrogel roughness. The governing equations—the momentum conservation (Navier-Stokes) and the continuity equations were solved for constant flow rate.

The slip velocity was controlled through the Navier slip boundary conditions:

$$u_s = l_x \frac{\partial u}{\partial y}\bigg|_{wall}, \quad w_s = l_z \frac{\partial w}{\partial y}\bigg|_{wall}, \tag{1}$$

which were applied symmetrically to both walls. Here u_s and w_s are the slip velocities in the streamwise and spanwise direction, respectively, and l_x and l_z are the slip lengths. The latter were varied in space via filtering functions:

$$f_x = \frac{1}{2}\left\{1 + \cos\left(\frac{2\pi}{\lambda_x}x\right)\right\}, f_z = \frac{1}{2}\left\{1 + \cos\left(\frac{2\pi}{\lambda_z}z\right)\right\} \tag{2}$$

$$l_x, l_z = f_x f_z l_0 \tag{3}$$

to account for a wavy hydrogel surface as the one presented on Fig. 1. Here l_0 is the reference slip length, corresponding to the depth of the valleys on the surface.

The numerical method was the same as in Ref. [11]. A computational domain of size $4\pi\delta \times 2\delta \times 2\pi\delta$ and $128 \times 150 \times 128$ grid points was used in the streamwise, wall-normal and spanwise directions, respectively. The simulations were performed at $Re = U_c\delta/v = 4200$ or $Re_\tau = u_{\tau_0}\delta/v \cong 180$, where U_c is the centerline velocity, δ is the channel half height, v is the kinematic viscosity and u_{τ_0} is the friction velocity for no slip channel flow. In order to investigate how the size and height of the dimples affect the turbulent vortex structure near the wall we

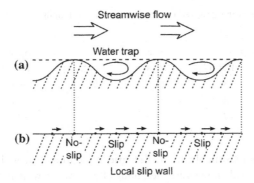

Fig. 1 Schematic of **a** the water trap effect by rough hydrogel surface and **b** the local slip model

Table 1 Parameters used in the DNS of turbulent channel flow

Parameter	Symbol	Value
Reference slip length	l_0 (l_0^+)	0.005 (0.9), 0.01 (1.8), 0.02 (3.6)
Streamwise wavelength	λ_x^+	275, 368, 550, 735, 1100
Spanwise wavelength	λ_z^+	50, 183

varied the reference slip length, and the streamwise and spanwise wavelengths as shown in Table 1. Hereafter, superscript "+" denotes quantities normalized by u_{τ_0} and v.

3 Results and Discussion

3.1 Code Validation

To validate the code we performed simulations with uniform slip lengths and compared them to the results by Min and Kim [3]. Three types of calculations were performed: (a) isotropic slip ($l_x = l_z = l_0$); (b) only streamwise slip ($l_x = l_0, l_z = 0$); (c) only spanwise slip ($l_x = 0, l_z = l_0$) for all the reference slip lengths used in the present study. The drag reduction ratio (DR) is calculated by:

$$\text{DR} = \frac{\left(-dp/dx|_{no-slip}\right) - \left(-dp/dx|_{slip}\right)}{\left(-dp/dx|_{no-slip}\right)} \times 100 \qquad (4)$$

where $-dp/dx|_{\text{no-slip}}$ and $-dp/dx|_{\text{slip}}$ are the mean pressure gradients under no-slip and slip conditions, respectively.

The comparison of the DR is presented in Table 2. It can be seen from the results that the values agree well qualitatively, although there are some noticeable differences. This is due to the fact, that in order to maintain the computational costs of the

Table 2 Comparison of the drag reduction by uniform slip length

	DR[%]					
	Isotropic slip		Streamwise slip		Spanwise slip	
l_0^+	Present	Ref. [3]	Present	Ref. [3]	Present	Ref. [3]
0.9	4	3	11	10	−9	−8
1.8	10	8	20	18	−18	−16
3.6	19	17	31	29	−31	−26

simulations within reasonable limit, the grid resolution had to be reduced. Even with this reduced grid, however, the main effects were well reproduced and the differences were relatively small. Therefore it was concluded that the present simulation conditions were sufficient to give the main effects of interest.

3.2 Effect of Spatially Varying Slip Length in Streamwise and Spanwise Directions

In the study the wavelengths of the wavy surface were chosen so that they correspond to the order of the size of the flow streak structures [12]. Figure 2 shows the effect of slip wavelength on the drag for the different slip lengths. The drag reduction ratio increases with the increase of the reference slip length, but the effect is considerably smaller than the respective uniform slip cases (Table 2). For $l_0^+ = 0.9$ the DR is about ~ 1 %, where it is about 4 % for the respective uniform isotropic slip case. For $l_0^+ = 1.8$ the DR is about ~ 2 % and for 3.6 it ranges between 3.3 and 5.2 %.

At lower reference slip lengths the choice of wavelengths strongly influences the skin friction and the difference in the DR between two wavelength combinations can be as high as ~ 1 %. As the reference slip length increases to 3.6 wall units the correlation of the wavelengths is mostly lost and the resultant drag reduction depends mainly on the size of the wavelengths. This is evidenced by the fact that in the cases with large slip the DR shows a general increase with the increase of the streamwise wavelength, where such is not observed in the respective low slip cases. A possible explanation for this phenomenon is that at high slip lengths the surface-turbulent vortex interaction is reduced, then the no-slip areas act only as obstacles that disturb the near-wall vortices with higher wavelengths corresponding to gentler, less frequent perturbances and lower friction.

Figure 3a shows the mean velocity profiles of the no slip and a slip case with $l_0^+ = 3.6$, $\lambda_x^+ = 550$, $\lambda_z^+ = 50$. As a result of a mean slip velocity on the surface, the mean velocity is increased everywhere in the channel. The effect, however, is more pronounced in the viscous region ($y^+ < 5$). The presence local of slip conditions influence the root mean square (rms) fluctuations mainly of the streamwise velocity component, u_{rms}^+, although some decrease is also observed for the wall-normal, v_{rms}^+,

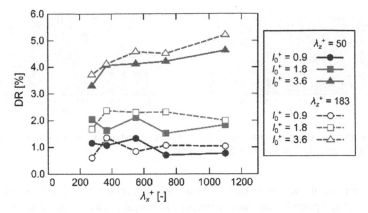

Fig. 2 Dependence of the drag reduction ratio on the slip length and slip wavelength combination. *Symbols* represent simulation results, the lines serve to guide the eye

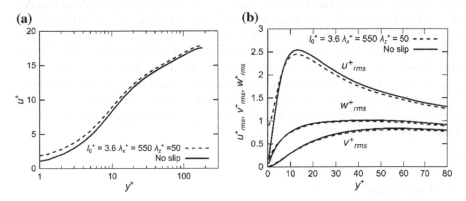

Fig. 3 Comparison of the **a** mean velocity profiles and **b** the profiles of the root mean square velocity fluctuations as function of the channel width, y^+, in the no slip case and case with $l_0^+ = 3.6$, $\lambda_x^+ = 550$, $\lambda_z^+ = 50$

and the spanwise velocity fluctuations, w_{rms}^+, (Fig. 3b). In case of u_{rms}^+ the peak is lower and in the near wall region the increase of the fluctuations is observed due to the fluctuations of the slip velocity. These results are similar to those for uniform slip length [3, 4], showing that the net effect of the local slip on the flow is of the same nature.

To investigate the effect of the surface patterning on the streak structures in Fig. 4 we compared the instantaneous streamwise velocity fluctuations in the horizontal plane at distance $y^+ \sim 6.1$. The differences between the structures in the two cases were small and required further investigation of the two-point correlations (data not shown). It was discovered that the size of the streak structures is not significantly influenced by the slip patterning, where in both the slip and no-slip

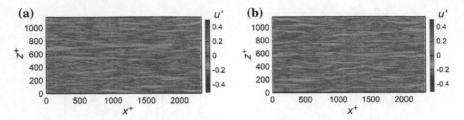

Fig. 4 Streamwise velocity fluctuations in the x–z plane at distance $y^+ \sim 6.1$ wall units from the wall. **a** No slip case. **b** $l_0^+ = 3.6$, $\lambda_x^+ = 550$, $\lambda_z^+ = 50$

cases the length of the streak structures in the streamwise direction was about 500 wall units. The spanwise distance between the streaks was about 50 wall units in good agreement with Ref. [12]. Therefore the spatially varying boundary conditions reduced drag without influencing the large coherent structures. Further studies at higher Reynolds number should be carried out for discussing the relation between the slip length and the wall-normal velocity gradient in the present hydrogel model.

As for the physical interpretation of the results, the increase of DR with slip length means that it might be more beneficial to have hydrogels with larger peak to valley roughness than smooth surfaces. However the simulation results can be trusted only to the extent that the trapped water layer hypothesis holds. It is expected that beyond a critical value the regime would break down and further increase in the roughness would not lead to drag reduction. This critical value should be obtained from experiments, e.g. by particle image velocimetry in turbulent water channel.

4 Conclusions

A three-dimensional simulation of turbulent channel flow over walls with modelled rough hydrogel surface was carried out. It was found that wavelengths of the order of the size of the streak structures are effective for drag reduction as drag was reduced in all cases. The reduction was proportional to the slip length reaching to up to 5 %. It was found that for low slip lengths the effect on drag depends strongly on the wavelength combination of the slip patterning of the wall. This dependency was mostly lost for the high slip lengths, where the effect of the wavelengths acted mostly through the wavelength size. For hydrogels this would mean that coatings, whose roughness constitutes of larger peak to valley height are beneficial for drag reduction.

References

1. Lindholdt, A., Dam-Johansen, K., Olsen, S.M., Yebra, D.M., Kiil, S.: Effects of biofouling development on drag forces of hull coatings for ocean-going ships: a review. J. Coat. Technol. Res. **12**, 415–444 (2015)
2. Yamamori, N., Shimada, M.: Verification of the reduction effect of new antifouling paint for ship bottoms "LF-Sea$_®$" on the friction resistance by utilizing actual ships (in Japanese). TECHNO-COSMOS **22**, 28–33 (2009)
3. Min, T., Kim, J.: Effects of hydrophobic surface on skin-friction drag. Phys. Fluids **16**, L55–L58 (2004)
4. Busse, A., Sandham, N.D.: Influence of an anisotropic slip-length boundary condition on turbulent channel flow. Phys. Fluids **24**, 055111 (2012)
5. Daniello, R.J., Waterhouse, N.E., Rothstein, J.P.: Drag reduction in turbulent flows over superhydrophobic surfaces. Phys. Fluids **21**, 085103 (2009)
6. Park, H., Sun, G., Kim, C.J.: Superhydrophobic turbulent drag reduction as function of surface grating parameters. J. Fluid Mech. **747**, 722–734 (2014)
7. Seo, J., Garcia-Mayoral, R., Mani, A.: Pressure fluctuations and robustness in turbulent flows over superhydrophobic surfaces. J. Fluid Mech. **783**, 448–473 (2015)
8. Bechert, D.W., Bruse, M., Hage, W., Van Der Hoeven, J.G.T., Hoppe, G.: Experiments on drag-reducing surfaces and their optimization with an adjustable geometry. J. Fluid Mech. **338**, 59–87 (1997)
9. Garcia-Mayoral, R., Jimenez, J.: Hydrophobic stability and breakdown of the viscous regime over riblets. J. Fluid Mech. **678**, 317–347 (2011)
10. Garcia-Mayoral, R., Jimenez, J.: Scaling of turbulent structures in riblet channels up to $Re_\tau \approx 550$. Phys. Fluids **24**, 105101 (2012)
11. Takagi, Y., Nakamoto, M., Okano, Y.: Numerical investigation of the drag reduction mechanism in turbulent channel flow with local slip velocity. In: 4th ASCHT, 0060-T01-1-A (2013)
12. Kim, J., Moin, P., Moser, R.: Turbulent statistics in fully developed channel flow at low Reynolds number. J. Fluid Mech. **177**, 133–166 (1987)

Development of High-Frequency Acoustic Source for Auditory Stimulated Magnetoencephalography

Anna Jodko-Władzińska, Michał Władziński,
Tadeusz Pałko and Tilmann Sander

Abstract Hearing can be tested objectively by measuring brain responses due to hearing of certain frequencies. Brain responses can be measured by means of electroencephalography or magnetoencephalography, but non-contact measurements of biomagnetic field allow avoiding difficulties associated with the use of electrodes. Furthermore, magnetoencephalography is sensitive to other brain current configurations than electroencephalography. Acoustic source for auditory evoked magnetoencephalography should deliver the stimulus signal directly to human's ear and should not interfere with MEG recordings of brain magnetic fields. Cost-efficient setup based on a piezoelectric transducer requires the use of polythene tube for delivering the sound to the subject, which attenuates the tones of frequencies above 9 kHz due to its length. Possible solutions to increase the range of frequencies delivered by auditory source were analyzed. Prototype setup was modified by changing waveguide medium to water and independently by reducing its dimensions, especially the tube length and diameter. Efficiency of acoustic sources was not sufficient and sound pressure level software correction was made. Acoustic properties of developed setups are presented.

Keywords Magnetoencephalography · MEG · Auditory evoked magnetic field · High-frequency acoustic source

A. Jodko-Władzińska (✉) · M. Władziński · T. Pałko
Faculty of Mechatronics, Warsaw University of Technology,
Św. Andrzeja Boboli 8, 02-525 Warsaw, Poland
e-mail: a.jodko@mchtr.pw.edu.pl

M. Władziński
e-mail: m.wladzinski@mchtr.pw.edu.pl

T. Pałko
e-mail: t.palko@mchtr.pw.edu.pl

T. Sander
Department of Biosignals, Physikalisch-Technische Bundesanstalt,
Abbestr. 2-12, 10587 Berlin, Germany
e-mail: Tilmann.Sander-Thoemmes@ptb.de

© Springer International Publishing AG 2017
R. Jabłoński and R. Szewczyk (eds.), *Recent Global Research and Education: Technological Challenges*, Advances in Intelligent Systems and Computing 519,
DOI 10.1007/978-3-319-46490-9_28

1   Introduction

In pure tone audiometry, hearing thresholds of frequencies from 125 Hz to 8 kHz
are measured. High-frequency (8–20 kHz) pure tone audiometry is not routinely
used for diagnostics and is only performed on groups of patients especially vul-
nerable to hearing loss as it can indicate damage to the inner ear before hearing loss
expands to the frequencies tested in conventional audiometry [1].

Hearing can be tested objectively by measuring brain responses due to hearing of
certain frequencies. Thus, the measurement of brain responses due to hearing of
high frequencies can be useful in prediction of hearing loss.

Brain responses can be measured by means of electroencephalography or
magnetoencephalography. Magnetoencephalographic and electroencephalographic
(MEG/EEG) signals contain fairly similar information due to the same source of
signals: activity of the cerebral cortex neurons. However, some components of EEG
signal cannot be detectable using MEG as magnetoencephalography is sensitive to
other brain current configurations than electroencephalography. Magnetic field
changes correspond to the tangential component of postsynaptic neuronal currents,
while the electric potentials relate to both tangential and radial components [2].

Nevertheless, non-contact measurements of biomagnetic field allow avoiding
difficulties related to the use of scalp electrodes, such as a poor electrode contact
with the skin and varied conductivity of the body tissues. These factors reduce the
strength of signals and the accuracy of EEG measurements. The magnetic field is
not influenced by the presence of skin, skull, cerebro-spinal fluid or grey-matter as
their magnetic permeability is actually equal [3, 4].

Measurement of magnetic brain responses due to hearing is a challenge because
of its extremely low value (10–13 T) and the influence of technical electromagnetic
sources and the Earth's magnetic field (5×10^{-5} T). Acoustic sources for auditory
evoked magnetoencephalography should deliver the sound directly to human's ear
and cannot cause any disturbance to the MEG recordings of brain magnetic fields
[5].

2   Development of Auditory Source for MEG

A cost-efficient acoustic source based on a piezoelectric transducer (KEMO L010,
https://www.luedeke-elektronic.de) and commonly available parts was designed for
auditory evoked magnetoencephalography as presented in Fig. 1 [5]. In the prototype setup the piezoelectric transducer was placed in a funnel (a), which was
connected to a 5 m—long polythene tube (b; $\varphi = 10$ mm) to avoid a disturbance of
the MEG signal from the currents flowing in the transducer. An ear insert (d) was
connected with the polythene tube via silicon tubes of smaller diameter (c).

The setup successfully delivered sounds at frequencies from 1 to 8 kHz. Sounds
at higher frequencies were strongly attenuated, which makes the acoustic source

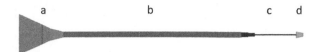

Fig. 1 Diagram of the acoustic source. **a** Funnel with piezoelectric transducer. **b** Polythene tube (*blue*). **c** Silicon tubes (*dark blue*). **d** Ear insert

unsuitable for high-frequency auditory stimulated MEG (Fig. 2). Possible solutions to increase the range of frequencies delivered by auditory source were analyzed.

2.1 Fluid as a Medium

Sound waves travel faster in liquids than in air and thus the length of sound wave of certain frequency is greater in liquids. The longer the wave is, the less disturbance according to reflections at the end of waveguide and interference of waves occurs. This leads to less disturbances of sound waves for the waveguide of certain length filled with fluid.

The 5 m—long polythene tube in the prototype auditory setup (b in Fig. 1) was filled with water, in which sound waves travel more than 4 times faster. One end of polythene tube was connected to the funnel holding piezoelectric transducer, which was contacting the fluid. The other end, connected to silicon tubes of smaller diameter (c in Fig. 1), was closed with an elastic membrane holding fluid. Silicon tubes with total length of 0.3 m, used to connect the polythene tube with the ear insert, were filled with air.

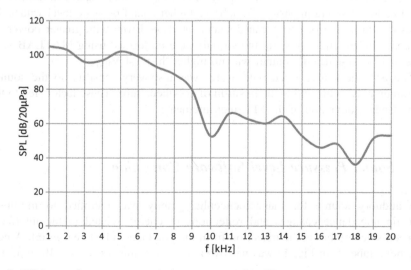

Fig. 2 Efficiency of prototype acoustic source presented in [5]

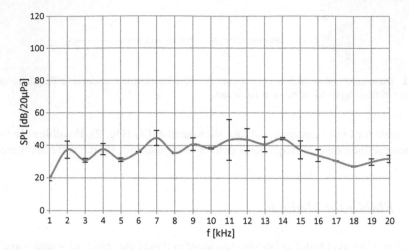

Fig. 3 Efficiency of acoustic source filled with fluid (water); *bars* represent standard deviation of data set obtained for each frequency

Efficiency of acoustic source was measured at Physikalisch-Technische Bundesanstalt Berlin using Brüel and Kjaer (http://www.bksv.com) type 4157 occluded ear simulator. The results are presented in Fig. 3.

Efficiency of acoustic source filled with fluid was much more flat comparing to the prototype acoustic source filled with air. Sound pressure level values of air-borne sounds were in the range of 80–100 dB for frequencies below 9 kHz and 40–60 dB for frequencies above 9 kHz. SPL values of water-borne sounds for almost all audible frequencies (except 1 kHz) were in the range of 30–45 dB.

Higher density of water comparing to air and higher sound speed in fluids both lead to a lower sound intensity. To achieve higher sound pressure level values the acoustic source (piezoelectric transducer) should be driven with greater power. In the cost-efficient auditory setup the stimuli were performed using MATLAB software and PC soundcard's output was no higher than 1,4 V_{RMS}.

The acoustic source filled with water was also very sensitive to the sounds surrounding the setup and to the position of polythene tube. These factors led to the high standard deviation of the SPL measurements.

2.2 Sound Pressure Level Software Correction

The auditory source for magnetoencephalography was modified to minimize acoustic impedance changes and reflections of energy related to the interfaces between silicon tubes at the transition from the polythene tube to ear insert. A new polythene tube (b in Fig. 1) was narrower (φ_{new} = 8 mm vs. φ_{old} = 10 mm) and

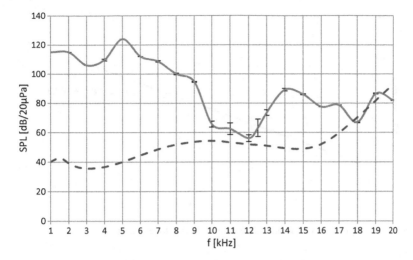

Fig. 4 Efficiency of designed acoustic source (*solid line*) and the equal-loudness curve of 40 phons due to ISO 226:2003 standard (*dashed curve*); *bars* represent standard deviation of data set obtained for each frequency

less connectors (c in Fig. 1) were needed. The new polythene tube was also shorter (L_{new} = 2.5 m vs. L_{old} = 5 m) to minimize attenuation of the high-frequency tones.

Due to the shortening of the polythene tube the piezoelectric transducer was not placed outside magnetically shielded room as in the prototype setup. Magnetic properties of the acoustic source had to be tested. The ear insert was placed in the MEG helmet and the transducer was driven by an electric signal. No disturbing signals due to the acoustic source could be detected. It can be concluded that the acoustic source will not interfere with the recordings of brain magnetic fields.

The efficiency of modified acoustic source was measured using Brüel & Kjaer type 4157 occluded ear simulator. The results are presented in Fig. 4 (solid line).

Pure tones of frequencies above 9 kHz were attenuated but not as strongly as in the prototype setup. The decrease of efficiency could be related to the tube geometry and reflections of sound waves.

Sound pressure level values of the tones were software corrected. For future use in magnetoencephalography, the SPL values of high-frequency sounds were set at perceived loudness level of 40 phons defined in the international standard ISO 226:2003 (dashed curve in Fig. 4).

Sound pressure level values for varied tones amplitude were measured using PC soundcard's standard output. Sound pressure level versus common logarithm for tone amplitude curves were calculated for the frequencies of interest: 8, 10, 12.5, 16 and 20 kHz. For all tested frequencies characteristics were linear with the coefficient of determination greater than 0.99 as presented in Fig. 5.

Amplitude of each tone was set according to SPL—lg(A) curve to provide SPL values defined in the international standard ISO 226:2003. SPL values for the

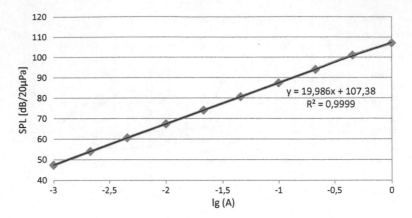

Fig. 5 SPL vs. lg(A) curve for pure tone of 8 kHz; A—tone amplitude

frequencies of interest (8, 10, 12.5, 16 and 20 kHz) were: 51.8, 54.3, 51.5, 52 and 92.8 dB respectively.

3 Conclusions

The developed acoustic source successfully delivers high-frequency sounds at perceived loudness level of 40 phons defined in the international standard ISO 226:2003. It meets the requirements for magnetoencephalography as it does not interfere with MEG recordings and can be used in measurements of brain responses due to high-frequency stimuli.

References

1. Fabijańska, A., Smurzyński, J., Kochanek, K., Skarżyński, H.: Audiometria wysokich częstotliwości u pacjentów z szumami usznymi i prawidłowym słuchem. Nowa Audiofonol. **3** (3), 17–23 (2014)
2. Huotilainen, M., Winkler, I., Alho, K., Escera, C., Virtanen, J., Ilmoniemi, R.J., Jääskeläinen, I. P., Pekkonen, E., Näätänen, R.: Combined mapping of human auditory EEG and MEG responses. Electroencephalogr. Clin. Neurophysiol. **108**(4), 370–379 (1998)
3. Bakker, L.N. (ed.): Brain Mapping Research Developments, Nova Science Publishers (2008)
4. Hansen, P.C., Kringelbach, M.L., Salmelin, R. (eds.): MEG: An Introduction to Methods, Oxford University Press (2010)
5. Jodko-Władzińska, A., Kühler, R., Hensel, J., Pałko, T., Sander, T.: Evaluation of cost-efficient auditory MEG stimulation, Advanced Mechatronics Solutions. In: Jabłoński R., Brezina T. (eds.) Advances in Intelligent Systems and Computing, pp. 153–158. Springer (2016)

Dynamic Promotion and Suppression Model for Plasmid Conjugal Transfer Under a Flow Condition

T. Watanabe and K. Takeda

Abstract Recipients (the plasmid-free bacteria) receive a copy of plasmid from donors (the plasmid-bearing bacteria) and become transconjugants. In case of the plasmid has a gene providing a drug resistant, recalcitrant substance and so on, these functions are transferred with plasmid carrying. Plasmid transfer makes a substantial contribution an ability to adjust to the environment. Plasmid transfer is mainly performed by conjugation, called conjugal transfer, of a pair of bacteria. Conjugal transfer need to approach the bacteria bodies. Meanwhile, in nature, bacteria lives in not only solid conditions such as soil but also liquid conditions such as rivers. Therefore dynamic analysis of plasmid in the liquid condition is very important. Conjugal transfer is affected by a stream of the liquid condition, because the behavior of bacteria is controlled by the stream. However, relationship between flow condition and the number of transconjugants isn't currently clarified. Therefore, in this study, we aim at analysis of plasmid dynamics under the flow condition. We develop a model with promotion and suppression term for relations between flow condition and the number of transconjugants, and identifying parameters by using the experimental data at agitating flask.

Notation

Z   Collision frequency [Hit]
D_{pv} Volume equivalent diameter [μ m]
$\aleph$ Volume number density [CFU/m^3]
H   Height of the liquid [m]
T   Temperature [K]
n_t Total molecular weight in the tank [mol]
r_c Individual rotation radius [m]
α Probability of valid collision [-]

T. Watanabe (✉) · K. Takeda
Graduate School of Integrated Science and Technology, Shizuoka
University, Shizuoka, Japan
e-mail: watanabe.toshinori.15@shizuoka.ac.jp

© Springer International Publishing AG 2017   203
R. Jabłoński and R. Szewczyk (eds.), *Recent Global Research and Education:
Technological Challenges*, Advances in Intelligent Systems and Computing 519,
DOI 10.1007/978-3-319-46490-9_29

y   Number of transoonjugants [CFU]
1   Number of donors [CFU]
γ   Probability of inhibition [CFU/Hit]
z   Collision frequency [Hit/s]

1   Introduction

Bacteria is present in every environment and adjust to environmental changes. The
key of this ability to the environment is gene transfer. There are various ways to
transfer genes such plasmid conjugal transfer, transformation, and transduction.
Plasmid conjugal transfer allow faster gene transfer than the others. Therefore, it is
thought the most important way to transfer genes [1]. Bacteria with plasmid, called
donor, can copy plasmid to plasmid-free cell, called recipient. In case of plasmid has
gene providing a drug resistant, recalcitrant substance and so on, plasmid allow
bacteria to get those functions. Bacteria transfer plasmid by conjugation, called
plasmid conjugal transfer, of pair of bacteria. Conjugation requires access of bacteria. By the way, bacteria live in not only solid condition such as soil but also liquid
such as rivers. Conjugal transfer is thought that affected by a stream of the liquid. In
the late 1970s, Levin et al. developed the mathematical plasmid transfer model [2].
As the beginning this study, many mathematical models have been constructed.
However, the relation between conjugal transfer and stream of the liquid still
couldn't be elucidated. Therefore, in this study, we aim to analyze the plasmid
dynamics under the flow condition. The relational equation between the number of
transconjugants and stream condition was constructed. The parameters of the
equation was identified by using with experimental data. We analyzed the plasmid
conjugal transfer dynamics with the parameter and experimental data.

2   Experimental Condition

We used a spinner flask as mix tank (Fig. 1) under the condition referred to
Table 1.

3   Plasmid Conjugal Transfer Model

In this study, conjugations that occur by mix and the other (we call dynamic and
static conjugal) was thought severe the logarithm each of the average value of the
transconjugants and the bacteria concentration shown that it was proportional

Fig. 1 The schematic
diagram of spinner

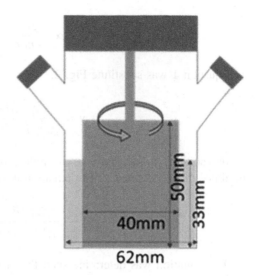

Table 1 Experimental
condition

Culture time	45 min
Temperature	30 °C
Liquid volume	100 mL
Initial bacteria concentration	10^5–10^8 CFU/mL
The ratio of donor to recipient	1:1
Agitation speed	0–200 rpm

relationship. Therefore, this relational expression is decided such a plasmid conjugal transfer model of static conjugal. Dynamics conjugal transfer model is based
on the collision of the cell body because the collision between cell bodies was
necessary for the joining. I simplified it boldly and evaluate as the theory of particle
collision in the turbulence was not established. It is assumed that the fluid was
turbulence and speed vector is isotropic in the mixing tank. Furthermore, it was
postulate that the bacteria accompanies a fluid. Therefore we calculated the collision
frequency by applying the gaseous molecules collision model. The equation of
bacterial collision, Z, was as follows.

$$Z = \frac{\sqrt{2}\pi^3}{6} D_{pv}^2 \aleph^2 HTnr_c^3 + \frac{5\sqrt{2}\pi^3}{12} D_{pv}^2 \aleph^2 HTnr_c^{1.8}$$
$$\times \left(r^{1.2} - r_c^{1.2} \right) \tag{1}$$

Conjugant transfer is performed only between the donor (or the transconjugant)
and the recipient, so that this combination was called the valid collision in this
study. The probability, α, that the collision occurring once is the valid collision was
considered, and the probability of valid collision was calculated.

$$\alpha = \frac{2(y - l_0)^2}{n(n - 1)} \qquad (2)$$

Equation 1 was substitute Fig. 2.

$$y = \frac{\frac{2}{n(n-1)} \beta z t l_0^2}{\frac{2}{n(n-1)} \beta z t (-l_0) + 1} \qquad (3)$$

In addition, it behooves us to reflect on probability that conjugation was inhibited. It was defined γ. The equation incorporated γ was shown as follows.

$$y = \frac{\frac{2}{n(n-1)} (\beta - \gamma) z t l_0^2}{\frac{2}{n(n-1)} (\beta - \gamma) z t (-l_0) + 1} \qquad (4)$$

This equation was determine such the plasmid conjugal transfer model in this study.

3.1 Probability Identification of Conjugal Transfer and Inhibitions of Conjugal

β was defined probability that conjugal transfer occur by one ideal collision, and γ was defined probability of inhibitions of conjugal occur by some kind of inhibition factor. The value of β and γ was identified from experimental data. When the bacteria concentration is low and the agitating speed is also low, it is assumed that conjugation isn't inhibited. We identify β using the slope of the approximation of the number of transconjugants versus agitating speed at the experiment that the

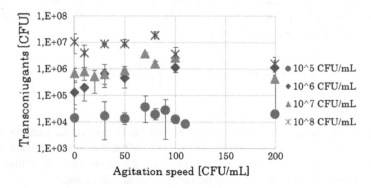

Fig. 2 The experimental data of number of transconjugants

bacteria concentration was 10^5 CFU/mL (the lowest bacteria concentration in this study) and the agitation speed was up to 90 rpm.

If the experimental data was below the Eq. 4 that is $\gamma = 0$ at exclusive of experimental condition to identify β, it was assumed that inhibitions of conjugal occurred. We identify γ through fitting the each plot by varying value of γ of Eq. 4.

4 Results and Discussion

Figure 2 shows the experimental data when the agitating speed was varied at each bacteria concentration. When the bacteria concentration is 105 or 106 CFU/mL, it was judged that the number of transconjugant is increase with increasing of the agitating speed. However, when the agitation speed is more than 110 rpm at 105 CFU/mL, it was shown that the number of transconjugants decreased. The reason for this is thought that the inhibition factor of conjugation occurred. At 107 or 108 CFU/mL, the number of transconjugant decreased with the increasing of the agitation speed. These showed the decline at not high agitation. From the above, it is suspected that the inhibition factor of conjugation which is effected by the bacteria concentration level exist.

Figure 3 shows the logarithm each of the average value of the transconjugant and bacteria concentration. The logarithm each of the average value of the transconjugant and bacteria concentration shown that it was proportional relationship. The approximate expression of this graph was identified as the plasmid conjugal transfer model of static conjugal. It is represented by the following equation.

$$Log_{10}y_0(x) = 0.943Log_{10}x - 0.521 \qquad (5)$$

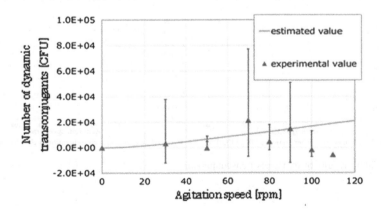

Fig. 3 Identifying the probability of conjugal transfer β

The probability of the conjugal transfer, β, was identified as the value when Eq. 3 correspond with the slop of the approximation straight line of experimental data.

$$\beta = 1.15 \times 10^{-4} \tag{6}$$

4.1 Identified the Probability of Conjugal

In case of the value of y in Eq. 4 was less than experimental data, γ in Eq. 4 is varied to conduct fitting to the experimental data. Figure 4 shows the results of fitting about dynamics transconjugants.

Figure 5 shows relation between γ, the agitation speed and the bacteria concentration. As it can be seen Fig. 2, transconjugants has a tendency to decrease over specific agitation speed. Due to in case of the bacteria concentration is higher, this tendency is stronger, γ was predicted that there is a relationship with the agitation speed and the bacteria concentration. Figure 5a shows that the increase tendency of γ with increasing concentration. However, Fig. 5b shows that it couldn't describe relationship between γ and the agitation speed.

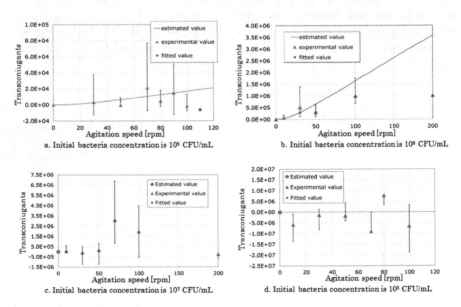

Fig. 4 Fitting results of dynamics transconjugants

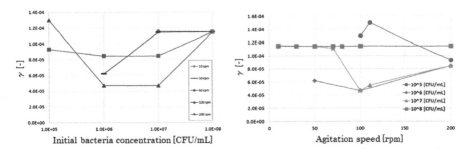

Fig. 5 The relationship between γ, agitation speed and bacteria concentration

5   Conclusion

In this study, conjugal was considered as static conjugal and dynamic conjugal, and the plasmid conjugal transfer model was constructed at agitation flask for each. Dynamic conjugal was attempted to explain by using bacteria collision, probability of conjugal transfer (β), and probability of conjugal inhibition (γ). Collision frequency was estimated by applying the gaseous molecules collision model. Due to only collisions between the donor (or the transconjugant) and the recipient allow conjugal transfer, this combination was defined the valid collision in this study. The probability occurring valid collision at one collision was derived, and it was incorporated into the equation as α. It was assumed that the inhibition of conjugal don't occur at low agitation speed. The estimation of the number of transconjugants could be performed well at low agitation speed. It was suggested that γ depend on initial concentration, but the dependence to the agitation speed of γ couldn't explained. As the possible reason for that there is a discrepancy between collision frequency estimated by applying the gaseous molecules collision model and actual. Therefore, we will attempt to estimate the collision frequency with more precision by using Computational Fluid Dynamics. In addition, we will simulate the plasmid conjugal transfer model by using new tank that more simpler stream can be created than mix tank.

References

1. Sørensen, S.J., Kroer, N., Bailey, M.J., Wuertz, S.: Studying plasmid horizontal transfer in situ: a critical review. Nat. Rev. Microbiol. Oct 2005
2. Levin, B.R., Stewart, F.M., Rice, V.A.: The kinetics of conjugative plasmid transmission: fit of a simple mass action model. Plasmid **2**, 247–260 (1979)

Impedance Spectroscopy as a Method for the Measurement of Calibrated Glucose Solutions with Concentration Occurring in Human Blood

Izabela Osiecka, Tadeusz Pałko, Włodzimierz Łukasik,
Dorota Pijanowska and Konrad Dudziński

Abstract This report presents the variation of impedance values in glucose-sodium chloride (0.9 %) and glucose-bovine plasma solutions at different concentrations, ranging from 50 to 400 mg/dl, with use of sensor model based on a tetrapolar current method. It is worth noticing, that changes of glucose concentration in all frequency range (5 kHz–2 MHz) are directly affected on the impedance modulus of each sample. Also, it is found that in reported method, even the smallest changes in impedance variations were clearly measurable. These findings may be the basis for possible development of a new approach, based on impedance technology, for the noninvasive monitoring of glycaemia.

Keywords Impedance spectroscopy · Glucose sensor · Four-terminal sensing

I. Osiecka (✉) · T. Pałko · W. Łukasik
Institute of Metrology and Biomedical Engineering,
Warsaw University of Technology, Warsaw, Poland
e-mail: i.osiecka@mchtr.pw.edu.pl

T. Pałko
e-mail: palkot@mchtr.pw.edu.pl

W. Łukasik
e-mail: lukasikw@mchtr.pw.edu.pl

D. Pijanowska · K. Dudziński
Institute of Biocybernetics and Biomedical Engineering,
Polish Academy of Science, Warsaw, Poland
e-mail: dpijanowska@ibib.waw.pl

K. Dudziński
e-mail: kdudzinski@ibib.waw.pl

© Springer International Publishing AG 2017   211
R. Jabłoński and R. Szewczyk (eds.), *Recent Global Research and Education:*
Technological Challenges, Advances in Intelligent Systems and Computing 519,
DOI 10.1007/978-3-319-46490-9_30

1 Introduction

Nowadays almost 500 mln people suffer from diabetes [1]. It is a very insidious disease, where one of the basic symptom is increased level of glucose concentration in blood. To control this concentration, diabetes have to puncture fingers several times a day, which are both painful and costly. Currently available invasive [2] glucometers do not address this requirement. Literature presents a lot of researches finding new, comfort, fast and highly sensitive approaches, where a promising spectroscopic technique is the impedance spectroscopy (IS) [3, 4]. In recent years, this spectroscopic technique was suggested as a new, possible method for noninvasive monitoring of glucose concentration in human blood [5].

Impedance spectroscopy is a technique based on the measurement of complex electrical impedance—the module and phase angle—as a function of frequency. It is widely used as a diagnostic method in electrochemical investigations to measure dielectric properties of biological tissues and is defined by following equations:

$$Z(\omega) = \frac{U(\omega)}{I(\omega)} = |Z(\omega)e^{j\varphi(\omega)}, \tag{1}$$

$$Z(\omega) = ReZ + jImZ, \tag{2}$$

where Z is impedance, U is voltage, I is current, ω is the radial frequency (the relationship between radial frequency ω (expressed in radians/second) and frequency f (expressed in hertz) is: $\omega = 2\pi$ f, φ-angle phase, ReZ and ImZ are respectively the real and imaginary part of the impedance.

Usually, to measure the electrical impedance of tissue electrode technique is used. From the point of the quantity of the electrodes, there are two types of possible methods: bipolar and tetrapolar. Among bipolar methods to measure impedance value there are [6]:

- Bridge method- realization of the Wien bridge; *limitation*: difficult balancing
- Voltage method—voltage input from the generator to the test area, a measure of impedance is measured current; *limitation:* dependence of current to measure impedance
- Current method—in the study area the current flow with constant amplitude is forced, a measure of impedance is the value of the voltage amplitude

The most advantageous for biomedical applications is tetrapolar current method, where sensing electrodes are separated from stimulation electrodes. This solution allows for elimination of the influence of stimulation electrodes impedance for the measurement result and getting in the study area more equal distribution of current density, which increase accuracy and repeatability.

Impedimetric glucose sensors are specially sensitive for intermolecular changes of electromagnetic properties, what allows monitoring glycaemia in wide range of frequency. It is known that the increase of the glucose concentration causes growth of the impedance in human blood [7, 8].

The main aim of this study was to more deeply analyze and examine variations of impedance on lower frequency range in sodium chloride solutions and bovine calf plasma solutions contains cellular components at different glucose concentration range.

2 Materials and Methods

2.1 Preparation of Samples

Concentration of glucose in normal human blood range from 70 to 110 mg/dl. We prepared 4 samples with glucose concentration spanning from hypoglycemia to hyperglycemia. For the preparation of this set of sample we use sodium chloride (0.9 %) and glucose. First, we prepared our stock solution by diluted 400 mg of glucose in 100 ml of water. Then, by using diluting method we obtained solution with following concentrations: 50, 100, 200 and 400 mg/dl. The second set with the same concentrations and method was prepared by diluting glucose in bovine calf plasma (Sigma-Aldrich), which was stored in refrigerator in 4 °C.

2.2 Impedance Measurement

Within 24 h from the sample preparation we performed the impedance measures through a own made impedance analyzer. It allows to measure the impedance of the sample by tetrapolar current method using sinusoidal current [9]. Range of frequency changes of application current was from 5 kHz to 2 MHz. The impedance modulus of the sample was measured in 100 frequency points for each decade.

As a measurement cell we use cylindrical probe made of polystyrene posted in Fig. 1a, b. It is characterized by four platinum electrodes to possible separation between two sensing and two stimulation parts to avoid any secondary effects (1, 2). The test were performed by applying 1 mA current to the sample through the external electrodes. For each sample studied, we carried out two independent measures: after the first measure the probe was cleaned before immersing it again into the sample. The impedance values presented for each sample are the average between the two measures. All the measurements were performed at ambient temperature (23 °C).

(a)   **(b)**

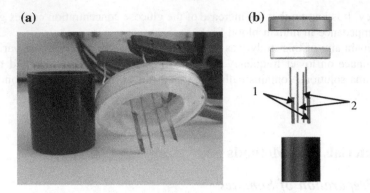

Fig. 1 Measurement probe with four platinum electrodes (**a**, **b**). *1* stimulation electrodes
(0.1 × 3 mm), *2* sensing electrodes (ø 0.6 mm)

3 Results and Discussion

The impedance modulus of studied samples is presented in Figs. 2 and 3. It is
shown that it increases with increasing glucose concentration value in all frequency
range. Due to the limit of the spectrometer capabilities, the response found at the
region of the top frequency in all samples carries some mistakes. Moreover,
measurements of the same glucose sample for three consecutive days demonstrate
that with respect to the first measurement on the fresh solution there is a visible
increase in resistance of the sample. To avoid this parameter changes, we performed
our investigation for all of the samples only for 24 h.

Variation in the impedance parameters shown in Fig. 2 were clear and modulus
increased for increasing glucose concentrations in the whole consider frequency
range. Also, the difference in impedance modulus between the first (50 mg/dl) and
the last sample (400 mg/dl) is significant. Only variation between 100 mg/dl and
200 mg/dl samples were less distinguishable.

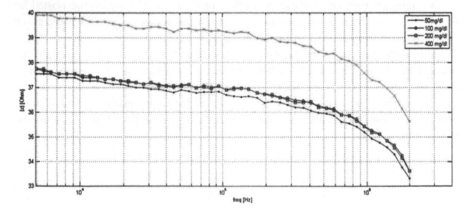

Fig. 2 Impedance modulus for glucose-sodium chloride samples

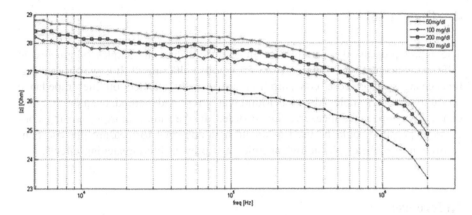

Fig. 3 Impedance modulus for glucose-bovine plasma samples

Similarly to above mentioned samples, the modulus of glucose-bovine plasma solutions (Fig. 3) increased for increasing glucose concentrations in the whole range. However, we observed better impedance distinguishability than in glucose-sodium chloride samples, specially between 100 and 200 mg/dl solutions.

Changes of dielectric properties of glucose solutions measured by exlectroimpedance spectroscopy are mostly affected by variable plasma sodium concentration and are investigated for many years [10–13]. Numerous reports in the literature concern studies with use of both aqueous and with similar properties to the blood (sodium chloride or plasma) solutions of glucose.

The aim of our study was to reproduced reported in literature experiments and deeply analyze changes in impedance value in glucose-sodium chloride and glucose-bovine plasma solutions at frequency range from 5 kHz to 2 MHz. The main cause of choice for the applied solutions is due to the fact that both have an osmotic pressure extremely close to that of blood. Moreover, both sodium chloride (0.9 %) and bovine plasma also have conductivity similar to that of blood. The advantage of focusing on frequency values below the MHz for possible clinical applications may consist in a lower sensitivity the electromagnetic noise in the environment.

Our results showed, that the impedance modulus of each sample was affected by the glucose concentration at all investigated frequency range [7]. Thus confirm the fact, that changes in glucose concentration, even in the physiological range (from 70 to 100 mg/dl) influence the dielectric properties of solutions. Still, those variations in dielectric properties were small (changes of tenths of Ohm, especially in glucose-sodium chloride solutions) at all tested samples.

It should be noted, that there is a possibility, that presented results can be affected by electrode polarization. However, it is well known, that this effect is more relevant at low frequency values (below 100 kHz). As it shown in Figs. 2 and 3. changes in dielectric properties are clear in all frequency range. However, to avoid this polarization phenomena we used platinum electrodes, which despite relatively high cost, are more resistant compare to other metals [14].

4　Conclusions

The report presents the variation of impedance values in samples of glucose at different concentrations, ranging from 50 mg/dl to 400 mg/dl. It is worth noticing, that changes in glucose concentration are directly affected on the impedance modulus of each sample. Also, it is found that in reported method, even the smallest changes in impedance measures are recorded. In the future, those promising conclusions possible could be used to create new approach to measure glucose concentration in diabetes physiological fluids and prevent life-threatening effects.

References

1. twojacukrzyca.pl/statystyki-cukrzycy-w-polsce-i-na-swiecie/
2. Biocybernetyka i Inżynieria Biomedyczna 2000, red. Nałęcz, M., tom 2, Biopomiary, AOW EXIT 2001, cz.IV roz. 27–30, 587–685
3. Pradhan, R., Mitra, A., Das, S.: Quantitative evaluation of blood glucose concentration using impedance sensing devices. J. Electr. Bioimp. **4**, 73–77 (2013)
4. Nazareth, I.A., Vernekar, S.R., Gad, R.S., Naik, G.M.: Analysis of blood glucose using impedance technique. Int. J. Innovative Res. Electr. Electron. Instrumen. Control Eng. **1**(8), 413–417 (2013)
5. Losoya-Leal, A., Camacho-Leon, S., Dieck-Assad, G.: State of the art and new perspectives in non-invasive glucose sensors. Revista Mexicana de Ingeniería Biomédica **33**(1), 41–52 (2012)
6. Palko, T., Galwas, B.: Electrical and dielectric properties of biological tissue, Invited paper. In: Proceedings Of 11th International Microwave Conference MIKON'96, Workshop Biomedical Applications of microwaves, Warsaw (1996)
7. Sbrignadello, S., Tura,A., Ravazzani, P.: Electroimpedance spectroscopy for the measurement of the dielectric properties of sodium chloride solutions at different glucose concentrations. J. Spectrosc. (2013)
8. Tura, A., Sbrignadello, S., Barison, S., Conti, S., Pacini, G.: Impedance spectroscopy of solutions at physiological glucose concentrations. Biophys. Chem. **129**, 235–241 (2007)
9. Pawlicki, G., Pałko, T., Golnik, N., Gwiazdowska, B., Królicki, L.: Fizyka Medyczna, Akademicka Oficyna Wydawnicza, Warszawa (2002)
10. Caduff, A., Hirt, E., Feldman, Y., Ali, Z., Heinemann, L.: First humanexperiments with a novel non-invasive, non-optical continuous glucose monitoring system. Biosens. Bioelectron. **19**, 209–217 (2003)
11. Kashyap, A.S.: Hyperglycemia-induced hyponatremia: is it time to correct the correction factor? Arch. Intern. Med. **159**, 2745–2746 (1999)
12. Kyle, U.G., Bosaeus, I., De Lorenzo, A.D., et al.: Bioelectrical impedance analysis—part I: review of principles and methods. Clinical Nutrition **23**(5), 1226–1243 (2004)
13. Mattsson, S., Thomas, B.J.: Development of methods for body composition studies. Phys. Med. Biol. **51**(13), R203–R228 (2006)
14. Ragheb, T., Geddes, L.A.: The polarization impedance of common electrode metals operated at low current density. Ann. Biomed. Eng. **19**(2), 151–163 (1991)

Physical Breast Model Design for Contact Thermography

Joanna Małyska, Michał Biernat, Włodzimierz Łukasik
and Tadeusz Pałko

Abstract Researches prove that contact thermography can be used in medical imaging of breast pathologies. Pathological lesions suspected of being malignancy are morphologically characteristic, showing higher temperatures than healthy places do. For validation and verification purposes of contact thermography applications as well as thermograms interpretation algorithms development, a simplified physical breast model was designed and constructed. Model provides different thermal surface reaction based on activation particular local heat sources which can be observed on the surface of liquid crystal foils as a specific colorful maps. The breast model is able to simulate different breast types and several breast pathologies as well.

Keywords Thermal modeling · Breast model · Cancer simulation · Breast thermal imaging · Contact thermography

1 Introduction

Researches and clinical trials prove that the differences between the characteristics of rhythmic changes in skin temperature of clinically healthy and cancerous breast are real and measurable [1]. Tumors generally have an increase in blood supply and

J. Małyska (✉) · W. Łukasik · T. Pałko
Faculty of Mechatronics, Institute of Metrology and Biomedical Engineering,
Warsaw University of Technology, Św. Andrzeja Boboli 8, 02-525 Warsaw, Poland
e-mail: j.malyska@mchtr.pw.edu.pl

W. Łukasik
e-mail: w.lukasik@mchtr.pw.edu.pl

T. Pałko
e-mail: t.palko@mchtr.pw.edu.pl

M. Biernat
BRASTER S.A, The National Stadium, Al.Ks.J.Poniatowskiego 1, 03-901 Warsaw, Poland
e-mail: m.biernat@braster.eu

© Springer International Publishing AG 2017
R. Jabłoński and R. Szewczyk (eds.), *Recent Global Research and Education:
Technological Challenges*, Advances in Intelligent Systems and Computing 519,
DOI 10.1007/978-3-319-46490-9_31

angiogenesis, as well as an increased metabolic rate, which in turn translates into increased temperature gradients compared to surrounding normal tissue [2]. Asymmetrical temperature distributions, as well as hot or cold spots, are known to be strong indicators of an underlying dysfunction including tumors. Detecting these "hotspots" and gradients can thereby help to identify and diagnose malignancy [2, 3]. Due to so-called 'dermothermal effect', contact thermography can be used in medical imaging of breast pathologies.

Diagnostic breast contact thermography is a non-invasive imaging technique based on the thermal conductivity between a body and a cholesteric liquid crystal foil placed directly on the skin [4, 5]. Thermography does not provide information about the morphological characteristics of the breast, rather it provides functional information on thermal and vascular conditions of the tissue [4, 6]. Contact thermography developed by BRASTER is based on chiral liquid-crystal compounds. These compounds can selectively reflect light depending on temperature (liquid crystals reflect unpolarized white light over a range of wavelengths). Once liquid-crystal compositions have been properly formulated and secured in film-forming materials, this capability makes it possible to create the thermographic film. When applied to the examined (breast) surface, such matrices provide a colorful imaging of temperature distribution across the three colors: red, green and blue [5].

The purpose of this work is to present the simplified physical breast model for validation and verification purposes of contact thermography applications. This model is able to simulate a temperature distribution of different breast types and several breast pathologies as well, according to thermal effects on the surface of the breast skin.

2 Breast Model Design

The transport of thermal energy in living tissue is a complex process involving multiple phenomenological mechanisms including conduction, convection, radiation, metabolism, evaporation and phase change [7]. Design of a physical breast model which is all-purpose and also include all this processes is hardly possible. For this reason, the main aim of this work was to design and develop a breast model able to simulate a temperature distribution of a natural breast which provide a colorful response on a BRASTER liquid crystal foil. Functionality of the model should enable simulation of both healthy and pathological breast based on fact that pathological lesions suspected of being malignancy can be seen on thermograms as distinct color areas (Fig. 1). The methodological flow consists of two steps: the design of the breast phantom and verification of the breast model.

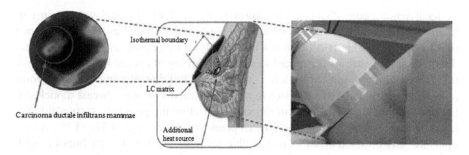

Fig. 1 The mechanism of a thermogram acquiring

2.1 Design and Development of the Physical Breast Model

The proposed idea of a physical breast phantom is based on resistors treated as a local heat sources. The design of the breast model and its main functional modules are shown in Fig. 2.

The main functional module of the breast model is a system of local temperature changes simulation, based on PCB with matrix of resistors (10×10), which generate areas of temperature higher than surrounding part of phantom within range from 0.25 to 2 °C. Proposed idea is based on conversion an electrical energy into thermal energy in controlled way dependent on a value of current flow. The system

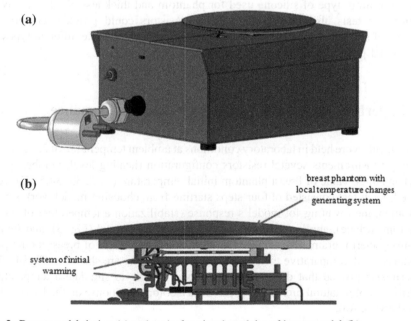

Fig. 2 Breast model design (**a**) and main functional modules of breast model (**b**)

of local temperature changes simulation is covered by a thin layer (2 mm) of additive silicone which thermal conductivity is close to fat tissue of natural breast (0,22 W/(m K)). This part of model is called breast phantom and it provides heat distribution and enables the model to contact with liquid crystal foil without a risk of damage. Dimensions of phantom is matched to foil diameter in order to perform tests by applying BRASTER tester to the breast model at once. Breast model also includes a system of initial warming, situated below the phantom, responsible for initial temperature provision of the breast phantom in the range of 31–35 °C. A user interface (UI) installed on a PC, responsible for indirect control of the breast model, enable to choose one of several model's configuration. The control cooperation of this four modules provides functionality of the breast model by possibility of getting different thermal surface reaction based on activation particular local heat sources (single or multiple resistors).

2.2 Verification of the Breast Model

The idea of breast model has been preliminary tested using infrared camera Optris PI200—several combinations of local temperature changes simulation system and materials used for breast phantom were compared. Initial tests enabled to choose optimal setup for each module, especially for type of resistors (150 O), distance between resistors on the matrix (6 mm), dependency characteristic between value of phantom surface temperature and value of temperature generated by system of initial warming, type of silicone used for phantom and thickness of silicone layer. Preliminary tests also proved that heating resistors could provide temperature changes of specific area of phantom and they are able to simulate different types of breasts and pathological changes of mammary gland.

3 Experimental Results of Breast Model Performance

Experiments were held in laboratory conditions at ambient temperature of 23 ± 1 °C. During measurements several resistors configuration (heating level, number, position) and three value of breast phantom initial temperature (31.5, 33, 34.5 °C) were treated. Each test consisted of four steps starting from choosing model work's configuration, then waiting for model's response (stabilization a temperature of breast phantom surface), applying BRASTER tester to breast model (Fig. 3), and finally acquiring thermograms. A sample of the thermogram, a result of breast model performance, and comparative thermogram of natural breast are shown in Fig. 4. The experiments proved that designed breast model is able to simulate a temperature distribution of a natural breast which provides a colorful response on the BRASTER liquid crystal foil.

Fig. 3 Experiments of breast
model performance—
complete breast model setup

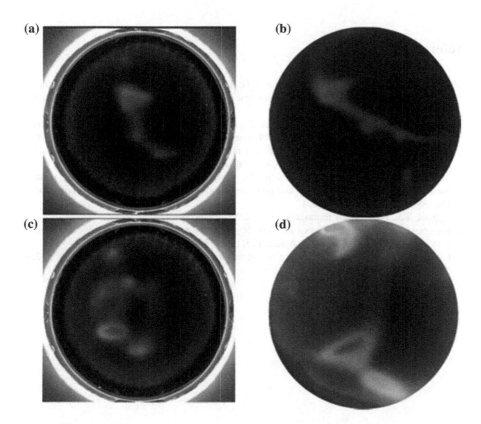

Fig. 4 Sample results of breast model working (**a**, **c**), thermograms of natural breast (**b**, **d**)

4 Discussion and Conclusions

The idea and construction of physical breast model for contact thermography have been described. The preliminary experiments of the designed model using the BRASTER tester (liquid crystal foil with an image acquisition system) proved its ability to simulate a surface temperature distribution of natural breast. Additionally the local heat sources enable to simulate the areas showing higher temperature than surrounding places (thermal hotspots) which provides the blood vessels and breast pathologies simulation. We expect that the breast model will be able to generate the temperature distribution alike any thermogram of natural breast. Simulation based on input thermogram is a next task of further research and development investigations. In the future, the breast model can be used by researchers to test refinements in their new hardware and software solutions as well as thermograms interpretation algorithms development.

References

1. Ng, E.Y.K., Rajendra Acharya, U., Keith, L.G., Lockwood, S.: Detection and differentiation of breast cancer using neural classifiers with first warning thermal sensors. J. Inf. Sci. **177**(20), 4526–4538 (2007)
2. Arora, N., Martins, D., Ruggerio, D., et al.: Effectiveness of a noninvasive digital infrared thermal imaging system in the detection of breast cancer. Am. J. Surg. **196**(4), 523–526 (2008)
3. Bartosz, Krawczyk, Gerald, Schaefer: Breast thermogram analysis using classifier ensembles and image symmetry features. IEEE Syst. J. **8**(3), 921–928 (2013)
4. Montruccoli, G.C., Montruccoli Salmi, D., Casali, F.: A new type of breast contact thermography plate: a preliminary and qualitative investigation of its potentially on phantoms. Phys. Med. **20**(1), 27–31 (2004)
5. Popiela, T.J., et al.: Atlas termograficzny. PZWL, Warszawa (2015)
6. Fitzgerald, A., Berentson-Shaw, J.: Thermography as a screening and diagnostic tool: a systematic review. J. N. Z. Med. Assoc. **125**(1350), 80–91 (2012)
7. González, F.J.: Thermal simulation of breast tumors. Revista Mexicana De Física **53**(4), 323–326 (2007)

Numerical Study of the PDMS Membrane Designed for New Chamber Stapes Prosthesis

Katarzyna Banasik and Monika Kwacz

Abstract New chamber stapes prosthesis (ChSP) is intended to restore hearing in patients with conductive hearing loss caused by otosclerosis. The proper selection of the membrane stiffness determines the effective prosthesis functioning. In this study, finite element (FE) model and simulation results of the poli-dy-methylo-siloxane (PDMS) membrane thickness are presented. The FE model was created using Abaqus 6.13 software. During simulations, the membrane thickness was adjusted so that its stiffness was similar to the stiffness of the normal stapes annular ligament. The simulation results allow to build a series of ChSP prototypes for preclinical experiments.

Keywords Chamber stapes prosthesis · PDMS membrane · Otosclerosis · Stapedotomy

1 Introduction

The most common method for hearing restoration in patients with conductive hearing loss, caused by otosclerosis resulting in reduction of stapes mobility, is stapedotomy surgery with piston stapes prosthesis [1–3]. During stapedotomy, the suprastructure of the otosclerotic stapes is removed and a small hole (diameter ~0.4–0.6 mm) in the stapes footplate (SF) is created. The piston of the stapes prosthesis is then placed into the hole and an prosthesis attachment is fixed to the long process of the incus. The postoperative hearing results are considered good but only for low and medium frequencies (0.5–3 kHz) [2, 4, 5]. To increase the stimulation of the inner ear, to improve hydrodynamics of the perilymph, and to

K. Banasik (✉) · M. Kwacz
Institute of Micromechanics and Photonics, Warsaw University of Technology, św. Andrzeja Boboli 8, 02-525 Warsaw, Poland
e-mail: kbanasik@mchtr.pw.edu.pl

M. Kwacz
e-mail: m.kwacz@mchtr.pw.edu.pl

© Springer International Publishing AG 2017
R. Jabłoński and R. Szewczyk (eds.), *Recent Global Research and Education: Technological Challenges*, Advances in Intelligent Systems and Computing 519, DOI 10.1007/978-3-319-46490-9_32

223

reduce disadvantages of the piston prostheses, new chamber stapes prosthesis (ChSP) is proposed [6, 7]. The ChSP (Fig. 1a) consists of: (1) a conical chamber ending with a thin tube, (2) a flexible membrane, and (3) a rigid plate with an attachment. The thin tube is designed to be placed into the hole made in the SF. The chamber is filled with fluid and covered with the flexible membrane. The rigid plate is fixed to the membrane. After attaching the plate to the long process of the incus, vibrations are being transmitted to the fluid and to the inner ear afterwards. The vibration transmission is possible due to the membrane compliance. The membrane in the ChSP acts as the annular ligament of the stapes in healthy ear. The ChSP functioning was numerically simulated [7] and experimentally confirmed [8].

The stiffness of the membrane determines the proper ChSP functioning and provides the inner ear stimulation at the level compared to physiological ear. This stiffness depends on both the material and the geometrical parameters of the membrane. The aim of this study is to determine the relationship between the area of the rigid plate and the optimal thickness of the membrane made from PDMS.

2 Methods

The 3D model of ChSP parts was build in CAD software (Fig. 1a). The ChSP shape and dimensions were established based on anthropometric data published in the literature [9, 10]. To determine the optimal thickness of the membrane, a finite-element (FE) model of the ChSP (Fig. 1b) was created using Abaqus 6.13.

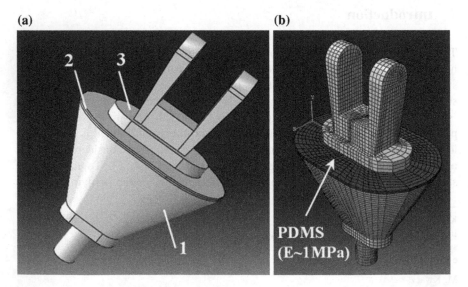

Fig. 1 New chamber stapes prosthesis. **a** CAD model. **b** FE model. *1* Chamber ending with thin tube, *2* membrane, *3* rigid plate with attachment

Table 1 Material parameters and number of elements assumed in the FE model

Part	Material	Young modulus, E (MPa)	Poisson's ratio (–)	Number of elements
Chamber	Titanium (Ti64)	115,000	0.34	42,364
Membrane	Poly-di-methylo-siloxane (PDMS)	1	0.4	575
Rigid plate	Titanium (Ti64)	115,000	0.34	3170

In the FE model, it was assumed that the membrane is made of PDMS and both the chamber and the rigid plate is made of titanium (Ti64). The chamber's upper surface area was accepted as 5.9 mm^2. The simulation were performed for different areas of the rigid plate, namely: 2.0, 2.3, 2.6, and 3.0 mm^2. The 3D model was meshed with C3D8R hexahedral 8-node elements. The material parameters and the number of elements for each part are listed in Table 1.

Load (0–20 µN) was applied to the plate and the membrane displacement was calculated afterwards. During simulations, the membrane thickness was adjusted so that the membrane stiffness was similar to the stiffness of the normal stapes annular ligament (SAL) in the healthy ear (120 N/m, as given in [11]). The membrane stiffness was calculated by dividing the load by the maximal membrane displacement. To compare the PDMS membrane stiffness with the stiffness of the SAL, linear regression analysis was used. It was assumed that the membrane thickness is properly adjusted when the determination coefficient is at least 0.99.

3 Results

The simulation results showed that the optimal thickness of the PDMS membrane depends on the surface area of the rigid plate. The course of this relationship is shown in Fig. 2.

The optimal thickness decreases from 0.2 to 0.13 mm with increase in the rigid plate's area. This optimal thickness was determined for the chamber upper surface area of 5.9 mm^2, which is close to the area of the oval window opening in the human middle ear. Furthermore, the maximal membrane displacement was always in the range of physiological displacement of the SAL (i.e. no greater than 100 nm).

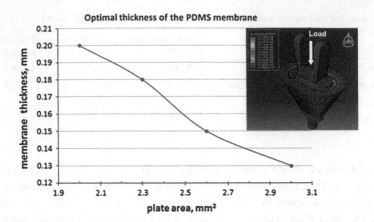

Fig. 2 Simulation results of the PDMS membrane thickness for different surface area of the rigid plate

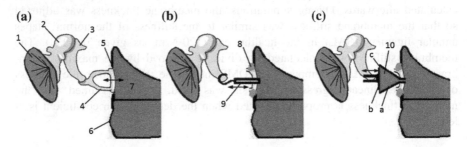

Fig. 3 Middle-ear biomechanics. **a** In normal state (healthy ear), **b** after piston-stapedotomy, and **c** after stapedotomy with new chamber stapes prosthesis. *Arrows* indicate vibration in the middle-inner ear interface. *1* Tympanic membrane, *2* malleus, *3* incus, *4* stapes, *5* stapes annular ligament (SAL), *6* round window membrane, *7* inner ear filled with fluid, *8* otosclerosis, *9* piston stapes prosthesis, *10* chamber stapes prosthesis (ChSP): *a* chamber, *b* rigid plate, *c* membrane

4 Discussion

New ChSP is currently under investigation at Warsaw University of Technology. Design and functioning of the ChSP (Fig. 3c) significantly differs from piston stapes prosthesis (Fig. 3b) that is presently only available for clinical use in stapedotomy surgery.

In normal healthy ear (Fig. 3a), sound transmission to the inner ear strongly depends on the SAL (5) stiffness, which determines the stapes mobility. According to measurements in human temporal bone specimens, the SAL stiffness in healthy ear is ∼120 N/m [11]. Otosclerosis (8) significantly reduces the stapes mobility due to the SAL stiffening.

The hearing results after piston-stapedotomy (Fig. 3b) are closely related to the smaller piston area (0.12–0.28 mm^2) compared with the area of the normal SF (~ 3 mm^2). The reduced area leads to decreased fluid volume displacement (VD) at the oval window. Predictions using a simple lumped element model [12] showed that the smaller the piston diameter the greater the residual air bone gap (ABG) and pore postoperative hearing results. Also, (1) clinical observations [13], (2) finite element modeling [7, 14], and (3) experiments on cadaver temporal bones [15, 16] showed that the piston-stapedotomy significantly changes middle-ear biomechanics and affects the residual ABG.

The ChSP functioning (Fig. 3c) mimics the healthy ear. The rigid plate (b) acts like the stapes and the membrane (c) like the SAL. The chamber (a) is a part of the fluid-filled inner ear. The proper choice of both the plate area and the membrane stiffness restores the physiological fluid VD and ensures the inner-ear stimulation at the physiological level. In this study, the plate area was assumed between 2.0 and 3.0 mm^2 (close to the SF area) and the membrane stiffness of 120 N/m (close to the SAL stiffness). The membrane should be made from biocompatible material. Thus, in our simulation, the PDMS was chosen, which is a very compliant silicon rubber often used in clinical applications. To achieve the desired membrane stiffness, the necessary thickness of the PDMS membrane was calculated (Fig. 2).

5 Conclusion

New ChSP is designed for patient with otosclerosis resulting in a significant reduction in stapes mobility. Design and functioning of the ChSP is similar to the healthy ear. Therefore, hearing results after the ChSP-stapedotomy will be better than after the piston-stapedotomy. The proper thickness of the ChSP membrane provides the same fluid volume displacement as the normal SF. The results presented in this study are crucial to build a series of ChSP prototypes for pre- and clinical experiments.

References

1. Shea, J.J.: Fenestration of the oval window. Ann. Otol. Rhinol. Laryngol. **67**, 932–951 (1958)
2. Häusler, R.: Advances in stapes surgery. In: Jahnke, K. (ed.) Middle Ear Surgery, pp. 105–119. Thieme, New York (2004)
3. Møller, P.: Stapedectomy versus Stapedotomy. Adv. Otorhinolaryngol. **651**, 69–73 (2007)
4. Vincent, R., Sperling, N.M., Oates, J., Jindal, M.: Surgical findings and long term hearing results in 3050 stapedotomies for primary otosclerosis: a prospective study with the otology-neurotology database. Otol. Neurotol. **27**, 25–47 (2006)
5. Bagger-Sjöbäck, D., Strömbäck, K., Hultcrantz, M., et al.: High-frequency hearing, tinnitus, and patient satisfaction with stapedotomy: a randomized prospective study. Sci. Rep. **5**, 13341 (2015). doi:10.1038/srep13341

228 K. Banasik and M. Kwacz

6. Gambin, W., Kwacz, M., Mrówka, M.: Komorowa protezka ucha środkowego [Chamber middle-ear prosthesis], Patent PL 217562B1 (2014)
7. Kwacz, M., Marek, P., Borkowski, P., Gambin, W.: Effect of different stapes prostheses on the passive vibration of the basilar membrane. Hear. Res. **310**, 13–26 (2014)
8. Sołyga, M., Mrówka, M., Gambin, W., Kwacz, M.: Round window vibration induced by new chamber stapes prosthesis: preliminary results of experimental investigation. In: 7th International Symposium on Middle Ear Mechanics in Research and Otology MEMRO'2015, Alborg, Denmark, 1–5 July 2015
9. Wengen, D.F., Nishihara, S., Kurokawa, H., Goode, R.L.: Measurements of the stapes superstructure. Ann. Otol. Rhinol. Laryngol. **104**, 311–316 (1995)
10. Kwacz, M., Wysocki, J., Krakowian, P.: Reconstruction of the 3D geometry of the ossicular chain based on micro-CT imaging. Biocybern. Biomed. Eng. **32**(1), 4–27 (2012)
11. Kwacz, M., Rymuza, Z., Michałowski, M., Wysocki, J.: Elastic properties of the annular ligament of the human stapes—AFM measurement. J. Assoc. Res. Otol. **16**(4), 433–446 (2015)
12. Rosowski, J.J., Merchant, S.N.: Mechanical and acoustic analysis of middle ear reconstruction. Am. J. Otol. **16**, 486–497 (1995)
13. Marchese, M.R., Cianfrone, F., Passali, G.C., Paludetti, G.: Hearing results after stapedotomy: role of the prosthesis diameter. Audiol. Neurootol. **12**, 221–225 (2007)
14. Koike, T., Wada, H., Goode, R.L.: Finite-element method analysis of transfer function of middle ear reconstructed using stapes prosthesis. In: Association for Research in Otolaryngology (ARO) Abstracts of the 24th Midwinter Meeting, p. 221. Mt. Royal, (NJ), Association for Research in Otolaryngology (2001)
15. Wysocki, J., Kwacz, M., Mrówka, M., Skarżyński, H.: Comparison of round window membrane mechanics before and after experimental stapedotomy. Laryngoscope **121**, 1958–1964 (2011)
16. Sim, J.H., Chatzimichalis, M., Röösli, C., Laske, R.D., Huber, A.M.: Objective assessment of stapedotomy surgery from round window motion measurement. Ear Hear. **33**(5), e24–e31 (2012)

Part IV
Plasma Physics

Optical Fibre Probing for Bubble/Droplet Measurement, and Its Possibility of the Application to Biotechnology

Takayuki Saito

Abstract Optical Fibre Probing (OFP) is a very useful technique to measure bubble/droplet (diameter, velocity and number density) in gas-liquid two-phase flows; furthermore the OFP is hopeful in other research fields. Its basic measurement principle is very simple: detection of change in the refraction indices. Accurate information about the penetration of a bubble/droplet by the probe is very difficult to obtain. Since the probing signals involve various kinds of optical information, no researcher succeeded in extracting physical meanings of the OFP signals. To improve the OFP measurement accuracy, deep understanding of the relationship between the probing signals and probe-bubble/droplet contacting process is needed. Through a newly developed lay-tracing simulator, the author analysed the optical signals and revealed the relationship. The objective of the present study is to improve the accuracy of the OFP, based on experimental results and the numerical results. Furthermore, the application of OFP to biotechnology will be proposed.

Keywords Optical fibre probing · Bubble · Droplet · Multiphase flow · Ray tracing simulation

1 Introduction

Gas-liquid flows are widely used in practical industrial processes; furthermore in order to understand nature, e.g. mass and heat transfer between the atmosphere and the ocean, the flows are essential. In the flows, usually bubbles or droplets are dispersed in the continuous phase. Visualizing deep inside of the flows is basically difficult. Hence, many researchers have been employing optical fibre probing inserted in a multiphase flow. Although this probing is a typical intrusive measurement method, well-designed optical fibre probe is able to simultaneously

T. Saito (✉)
Graduate School of Science and Technology, Shizuoka University, Shizuoka, Japan
e-mail: saito.takayuki@shizuoka.ac.jp

© Springer International Publishing AG 2017
R. Jabłoński and R. Szewczyk (eds.), *Recent Global Research and Education: Technological Challenges*, Advances in Intelligent Systems and Computing 519,
DOI 10.1007/978-3-319-46490-9_33

231

measure diameter, velocity and number density of the bubbles or droplets in apparatus of size from laboratory [1, 2] to practical field [3]. Its measurement principle is, basically, detection of change of refractive indices [4]. From this measurement principle, most of researchers have been imprudently thinking that at least two sets of optical fibres were needed to measure the diameter and velocity. Thus, multi-tip probes (e.g. two-tip probe or four-tip probe) were used. Since, unfortunately, the multi-tip probes are inevitably larger than 1 mm in probe-bundle's outer diameter, the multi-tip probes are too large to measure bubbles/droplets smaller than about 1.5 mm. The bubbles from 1 mm to 3 mm in diameter are always used in industrial plants due to their highest mass transfer coefficient. In order to measurement such small bubbles/droplets, a very few researchers proposed a single-tip optical fibre probes (S-TOPs) [5, 6]. Cartellier [5], who was a pioneer of single-tip optical fibre probe, developed an S-TOP named "mono-tip optical fibre probe" with a corn-shaped primary sensing tip and a secondary sensing tip etched through a strong acid. Saito [6] developed an S-TOP (S-TOPW) with a wedge-shape primary sensing tip and a secondary sensing tip micro-fabricated through femtosecond laser pulses. These S-TOPs are adequate for measurement of the small bubbles/droplets. However, these have a difficult problem to solve for accurate measurement; i.e. penetration position where the probe penetrates the bubble/droplet, and penetration angle between the bubble/droplet motion direction and the probe axis. When the probe penetrate the centre of the bubble/droplet, the measured chord length is nearly equal to its minor axis; when it does the edge of the bubble/droplet, the measured chord length is shorter than the minor axis. When the probe penetrates the bubble/droplet at the centre as the probe axis is parallel to the bubble/droplet motion direction, the measurement chord length well accords with the minor axis.

The author and his research group have solved this problem [7–9] on the basis of careful experimental results and well-designed computational simulation. In the present study, the author summarises how to solve this problem. In addition, the author proposes application of his single-tip optical fibre probe to biotechnology.

2 Measurement Principle of Optical Fibre Probing, and the Problem to Solve

2.1 Measurement Principle

As shown in Fig. 1a, incident light into an end (incident-side tip) of optical fibre probe propagates in the fibre core as repeating reflection at the interface between the fibre core and clad, and some of the incident light reaches the other end (sensing tip). When the sensing tip is covered with a gas (for instance air), since the difference in refractive indices between the core and the gas phase is large, most of the light reaching the sensing tip is reflected at the sensing surface in accordance with

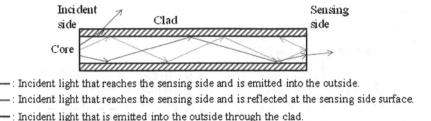

------- : Incident light that reaches the sensing side and is emitted into the outside.
------- : Incident light that reaches the sensing side and is reflected at the sensing side surface.
------- : Incident light that is emitted into the outside through the clad.

Fig. 1 Model patterns of incident light propagation in an optical fiber

Snell's law. Further, some of the reflected light propagates in the core, then reaches the incident-side tip. When the sensing tip is covered with a liquid (for instance water; this principle is approved against opaque liquids), most of the light reaching the sensing tip is emitted from the sensing surface into the liquid phase. Hence, the intensity of the returned light is small. Thus, at the incident-side tip, bright and dark On-Off signals are obtained; i.e. the bright signals correspond with the situations where the sensing tip is covered with the gas-phase, and the dark signals correspond with the situations where the sensing tip is covered with the liquid-phase. As stated above, the measurement principle is repetitive phase detection on the basis of these On-Off signals.

A set of two optical fibre can measure velocity of a bubble/droplet from a time difference (T1 in Fig. 2a, b) between the signals of the two probes, and can measure the bubble/droplet diameter from the above velocity and the gap (L1 in Fig. 2a, b) between the two sensing points.

2.2 Velocity Measurement Through a Single Tip Optical Fibre Probe with a Wedge-Shape Tip

The author and his group realized simultaneous measurement of bubble/droplet diameter and velocity in a different way from Cartellier; i.e. by using a single-tip optical fibre probe with a wedge-shape tip (S-TOPW) [6]. As shown in Fig. 3, the wedge-shape tip is gradually covered with a gas-phase (bubble) or a liquid-phase (droplet). In bubble measurement, when the summit of the wedge-shape tip touches the frontal surface of the bubble, the intensity of the returned light increases, furthermore, as the tip penetrates the bubble's frontal surface more deeply, the increase in the area covered with the gas-phase brings increase in the returned light intensity. The increase rate (dIR/dt) of the returned light intensity is proportional to the bubble velocity UB (in a strict sense, the bubble's frontal surface velocity),

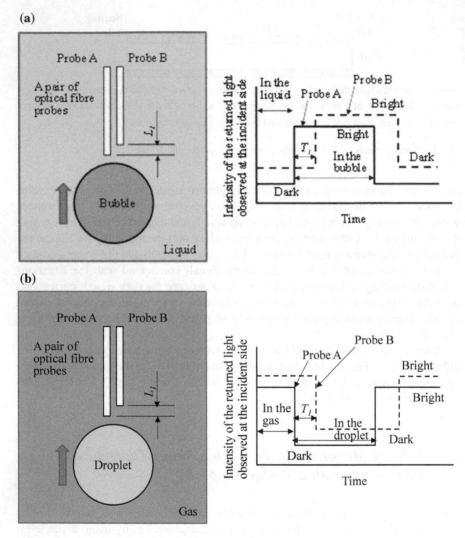

Fig. 2 **a** Optical signals of a pair of optical fiber probes in bubble measurement. **b** Optical signals of a pair of optical fiber probes in droplet measurement

$$U_B \propto \alpha_1 dI_R/dt, \tag{1}$$

where α1 is a proportional coefficient decided by surface tension and wettability [6].
    Hence UB is expressed by the other form as below,

$$U_B \propto \alpha_2 dV_m/dt, \tag{2}$$

where α_2 is a proportional coefficient [6].

Fig. 3 Principle of
bubble/droplet velocity
measurement through the
S-TOPW

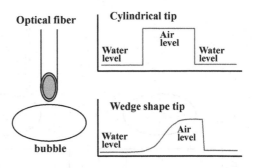

In addition, the signals in droplet measurement are inverted against the signals in bubble measurement.

2.3 The Problems to Solve Bubble/Droplet Measurement Through a Single Tip Optical Fibre Probe

Researchers using an optical fibre probe have been struggling with the following problems: penetration position and penetration angle.

1. **The problem of penetration position**.
   A measured chord length LC when a single-tip optical fibre probe penetrates the centre area of a bubble is almost the same as the bubble's minor axis, and LC is longer than LS that is measured chord length when the single-tip optical fibre probe penetrates the side edge area of a bubble [6, 9]. As illustrated in Fig. 4a, an output signal when the single-tip optical fibre probe penetrates the side-edge area of a bubble is different from the one when it penetrates the centre area of a bubble, even if the bubbles are the same size. This problem brings randomness in the measured chord length.

2. **The problem of penetration angle**.
   As illustrated in Fig. 4b, a penetration angle also brings randomness in the measured chord length [8, 9]. By using a single-tip optical fibre probe, only when the probe penetrates the centre area of a bubble in parallel to the bubble's motion, we are able to accurately measure bubble's/droplet's minor axis.

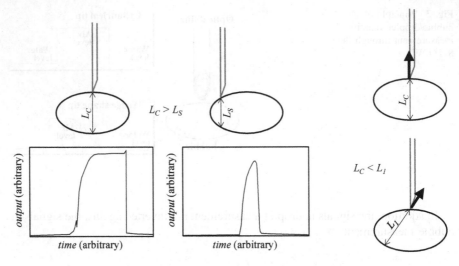

Fig. 4 Relationship between a chord length and a penetration position/angle through the S-TOPW

3 Typical Optical Configuration for a Single-Tip Optical Fibre Probe

3.1 Plane Optical Fibre

A quartz or plastic optical fibre of multi-mode type is usually used for optical fibre probes. Both optical fibres are composed of a core (: waveguide) and a clad (: reflection layer); the refractive index of the core is slightly larger than that of the clad. The quartz optical fibre is better than the plastic one because of easy-processed, heat-resistant and chemical-resistant properties. For instance, the quartz fibre is easily fine-drawn by heating (Fig. 5). Limitation of the sensing tip diameter is about 5–10 μm, because the clad thickness should be larger than wavelength of incident light.

3.2 Optical Configuration

Figure 6 diagrams typical optical configuration for the S-TOPW. The system consists of a LD (e.g. 635 nm laser), half mirror, objective lens, S-TOPW, polarizer, photo detector (e.g. photo multiplier), and so on. The polarizer cuts noise by using change of polarization plane of the laser beams).

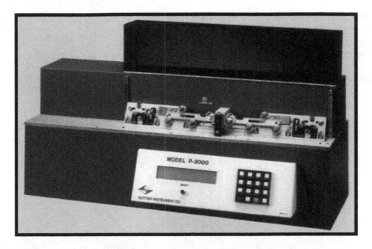

Fig. 5 Apparatus for fine-drawing a plane quartz optical fibre (Micro pipette puller)

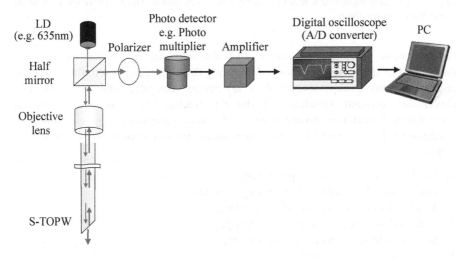

Fig. 6 Typical optical configuration for the S-TOPW

4 Ray-Tracing Computational Simulation

4.1 Outline of Ray-Tracing Computational Simulation

The system composed of S-TOPW, a bubble (or droplet) and a liquid-phase (or a gas-phase) is apparently simple, however optical signals generated in this simple system are very complex due to reflection at a moving and deforming gas-liquid interface. In particular, noise including the optical signals is indeed various,

Table 1 Reflectivity and transmissivity (Fresnel law)

	P-polarization	S-polarization
Reflectivity R	$R_P = \frac{\tan^2(\theta_i - \lvert\theta_t\rvert)}{\tan^2(\theta_i + \lvert\theta_t\rvert)}$	$R_S = \frac{\sin^2(\theta_i - \lvert\theta_t\rvert)}{\sin^2(\theta_i + \lvert\theta_t\rvert)}$
Transmissivity T	$T_P = \frac{\sin(2\theta_i)\sin(2\lvert\theta_t\rvert)}{\sin^2(\theta_i + \lvert\theta_t\rvert)\cos^2(\theta_i - \lvert\theta_t\rvert)}$	$T_S = \frac{\sin(2\theta_i)\sin(2\lvert\theta_t\rvert)}{\sin^2(\theta_i + \lvert\theta_t\rvert)}$

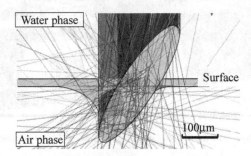

Fig. 7 A typical computational result of the ray-tracing simulation. The S-TOPW enters in an air phase from a water phase

furthermore no previous researcher was not able to reveal the physical meanings of the noise. For the specific purpose of revealing the hidden physics in the noise, the author and his group developed a ray-tracing computational simulator [7]. Developing original simulator of the ray tracing, they successfully traced ray-segment trajectories geometrically and render complicated optical boundary conditions [7]. In the simulator, ray segments are three-dimensionally calculated as follows;

1. Incident ray segments are generated,
   Directions: Considering N.A. of objective lens,
2. Total reflection at the core-clad interface,
3. Reflected or refracted at the sensing-tip,
   Some of the reflected rays are returned,
4. Returning to the inlet tip,
5. Energy of the returned rays are detected at the incident surface in consideration of P- and S-polarization listed in Table 1.
   A typical computational result is shown in Fig. 7.

4.2 Computational Results

A typical computational result in virtual measurement of a bubble by the S-TOPW is shown in Fig. 8 [7, 9]. The computed bubble is categorized in an oblate

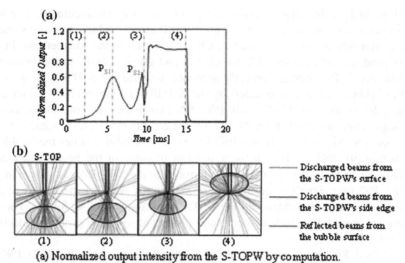

(a) Normalized output intensity from the S-TOPW by computation.
(b) Geometric relationship between the bubble and the S-TOPW, and destinies of the beams.
(A) Computational results of output signal.

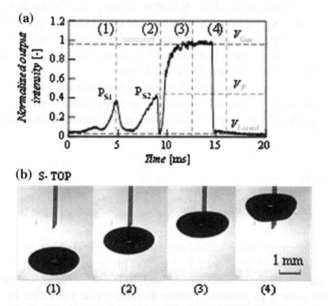

(a) Normalized output intensity from the S-TOPW by experiments.
(b) Geometric relationship between the bubble and the S-TOPW.
(B) Computational results of output signal.

Fig. 8 Comparison of typical output signals between computational and experimental results

ellipsoidal bubble that ascends zigzagging in a rest water column. The bubble's diameter was set at 2.9 mm in equivalent diameter, which was the same as the experimental bubble's diameter in Chapter "Study on the Magnetizing Frequency Dependence of Magnetic Characteristics and Power Losses in the Ferromagnetic Materials". The bubble vertically ascended, touched the S-TOPW at the centre of the bubble, and was penetrated by the S-TOPW as the bubble's minor axis was parallel to the S-TOPW's axis (Fig. 8A (b)). Figure 8A (a) shows normalized output intensity from the S-TOPW by the computation. The normalized intensity possesses two peaks, PS1 and PS2 before the S-TOPW touches the bubble frontal surface [9]. PS1 is brought by re-incident of some of the beams reflected at the bubble rear interface (as a concave mirror) into the S-TOPW, as shown in Fig. 8A (b). PS2 (pre-signal) is brought by re-incident of some of the beams reflected at the bubble frontal surface into the S-TOPW, as shown in Fig. 8A (b). These peaks are not noise but include a very useful physics to solve the problems mentioned in Sect. 2.3.

Figure 8B (a) shows normalized output intensity from the S-TOPW by the experiments mentioned in Chapter "Frequency Resolution and Accuracy Improvement of a GaP CW THz Spectrometer". As shown in Fig. 8B (b), the bubble vertically ascended, touched the S-TOPW at the centre of the bubble, and was penetrated by the S-TOPW as the bubble's minor axis was parallel to the S-TOPW's axis. The experimentally-obtained normalized output intensity is very similar to the computational one, furthermore possesses two peaks, P_{S1} and P_{S2} before the S-TOPW touches the bubble frontal surface.

As mentioned above, the computational results well accorded with experimental ones. The computational simulator developed by the author and his group is able to accurately reveal physical meanings hidden in noise that should be rejected.

5 Experiments

5.1 Bubble Measurement

The author and his group made an experimental setup for bubble measurement through the S-TOPW as diagrammed in Fig. 9. This setup consisted of a very unique bubble launch device, water vessel and S-TOPW system, as well as a high-speed video camera to measure bubble diameters and velocities by visualization to obtain reference data. The bubble launch device launched bubbles uniform in size, and controlled the launching interval arbitrarily [10]. Hence, the author conducted bubble experiments under ideal conditions to acquire pure influences of penetration position and penetration angle on measurement results, without any effects of bubble-diameter difference on measurement results.

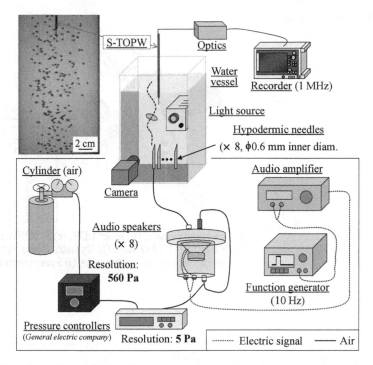

Fig. 9 Experimental setup for bubble measurement

5.2 Droplet Measurement

The author and his group made experimental setup for droplet measurement through the S-TOPW as diagrammed in Fig. 10. This setup consisted of an injection nozzle, nozzle controller, S-TOPW system. In addition, in order to obtain reference data against the S-TOPW results, a YAG laser system and high-speed video camera were used for glare point technique that measured droplet size, and a smart LDV was used to measure droplet velocity.

5.3 Experimental Results

1. **Basic results of bubble measurement through S-TOPW.**
   Figure 11 shows a typical signal obtained in the bubble measurement through the S-TOPW. A bubble burst signal with a clear pre-signal is observed. Although the bubble's diameter was uniform, the bubble burst signals are different due to randomness of penetration position and penetration angle [9]. The

Fig. 10 Experimental setup
for droplet measurement

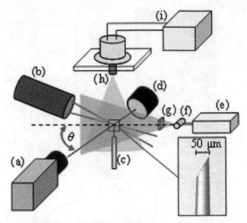

(a) High speed camera, (b) smart LDV, (c) S-TOPW system,
(d) Back light, (e) YAG laser, (f) Cylindrical lens, (g) Plane-
convex lens, (h) Injection nozzle, (i) Spray controller

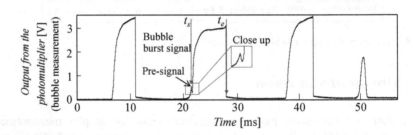

Fig. 11 A typical output signal in the bubble measurement (experiments)

bubble's frontal surface touched the S-TOPW at t_s, and the bubble rear interface
touched S-TOPW at t_e. The measured chord length of the bubble is calculated
from the residential time $(t_s - t_e)$ and the bubble velocity calculated from
Eq. (1).

2. **Basic results of droplet measurement through S-TOPW.**
   Figure 12 shows a typical signal obtained in the droplet measurement through
   the S-TOPW. A droplet burst signal with a clear post-signal is observed.
   Although the droplet's diameter was uniform, the bubble burst signals are different due to randomness of penetration position and penetration angle. The
   chord length is calculated via the similar way mentioned above.

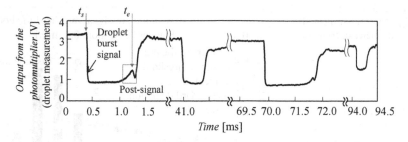

Fig. 12 A typical output signal in the droplet measurement (experiments)

6 Newly Developed Methods for Accurate Measurement Through S-TOPW

6.1 Pre-signal Threshold Method (For Bubble Measurement)

As analysing optical signals via newly developed ray-tracing computational simulator in Sect. 4.2, the pre-signal is very hopeful to solve the difficult problems in bubble measurement via the single-tip optical fibre probe (e.g. S-TOPW). As described in Sect. 5.1, when the S-TOPW penetrate the centre area of a bubble in parallel to the bubble's motion direction, the bubble burst signal with a clear pre-signal were successfully obtained. The author and his group have developed "Pre-signal threshold method" for the particular purpose of accurate and practical measurement of bubbles, using this characteristics of the pre-signal [9].

Furthermore, by analysing optical signals for the droplet measurement via the ray-tracing computational simulator, it was revealed that the post-signal indicated in Fig. 12 clearly appeared when the S-TOPW penetrate the centre area of a droplet in parallel to the droplet's motion direction. As shown in Fig. 12, the droplet burst signals with a clear pre-signal were successfully obtained experimentally [9]. The author and his group have developed "Post-signal threshold method" for the particular purpose of accurate and practical measurement of droplet, using this characteristics of the post-signal.

6.2 Demonstration of the Pre-signal Threshold Method

The standard deviation of the former was 0.2 mm; on the other hand that of the latter result was 0.5 mm. The diameter distribution of the former is very similar to that obtained by visualization. Thus, the measured results are significantly improved by processing with the pre-signal threshold method [9].

Fig. 13 Effects of the
pre-signal threshold method.
Bubble diameter distribution

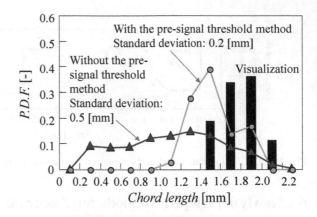

The results of the velocity are also significantly improved by processing with the pre-signal threshold method (Fig. 13).

7 Application of a Single-Tip Optical Fibre Probe to Biotechnology

The author and his group have developed other type of S-TOPW named Fs-TOP that was micro-fabricate by femtosecond laser pulses [11]. As shown in Fig. 14, it possesses a primary sensing tip of wedge shape, and a secondary sensing part like a groove on the clad; this groove was micro-fabricated by femtosecond laser pulses. They can design and make Fs-TOPs according to their wishes, on the basis of the computational simulation.

For instance, a red blood cell (Fig. 15a) is about 10–15 μm in diameter, a white blood cell is about 6–30 μm in diameter, and a diseased cell (Fig. 15b) is about 10 μm in diameter. The Fs-TOP can be considered to be applied to in situ measure for these cells, under some innovative ideas.

Fig. 14 A sample of Fs-TOP

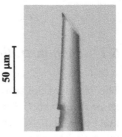

(a)   (b)

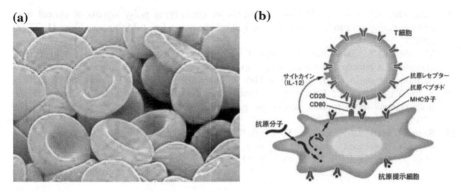

Fig. 15 a Red blood cells. **b** Diseased cells

8 Conclusions

In the present study, the optical fibre probing was outlined on the basis of experiments and computational simulation. The optical fibre probing is usually employed in multi-phase flow measurement: bubble/droplet diameter, velocity and number density.

By micro-fabricating a plane optical fibre (Fs-TOP) through femtosecond laser pulses, the Fs-TOP is considered to apply to other research field such as biotechnology.

References

1. Mudde, R.F., Saito, T.: Hydrodynamical similarities between bubble column and bubbly pipe flow. J. Fluid Mech. **437**, 203–228 (2001)
2. Juliá, J.E., Harteveld, W.K., Mudde, R.F., Van Den Akker, H.E.A.: On the accuracy of the void fraction measurements using optical probes in bubbly flows. Rev. Sci. Instrum. **76**, 035103 (2005)
3. Higuchi, M., Saito, T.: Quantitative characterizations of long-period fluctuations in a large-diameter bubble column based on point-wise void fraction measurements. Chem. Eng. J. **160**, 284–292 (2010)
4. Abuaf, N., Jones, O.C. Jr., Zimmer, A.: Optical probe for local void fraction and interface velocity measurements. Rev. Sci. Instrum. **49**, 1090–1094 (1978)
5. Cartellier, A.: Simultaneous void fraction measurement, bubble velocity, and size estimate using a single optical probe in gas-liquid two-phase flows. Rev. Sci. Instrum. **63**, 5442–5452 (1992)
6. Mizushima, Y., Saito, T.: Detection method of a position pierced by a single-tip optical fibre probe in bubble measurement. Measure. Sci. Technol. **23**, 085308 (2012)
7. Sakamoto, A., Saito, T.: Analysis of optical fiber probing based on a ray tracing. Rev. Sci. Instrum. **83**, 075107 (2012)

8. Sakamoto, A., Saito, T.: Robust algorithms for quantifying noisy signals of optical fiber probes employed in industrial-scale practical bubbly flows. Int. J. Multiphase Flow **41**, 77–90 (2012)
9. Mizushima, Y., Sakamoto, A., Saito, T.: Measurement technique of bubble velocity and diameter in a bubble column via single-tip optical-fiber probing with judgment of the pierced position and angle. Chem. Eng. Sci. **100**, 98–104 (2013)
10. Saito, T., Toriu, M.: Effects of a bubble and the surrounding liquid motions on the instantaneous mass transfer across the gas-liquid interface. Chem. Eng. J. **265**, 164–175 (2015)
11. Saito, T., Matsuda, K., Ozawa, Y., Oishi, S., Aoshima, S.: Measurement of tiny droplets using a newly developed optical fibre probe micro-fabricated by a femtosecond pulse laser. Measure. Sci. Technol. **20**, 114002 (2009)

Fluorescence Analysis of Micro-scale Surface Modification Using Ultrafine Capillary Atmospheric Pressure Plasma Jet for Biochip Fabrication

Masaaki Nagatsu, Masahiro Kinpara and Tomy Abuzairi

Abstract In this study, we propose the micro-scale surface functionalization techniques using a maskless, versatile, simple tool, represented by a nano- or micro-capillary atmospheric pressure plasma jet (APPJs). We show the possibility of size-controlled surface functionalization on polymer substrate with amino groups or carboxyl groups by using fluorescent technique. Moreover, we prove the successful connection of biomolecules on the functionalized micro-scale patterns, indicating the possibility to use ultrafine capillary APPJs as versatile tools for biosensing, tissue engineering, and related biomedical applications. With use of this technology, we study the biomolecules patterning onto CNT dot array via ultrafine APPJ for biochip fabrication. An ultrafine APPJ with a micro-capillary was utilized to functionalize amino groups on designated CNT spots. Two-stage plasma treatments, pre-treatment and post-treatment, were conducted to functionalize CNT array by ultrafine APPJ. Biomolecules system, such as biotin-avidin system was employed to assess the feasibility of biomolecule immobilization on CNT. The possibility of this technique for micro-biochip applications was successfully demonstrated by patterning biomolecules onto CNT microarrays.

Keywords Surface modification · Ultrafine capillary atmospheric pressure plasma jet · Micro-scale patterning · Biochip sensor

M. Nagatsu (✉)
Research Institute of Electronics, Shizuoka University, Shizuoka, Japan
e-mail: nagatsu.masaaki@shizuoka.ac.jp

M. Nagatsu · M. Kinpara
Department of Engineering, Graduate School of Integrated Science
and Technology, Shizuoka University, Shizuoka, Japan

M. Nagatsu · T. Abuzairi
Graduate School of Science and Technology, Shizuoka University, Shizuoka, Japan

T. Abuzairi
Department of Electrical Engineering, Faculty of Engineering, Universitas Indonesia,
Shizuoka, Japan

© Springer International Publishing AG 2017
R. Jabłoński and R. Szewczyk (eds.), *Recent Global Research and Education:
Technological Challenges*, Advances in Intelligent Systems and Computing 519,
DOI 10.1007/978-3-319-46490-9_34

1 Introduction

For successful realization of miniaturized biochip devices, surface functionalization which provides sites to immobilize biomolecules onto the chip is one of the most important fabrication steps. So far several techniques with chemical patterning are extensively developed, requiring expensive devices and having limited use, or, colloidal lithography masks [1–5]. For the surface functionalization, plasma treatment is also used as one of effective methods compared to other conventional chemical methods, such as carboxylic acid, nitric acid, high temperature vapor, etc. Plasma treatment has advantages of low temperature treatment, little damaging effects, and providing a wide range of different functional groups depending on the plasma discharge conditions. For the development of multi-functional biochip device, a normal low pressure plasma processing might be difficult to modify the surface of substrate with different functional groups using a lithographic technique. Therefore, we propose here to use the atmospheric pressure plasma jet with a nano-capillary for micro-scale surface modification of the substrate [6–10].

In this work, a nano-size capillary atmospheric pressure plasma jet technique has been developed to functionalize amino or carboxyl groups selectively on the polymer or CNT dot-array substrate for fabricating biochip sensor for virus detection, as illustrated in Fig. 1. Selective surface functionalization has been demonstrated by atmospheric plasma pressure jet with two stages: pre-treatment stage by atmospheric pressure plasma jet with a negative biasing for activation and post-treatment stage for functional group modification. Our recent results of fluorescent microscopy with the fluorescent dye indicate that pre-treatment and

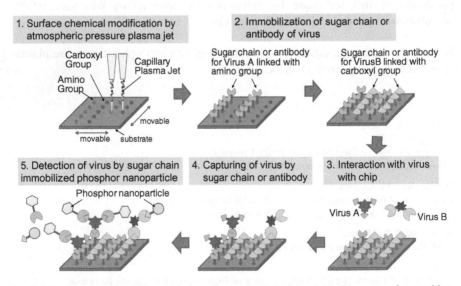

Fig. 1 Illustration of fabrication processes of the multi-functional biochip sensor for sensitive detection of various viruses

post-treatment period time strongly affect the fluorescent intensity and area of surface functionalization [10, 11].

2 Experimental Details

The schematic representation of the experimental device is shown in Fig. 2. The discharge was produced in a glass tube in two electrode configurations. Square pulsed voltage of about ±7.5 kV peak to peak value at a frequency of 5 kHz and 50 % duty ratio was applied between two copper electrodes placed on the discharge tube at a distance of 1 cm from each other. The driven electrode was placed on the discharge tube 6 cm away from its tip.

The aperture diameter of the capillary was 100 nm or 1 μm. The distance between the tip of the discharge tube and the substrate surface processed was adjusted to ∼25 for 100 nm aperture or 250 μm for 1 μm aperture, respectively. To prevent capillary damages by the high pressure gas flow inside the tube, the gas pressure was reduced by using leak holes at a tapered plastic tip between discharge tube and capillary tube.

For the first stage of the patterning procedure (pretreatment), −500 V bias was applied on a metallic mesh placed right under the substrate sample. The role of this voltage is to accelerate the positively charged species and intensify the effects of pretreatment. As a discharge gas in the pretreatment, only helium was used at a gas flow rate of about 500 sccm, while in the post-treatment step, a fractional amount of

Fig. 2 The schematic of the experimental set up of capillary APPJ

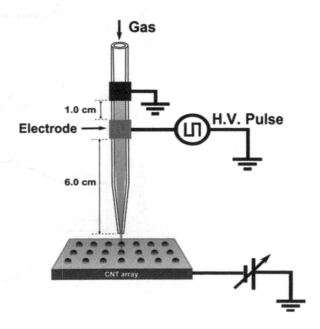

ammonia or oxygen of about 3 % was added to assure the radicals for amine or hydroxyl group functionalization. In order to demonstrate the functionalized ultrafine patterns on the substrate, a computer-controlled x-y stage was used.

3 Results and Discussion

Typical waveforms of applied voltage and current flow between two electrodes were shown in Fig. 3. It is seen that sharp current peaks having about 1 μs pulse width are observed in the current waveforms together with broad displacement current during rising and falling phases of the applied voltage. From our recent experiment with ICCD camera, it is found that these current peaks correspond to the dielectric barrier discharges between two copper electrodes through the quartz discharge tube [11]. During spiky current pulses observed about 10–12 μs after a sharp current peak, plasma bullets were ejected from APPJ and irradiated onto the surface of substrate.

In Fig. 4a–c, illustration of maskless surface modification of polymer such as photoresist film by capillary APPJ and their experimental results of maskless surface modification of photoresist film with amino groups by He/NH₃ APPJ and

Fig. 3 Typical waveforms of applied voltage and current between two electrodes of capillary APPJ

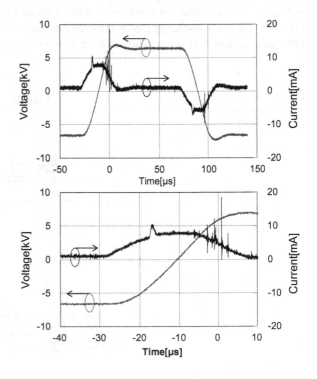

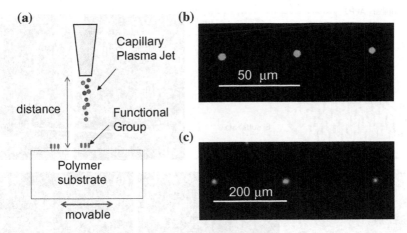

Fig. 4 **a** Illustration of maskless surface modification of photoresist film by capillary APPJs, and fluorescent microscope images of surface modification patterns analyzed by using fluorescence dyes; **b** Alexa Fluor 488 for amino group, and **c** CFDI for hydroxyl group

hydroxyl groups by He/O$_2$ APPJ, respectively. We performed the pretreatment by He APPJ with a bias voltage of -500 V for 0.1 s and He/NH$_3$ APPJ without bias for 3 s. Figure 4b shows the fluorescence pattern of Alexa Fluor 488 used as fluorescent dye, where typical size of fluorescence pattern is about 5 μm. As for hydroxyl group, we used He/O$_2$ APPJ for post-treatment for 1 s and analyzed with a fluorescent dye, 5(6) carboxyfluorescein diisobutyrate (CFDI).

We have also studied the immobilization of biomolecules onto CNT dot array functionalized by using an ultrafine APPJ. CNT dot arrays were synthesized by a combined thermal and plasma CVD devices [12] and constructed in dot array form for realizing the development of biochip device (see Fig. 5a). For patterning of surface modification onto CNTs, an APPJ with a micro-capillary of ~ 1 μm was utilized. The substrate was scanned automatically by computer-controlled stage for patterning hundred of CNT dot array, as shown in Fig. 5. The patterning of CNT dot-array as a microarray biosensor has been demonstrated by successful maskless functionalization of amino ($-$NH$_2$) and carboxyl ($-$COOH) groups onto CNTs by using an ultrafine APPJ.[8] The experimental results of chemical derivatization with the fluorescent dye (Alexa Fluor 488 for amino and AABD-SH for carboxyl groups) showed that the CNT dot-array was not only successfully functionalized with amino group and carboxyl group, but also was functionalized without any interference between functional groups. As shown in Fig. 5b, c, the success of maskless functionalization in the line pattern provides a means of a multi-functionalization CNT dot-array with direct implications for future application of a microarray biosensor.

The biomolecules immobilization was simulated by using PEG-biotin-avidin-fluorescein isothiocyanat (FTIC) immobilization in place of actual antibody and antigen reaction. Figure 6 shows comparison between the CNT dot array samples

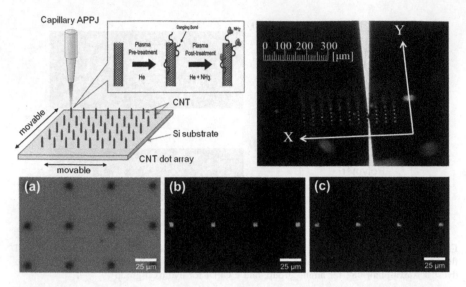

Fig. 5 Illustration and photo of maskless patterning of CNT dot array by using a ultrafine APPJ (*top*). Photos of **a** bright field image of CNT dot array, and the fluorescence line patterns of (**b**) Alexa Fluor 488 for amino (–NH₂) and (**c**) AABD-SH for carboxyl (–COOH) groups onto CNT dot array substrate analyzed with fluorescent microscope [8]

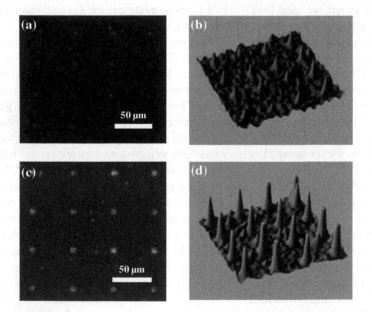

Fig. 6 Comparison between untreated and treated APPJ plasmas obtained from CNT dot array performed biomolecules immobilization. **a** and **c** Dark field image covered with Avidin-Biotin; **b** and **d**: 2D fluorescence intensity corresponding to **a** and **c** (color figure online)

having 4×50 μm spots, untreated (Fig. 6a, b) and treated (Fig. 6c, d) by APPJ plasmas. The black areas in Fig. 6a indicate that fluorescent dye of FTIC-avidin is not connected to amino group functionalized CNT dot array. On the other hand, the green areas in Fig. 6c correspond to fluorescent dye of avidin (FTIC-avidin) connected to biotin and amino group functionalized CNT dot array. Good spot uniformity of biomolecules immobilization was confirmed by 2D fluorescence intensity line profile across CNT dot-array of 4×50 μm shown in Fig. 6b, d.

4 Conclusion

We presented the micro-scale surface functionalization techniques using a mask-less, versatile, simple tool, represented by a nano- or micro-capillary atmospheric pressure plasma jet (APPJs). We showed the possibility of size-controlled surface functionalization on polymer substrate with amino groups or hydroxyl groups by using fluorescent technique. Moreover, we prove the successful connection of fluorescence dyes on the functionalized micro-scale patterns, indicating the possibility to use ultrafine capillary APPJs as versatile tools to modify the surface locally without any physical ask for biosensing and related biomedical applications. With use of this technology, we study the biomolecules patterning onto polymer or CNTs via ultrafine APPJ for biochip fabrication. An ultrafine APPJ with a micro-capillary was utilized to functionalize amino and carboxyl groups on designated CNT spots. Two-stage plasma treatments, pre-treatment and post-treatment, were conducted to functionalize CNT array by ultrafine APPJ. Biomolecules system, such as biotin-avidin system was employed to assess the feasibility of biomolecule immobilization on CNT. The possibility of this technique for micro-biochip applications was successfully demonstrated by patterning biomolecules onto CNT microarrays.

Acknowledgments This work has been supported in part by Grant-in-Aid for Scientific Research (No. 25246029) from the JSPS and the International Research Collaboration and Scientific Publication Grant (DIPA-23.04.1.673453/2015) from DGHE Indonesia.

References

1. Salaita, K., Wang, Y., Mirkin, C.A.: Nat. Nanotechnol. **2**, 145–155 (2007)
2. Tan, C.P., Cipriany, B.R., Lin, D.M., Craighead, H.G.: Nano Lett. **10**, 719–725 (2010)
3. Huo, F., Zheng, Z., Zheng, G., Giam, L.R., Zhang, H., Mirkin, C.A.: Science **321**, 1658–1660 (2008)
4. Coyer, S.R., Garcia, A.J., Delamarche, E.: Angew. Chem. Intl. Ed. **46**, 6837–6840 (2007)
5. Malmstrom, J., Christensen, B., Jakobsen, H.P., Lovmand, J., Foldbjerg, R., Sorensen, E.S., Sutherland, D.S.: Nano Lett. **10**, 686–694 (2010)
6. Kakei, R., Ogino, A., Iwata, F., Nagatsu, M.: Thin Solid Films **518**, 3457 (2010)
7. Motrescu, I., Ogino, A., Nagatsu, M.: J. Photopol. Sci. Technol. **25**, 529 (2012)

8. Abuzairi, T., Okada, M., Mochizuki, Y., Poespawati, N.R., Purnamaningsih, R.W., Nagatsu, M.: Carbon **89**, 208 (2015)
9. Ionut, T., Nagatsu, M.: Appl. Phys. Lett. 106, 054105 (5 p) (2015)
10. Motrescu, I., Nagatsu, M.: Nanocapillary atmospheric pressure plasma jet: a tool for ultrafine maskless surface modification at atmospheric pressure. ACS Appl. Mater. Interfaces **8**, 12528–12533 (2016)
11. Abuzairi, T., Okada, M., Purnamaningsih, R.W., Poespawati, N.R., Iwata, F., Nagatsu, M.: Appl. Phys. Lett. **109**, 23701(5 p.) (2016)
12. Matsuda, T., Mesko, M., Ogino, A., Nagatsu, M.: Diam. Related Mater. **17**, 772–775 (2008)

Cleaning of Silica Surfaces by Surface Dielectric Barrier Discharge Plasma

Lucel Sirghi, Florentina Samoila and Viorel Anita

Abstract This work presents a simple and easy method of surface cleaning of silica substrates used for thin film depositions and preparation of microscopy samples. To this purpose, we use a surface dielectric barrier discharge (SDBD) in flow of dried air (relative humidity below 2 %) at atmospheric pressure. The silica substrates and SDBD electrodes were mounted face to face with a small gap (2 mm) through which was flown the working gas. The cleaning effectiveness was evaluated by measurements of water contact angle of the silica surfaces, which showed a decrease from 60° to >30° in 20 min of treatment.

1 Introduction

Silica (glass or silicon wafers with native oxide top layers) substrates are the most common material used in thin film depositions and biologic sample preparation for optical and atomic force microscopy. In many cases the surfaces of silica substrates used for these proposes are not clean due to adsorption of airborne and package-released hydrophobic organic contaminants [1]. This surface contamination may decrease drastically the adhesion of deposited films on the substrates and may affect also the optical and atomic force microscopy measurements. Therefore, a simple and easy method of silica surface cleaning is very useful in preparation of silica substrates for thin film deposition or for preparation of microscopy samples. In previous works [1, 2], we have shown that the negative glow plasma of a glow discharge in air at low pressure is very efficient in removing hydrophobic contaminates from silica surfaces. However, low pressure plasma cleaning technique requires relatively expensive plasma setups and is not handy because requires substrate loading in a vacuum chamber, vacuuming the chamber, plasma cleaning and venting. Therefore, we have investigated the capability of surface dielectric

L. Sirghi (✉) · F. Samoila · V. Anita
Faculty of Physics, Iasi Plasma Advanced Research Center (IPARC),
Alexandru Ioan Cuza University of Iasi, Blvd. Carol I Nr. 11, Iasi 700506, Romania
e-mail: lsirghi@uaic.ro

© Springer International Publishing AG 2017   255
R. Jabłoński and R. Szewczyk (eds.), *Recent Global Research and Education:*
Technological Challenges, Advances in Intelligent Systems and Computing 519,
DOI 10.1007/978-3-319-46490-9_35

barrier discharge (SDBD) working in a closed Petri dish in atmospheric air (without flow) for cleaning of silicon surfaces [3]. We have found that the SDBD working in closed volume air at atmospheric pressure is less effective than the low-pressure discharge plasma for cleaning of silicon surfaces. In the present work, we have improved the SDBD capability for cleaning silica surfaces. To this purpose, the silica substrates and SDBD electrodes were mounted face to face with a small gap (around 2 mm). The working gas (dry air) was flown through the gap while the SDBD was powered by a high voltage sinusoidal wave generator. The cleaning effectiveness was evaluated by measurements of water contact angle of the silica surfaces after plasma treatment.

2 Materials and Methods

The as-supplied microscopy cover glass slides (coverglasses from Agar scientific, 13 mm in diameter and 0.1 mm in thickness) were loaded on the SDBD plasma cleaning device. A sketch of the SDBD cleaning device used in experiments is presented in Fig. 1. Dry and filtered air provided by a pure air circulator unit (Torlabs Inc.) was flown at a flow rate of about 1 L/min at room temperature and atmospheric pressure through a small gap (2 mm × 30 mm) formed by the SDBD electrodes and the sample holder with the glass sample facing the SDBD plasma. The SDBD device used in present study had a sandwich structure (area of 2.3 × 3 cm^2) formed by continuous bottom copper electrode, a polyamide dielectric layer (0.2 mm in thickness) and a top electrode structured as 18 stripes with the width of 0.2 mm and length of 30 mm. The glass sample holder was mounted on a poly-dimethylsiloxane (PDMS) O-ring to securely sealing the gap between the SDBD plasma and sample surface. The SDBD device was powered (around 2 W in average power) by a sinusoidal wave generator with voltage peak to peak amplitude around 5 kV and a frequency around 14 kHz. The time variation of the discharge current intensity through the SDBD resemble the time variation of the current

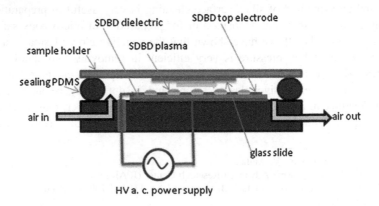

Fig. 1 Schematic representation of the experimental setup

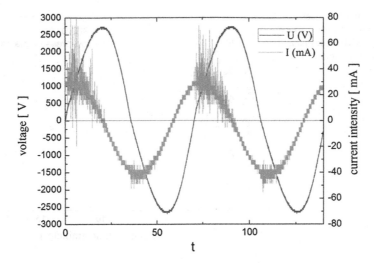

Fig. 2 Typical time variations of voltage (*blue line*) and current intensity (*red line*) during a period of SDBD

intensity through a capacitor with the difference of high frequency spikes corresponding to the SDBD working in multi filamentary regime [4]. Figure 2 presents typical time variations of SDBD voltage and current intensity during an oscillation period.

The water contact angle of the glass samples before and after various plasma treatment times was measured by analysis of profile images of small sessile water droplets (1–2 μL) on the sample surfaces.

3 Results and Discussion

Figure 3 shows comparatively the variations of water contact angle as result of treatment of glass surfaces by SDBD atmospheric plasma and a d. c. glow discharge plasma in air at low pressure (26 Pa).

The water contact angle for the as-supplied glass cover slides was around 60°. The d. c. glow discharge plasma in air at low pressure is very effective in turning the glass surface from less hydrophilic to super hydrophilic (water contact angle smaller than 10°). After 5 min of treatment in low-pressure plasma in air, the water contact angle of the glass surface decreased to very low values (around 2°) due to the cleaning and removing of the hydrophobic contaminant. However, cleaning alone can not explain the super hydrophilicity of glass substrates treated by low-pressure plasma. The X ray photoelectron spectroscopy measurements of O1s peak structure have shown formation of Si–OH functional groups that are responsible for the super-hydrophilic character of the glass surfaces treated in

Fig. 3 Change of water
contact angle of glass surface
as result of SDBD plasma and
low-pressure plasma
treatments

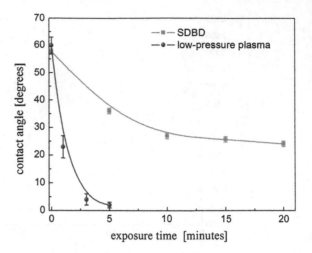

low-pressure plasma in air [5]. The SDBD plasma is much less effective in improving of the glass surface hydrophilicity. The SDBD plasma treatment of glass surface for about 20 min resulted in a noticeable decrease of the water contact angle, but the contact angle did not reach values lower than 24°, even for long treatment time. This result may be explained by considering that the SDBD plasma is efficient for removing of contaminant hydrophobic molecules adsorbed on the glass surface, but not efficient for generation of Si–OH functional groups responsible for the super hydrophilic character of silica surfaces.

4 Conclusion

The efficiency of the surface dielectric barrier discharge plasma in air flow at atmospheric pressure for cleaning and hydrophilization of glass surfaces has been investigated and compared to the efficiency of low-pressure plasma treatment in a d. c. discharge in air. It has been found that the surface DBD plasma treatment is much less efficient in hydrophilization of glass surfaces than the low-pressure plasma treatment. However, as resulted from the measurements of water contact angle values, the surface DBD plasma treatment is an easy, low-cost and efficient method for removing contaminant hydrophobic molecules adsorbed on glass surfaces.

Acknowledgments This work was supported by CNCSIS, IDEI Research Program of Romanian Research, Development and Integration National Plan II, Grant no. 267/2011.

References

1. Sirghi, L., Kylian, O., Gilliland, D., Cecone, G., Rossi, F.: Cleaning and hydrophilization of atomic force microscopy silicon probes. J. Phys. Chem. B **110**, 25975 (2006)
2. Sirghi, L.: Plasma cleaning of silicon surface of atomic force microscopy probes. Rom. J. Phys. **56**, 144 (2011)
3. Samoila, F., Anita, V., Sirghi, L.: Cleaning of silicon surface by surface dielectric barrier discharge. In: International Conference of physics of Ionized Gases (ICPIG) 32nd (26–31 July 2015) Iasi, Romania
4. Audier, P., Joussot, R., Rabat, H., Hong, D., Leroy, A.: ICCD imaging of plasma filament in a circular surface-dielectric-barrier-discharge arrangement. IEEE trans. Plasma Sci. **39**, 2180 (2011)
5. Apetrei, A., Sirghi, L.: Stochastic adhesion of hydroxylated atomic force microscopy tips to supported lipid bilayers. Langmuir **26**, 144 (2013)

Removal of Cs Ion from Aqueous Solution Using Prussian Blue-Carrying Magnetic Nanoparticles

Toshiya Takayanagi and Masaaki Nagatsu

Abstract In this study, we have carried out the experiment to immobilize Prussian blue (PB) onto magnetic nanoparticles (MNPs) to develop the novel adsorption materials to remove Cs ions efficiently from aqueous solution. An immobilization process with amino group modified PBs and carboxyl group modified MNPs was investigated using plasma surface functionalization and demonstrated to fabricate PB@MNPs. To functionalize the surface of these nanomaterials, we used an inductively-coupled radio frequency plasma device with Ar/H_2O gas mixture for carboxyl groups and Ar/NH_3 gas mixture for amino groups, respectively. Preliminary result of Cs ion removal property using these PB@MNPs adsorbent materials will be discussed.

Keywords Removal of Cs ion · Prussian Blue · Magnetic nanoparticle · Plasma surface functionalization

1 Introduction

Nowadays, the radioactive Cs leakage due to the nuclear power plant disaster at Fukushima in Japan is an urgent and serious problem to be solved. The effective removal of Cs^+ ions from radioactive nuclear waste solutions is crucial for public health. Therefore, an economic and high efficient sorbent for capture of radioactive Cs^+ ions from nuclear waste solutions is required. Since then, many researchers have been carrying out developing the efficient adsorption materials for radioactive Cs ions from polluted aqueous solution. Among them, nano-structured materials,

T. Takayanagi · M. Nagatsu (✉)
Graduate School of Integrated Science and Technology,
Shizuoka University, Hamamatsu, Japan
e-mail: nagatsu.masaaki@shizuoka.ac.jp

M. Nagatsu
Research Institute of Electronics, Shizuoka University, Hamamatsu, Japan

© Springer International Publishing AG 2017
R. Jabłoński and R. Szewczyk (eds.), *Recent Global Research and Education:
Technological Challenges*, Advances in Intelligent Systems and Computing 519,
DOI 10.1007/978-3-319-46490-9_36

261

Fig. 1 Illustration of
Prussian blue-immobilized
GEMNPs for Cs⁺ ion removal

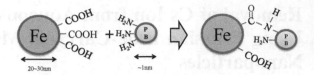

such as carbon nanotubes [1, 2], zeolite [3], bentonite [4, 5], etc., have received a lot
of attention in recent years due to their high specific surface areas.

So far, we have carried out the study on adsorption of metal ions from liquid by
using plasma surface-functionalized graphite-encapsulated magnetic nanoparticles
(GEMNPs).

To capture Cs ions on the surface of GEMNPs, however, specific adsorption
materials are essential. Prussian blue (PB) is well known materials as Cs adsorbent
materials, however, it is difficult to separate them from solution after adsorption
procedure. Hence, in this study, we propose the new type of adsorption materials,
that is, PB-immobilized magnetic nanoparticles (see Fig. 1). To immobilize PB onto
the surface of GEMNPs, we used an inductively coupled radio frequency (RF)
plasmas for surface modification of PB with amino groups and GEMNPs with
carboxyl groups, respectively [6, 7].

2 Experimental Setup

Figure 2 shows a schematic of an inductively-coupled RF plasma device. For
surface modification, we used two stages of plasma treatments; the first step is the
pretreatment by Ar plasma at pressure of 50 Pa, and power of 80 W for 10 min.
And then, we used Ar/H₂O plasma for carboxyl group introduction onto GEMNPs

Fig. 2 Experimental setup of
inductively coupled RF
plasma [6, 7]

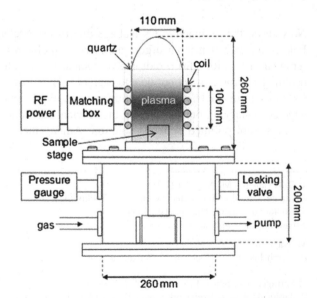

and Ar/NH$_3$ plasma for amino groups onto PBs, respectively. To evaluate the numbers of functional groups on to GEMNPs and PBs quantitatively, we used so-called chemical derivatization methods. The number of amino groups onto the PBs were quantified by a chemical derivatization method using sulfosuccinimidyl 6-[3'(2-pyridyl-dithio)-propionamido] hexanoate (sulfo-LC-SPDP), as shown in Fig. 3 [8, 9].

The treated PBs were suspended by bath sonication in 10 mM sulfo-LC-SPDP in phosphate buffer saline (PBS) and reacted for 30 min under light shielding conditions, repeating the ultrasonication every 5 min. The treated PBs were washed three times with PBS through ultrasonication and centrifugation and collected by centrifugal separator. The PBs with sulfo-LC-SPDP complexes were then reacted with 20 mM dithiothreitol (DTT) in PBS. After 15 min reaction, 5 min centrifugation at 20,400 g was performed and the cleavage product pyridine-2-thione (P2T) liberated from the sulfo-LC-SPDP present in the recovered supernatant liquid, was determined by spectrophotometry at 343 nm. The number of amino groups of the modified PBs was quantitatively determined from the calibration curve or by theoretical evaluation using the extinction coefficient of P2T at 343 nm.

As for carboxyl group analysis, we used Toluidine Blue O (TBO) assay [10]. Each carboxyl group on the surface GEMNP connects with TBO molecule. After several washing processes of GEMNPs, TBO molecules are separated by adding dodecyl sulfate (SDS) solution with low pH 1. The absorbance of separated TBO molecules was measured at 632 nm to analyze the number of carboxyl groups modified onto GEMNPs.

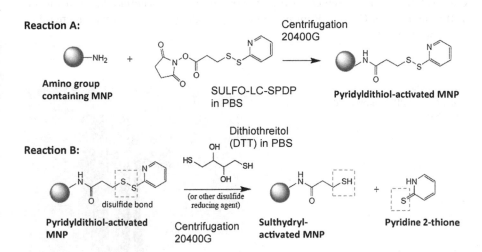

Fig. 3 Protocol of quantitative analysis of amino group population [8, 9]

3 Experimental Results

After surface modification of PB with amino groups and GEMNPs with carboxyl groups, these numbers of functional groups are analyzed using a chemical technique as described in the previous section. Optimizing the plasma pretreatment and post treatment conditions, we obtained the maximum number of carboxyl groups functionalized onto GEMNPs of roughly $1.7–3.4 \times 10^{19}$/g. This number roughly corresponds to the total numbers of carbon atoms onto the outmost surface of GEMNPs with ~ 20 nm average diameter. As for the number of amino groups introduced onto PB, we obtained about 1.18×10^{17}/g. The PB immobilized GEMNPs are prepared by the chemical reaction using cross-linking reagent. Figure 4 shows the photo of magnetic collection of PB immobilized GEMNPs by a magnet. It is clearly seen that the solution becomes transparent from dark blue color after magnetic separation.

Now, PB-immobilized GEMNPs are tested to remove the Cs ions from aqueous solution with different concentrations. Procedure to measure Cs ion removal property is as follows:

1. Treat 3.0 mg PB-immobilized GEMNPs with 4 ml CsCl solution with varying Cs concentration of 0–100 ppm.
2. After shaking it during 2 days, supernatant solution was taken by centrifugal separation process.
3. Measure the concentration of Cs ions remaining in supernatant solution by atomic absorption spectrometer (Thermo Solaar S4-AA).

The results of Cs ions removal property in liquid for different concentration of CsCl solution were presented in Fig. 5. For lower Cs concentration, we achieved 100 % removal of Cs ions. Removal efficiency gradually decreases with increase of Cs concentration.

Fig. 4 Photo of magnetic collection of PB-immobilized GEMNPs

Fig. 5 Cs removal efficiency
versus Cs concentration in
CsCl sollution

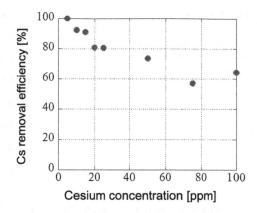

Figure 6 shows the Cs ions adsorption capacity of PB-immobilized GEMNPs having a weight of 3.0 mg as a function of equilibrium concentration of Cs ions. From this result, we can evaluated the maximum adsorption capacity of PB@MNPs as about 159.6 mg/g. It is noted that this value is higher than that of pure PB, 132 mg/g.

Figure 7 shows the comparison of the Cs^+ ion removal efficiencies between different samples with the same weight of 3 mg; graphite-encapsulated magnetic nanoparticles (MNPs) only, carboxyl group modified MNPs, and PB-immobilized MNPs. Here, the last sample was prepared by using cross-linking promoter to connect carboxyl-modified MNPs and aminated PB. The net weight of PB of the last sample was 2.1 mg. It is clearly found from Fig. 7 that the PB-immobilized MNPs showed a remarkable improvement of Cs removal property compared with other samples. When we tested the Cs removal using pure PBs of 2.1 mg, then we obtained about 60 % Cs ion removal under the same experimental conditions. Therefore, the present results suggest that the PB-immobilized MNPs have a better

Fig. 6 Relation between
adsorption capacity Qe versus
equillibrium concentration Ce

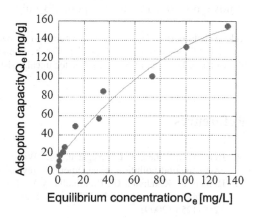

Fig. 7 Comparison of the
Cs$^+$ ion removal efficiencies
between different samples
with the same weight of 3 mg

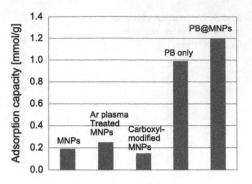

Cs removal property rather than pure PB only, which means that the PB-immobilized MNPs are promising adsorbents for practical radioactive Cs ion removal.

4 Conclusion

In this study, we have successfully immobilized Prussian blue onto magnetic nanoparticles to develop the novel adsorption materials to remove Cs ions efficiently from aqueous solution. An immobilization process with amino group modified PBs and carboxyl group modified MNPs was demonstrated by using an inductively-coupled radio frequency plasma device with Ar/H$_2$O gas mixture for carboxyl groups and Ar/NH$_3$ gas mixture for amino groups, respectively. Preliminary result of Cs ion removal property using these PB@MNPs adsorbent materials was presented and compared with those of other samples, such as untreated MNPs, Ar plasma treated MNPs, carboxyl—modified MNPs, and pure PB. The present results suggest that the PB-immobilized MNPs have a better Cs removal property rather than pure PB only.

Acknowledgments This work was supported in part by Grant-in-Aid for Scientific Research (No. 16K13709) from the Japan Society for the Promotion of Science (JSPS). The authors would like to thank Dr. Shubin Yang for her technical assistance of the present experiment.

References

1. Ren, X.M., Chen, C.L., Nagatsu, M., Wang, X.K.: Carbon nanotubes as adsorbents in environmental pollution management: A review. Chem. Eng. J. **170**, 395–410 (2011)
2. Belloni, F., Kütahyali, C., Rondinella, V.V., Carbol, P., Wiss, T., Mangione, A.: Can carbon nanotubes play a role in the field of nuclear waste management? Environ. Sci. Technol. **43**, 1250–1255 (2009)

3. Borai, E.H., Harjula, R., Malinen, L., Paajanen, A.: Efficient removal of cesium from low-level radioactive liquid waste using natural and impregnated zeolite minerals. J. Hazard. Mater. **172**, 416–422 (2009)
4. Yang, S., Han, C., Wang, X., Nagatsu, M.: Characteristics of cesium ion sorption from aqueous solution on bentonite- and carbon nanotube-based composites. J. Hazard. Mater. **274**, 46–52 (2014)
5. Yang, S., Okada, N., Nagatsu, M.: The highly effective removal of Cs^+ by low turbidity chitosan-grafted magnetic bentonite. J. Hazard. Mater. **301**, 8–16 (2016)
6. Saraswati, T.E., Ogino, A., Nagatsu, M.: RF plasma-activated immobilization of biomolecules onto graphite-encapsulated magnetic nanoparticles. Carbon **50**, 1253–1261 (2012)
7. Yang, E., Chou, H., Tsumura, S., Nagatsu, M.: Surface properties of plasma functionalized graphite-encapsulated gold nanoparticles prepared by direct current arc discharge method. J. Phys. D Appl. Phys. **49**(18), 185304 (2016)
8. Carlsson, J., Drevin, H., Axén, R.: Protein thiolation and reversible protein-protein conjugation. N-Succinimidyl 3-(2-pyridyldithio) propionate, a new heterobifunctional reagent. Biochem. J. **173**, 723–737 (1978)
9. Hermanson, G.T.: Bioconjugate techniques (Chap. 9). In: Fluorescent probes. Bioconjugate techniques (2008). doi:10.1016/B978-0-12-370501-3.00009-6
10. Zhan, X., Wu, J., Chen, Z., Hinds, B.J.: Single-step electrochemical functionalization of double-walled carbon nanotube (DWCNT) membranes and the demonstration of ionic rectification. Nanoscale Res. Lett. **8**, 279 (2013)

Low-Temperature Disinfection of Tea Powders Using Non-equilibrium Atmospheric Pressure Plasma

Syuhei Hamajima, Naohisa Kawamura and Masaaki Nagatsu

Abstract In this study, we aim at developing a new low-temperature disinfection technique for powdery foods, such as green tea powder, using non-equilibrium atmospheric pressure plasma. We have carried out experiment to inactivate *Escherichia coli* (*E. coli*) adhering to the tea leaves by using cylindrical dielectric-barrier discharge (DBD) non-equilibrium atmospheric pressure plasma. In addition, we have studied the effect of plasma treatment on flavor components of the tea leaves after plasma treatment. Preliminary results of disinfection experiments and flavor component analysis are presented and discussed.

1 Introduction

Recently, from the point of view of food safety, various seeds, such as soybean and granular, powdered foods, required sterilization process [1]. Up to now, the heat sterilization has been widely used as common methods for sterilizing various kinds of food materials. However, the heat sterilization method has a serious problem of deterioration of the food flavor and taste, especially of tea leaves and powder, wheat flour, and other seed powder. Compared with conventional methods including microwave heating [1, 2], chemical disinfection by sodium hypochlorite [3] or chlorine chemical solutions [4], low temperature atmospheric pressure plasma technology [5, 6] is an attractive method as alternated and has a number of advantages, for examples environmentally-friendly, low temperature, no need of vacuum pump system, and easy to use by a continuous processing.

S. Hamajima · M. Nagatsu (✉)
Department of Engineering, Graduate School of Integrated Science
and Technology, Shizuoka University, Hamamatsu, Japan
e-mail: nagatsu.masaaki@shizuoka.ac.jp

N. Kawamura
Kumeta Manufacturing Co. Ltd, Haibara-Gun, Japan

M. Nagatsu
Research Institute of Electronics, Shizuoka University, Hamamatsu, Japan

© Springer International Publishing AG 2017   269
R. Jabłoński and R. Szewczyk (eds.), *Recent Global Research and Education:*
Technological Challenges, Advances in Intelligent Systems and Computing 519,
DOI 10.1007/978-3-319-46490-9_37

In this study, we propose an inactivation of *Escherichia coli* (*E. coli*) attached to the tea powder by using a cylindrical dielectric-barrier discharge (DBD) at atmospheric pressure. In addition, we have studied the effect of plasma treatment on flavor components of the tea powder after plasma treatment. The preliminary experimental results of disinfection and flavor component analysis of green tea powder by using the cylindrical DBD plasma are presented and discussed.

2 Experimental Setup

Figure 1 shows a schematic of cylindrical DBD type atmospheric pressure plasma. Diameter and length of punched stainless steel cylinder with 2 mm diameter hole are 56 and 100 mm, respectively. To insulate two electrodes of inner punched stainless steel cylinder and aluminum tape, we used a Kapton® sheet with 0.125 mm thickness, as shown in Fig. 1. The tea powder sample was set on the stage inside the cylindrical DBD. To excite the atmospheric pressure plasma on the internal wall surface of cylindrical DBD, we used a high voltage, square wave power supply. Typical applied voltage is ±2.5 kV and pulse frequency is 5 kHz. Pictures of cylindrical DBD are also given in Fig. 1. It is seen that the plasma discharge are generated inside internal wall of cylindrical DBD electrode.

To study the disinfection of bacteria attached onto the green tea powder, we used *Escherichia coli* (*E. coli*). *E. coli* was grown overnight in Luria-Bertani (LB) broth medium in a shaking incubator at 150 rpm, 370 C. This bacterial culture was then serially diluted to get a desirable dilution. About 10^4 *E. coli* bacteria were pasted into 5 mg green tea powders on the stainless steel disk, as shown in Fig. 2. These samples were treated by DBD plasma under air circumstance and investigated by using colony counting method, as shown in Fig. 2.

The effect of plasma treatment on the flavor components of green tea powder was also investigated using a solid-phase microextraction-gas chromatography mass spectrometer (SPME-GC/MS) [7].

3 Experimental Results

Figure 3 shows typical waveforms of applied voltage and current of cylindrical DBD plasmas, where the right figure shows the magnified waveforms of left figure. It is seen that spiky current pulses having peak amplitude of about 4 A and full width at half maximum of 0.18–0.23 μs are observed during rising and falling phases of applied high voltage square-waves. From the voltage and current waveforms, we can calculate a net power of the DBD discharge as about 2.75 W per period. To avoid the electrical breakdown of Kapton® polyimide sheet [8], we operated the DBD plasmas at voltage amplitude less than 2.5 kV. The plasma treatment time was varied up to 300 s.

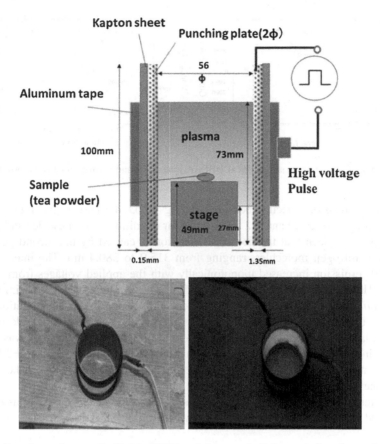

Fig. 1 Experimental setup of cylindrical DBD and photographs before and after plasma discharge

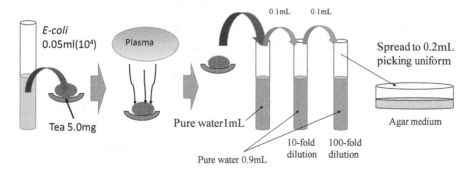

Fig. 2 Illustration of *E. coli* preparation and colony counting method

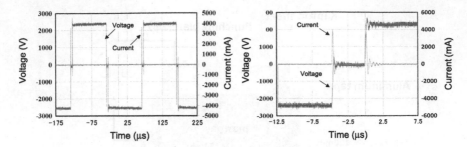

Fig. 3 Typical waveforms of cylindrical DBD plasmas (*left*) and its magnified waveform (*right*)

Optical emission spectra measured during DBD discharges generated at an applied voltage of ±2.0 and ±2.5 kV under air condition were presented in Fig. 4. It is apparently seen that the dominant emission is caused by the second positive system of nitrogen molecules, ranging from 315.9 to 380.4 nm. The intensity of these UV emission increased monotonically with the applied voltages from 1.5 to 2.5 kV. Hence, it is expected that the UV effect will be playing a role in disinfection of *E. coli*. Moreover, the DBD plasma generated ozone under air condition. To measure the ozone concentration, we installed the cylindrical DBD electrodes inside a larger chamber. Then the ozone concentration in the chamber was measured using an ozone analyzer. At 300 s after the DBD plasma ignition at ±2.5 kV, ozone concentration monotonically reached to roughly 400–500 ppm. Therefore, the ozone might play a critical role on the inactivation of *E. coli*.

The results of colony counting analysis of *E. coli* treated by DBD plasmas at ±2.0 and ±2.5 kV are shown in Fig. 5, where the initial numbers of *E. coli* are

Fig. 4 Typical optical emission spectra of cylindrical DBD plasmas at ±2.0 and ±2.5 kV

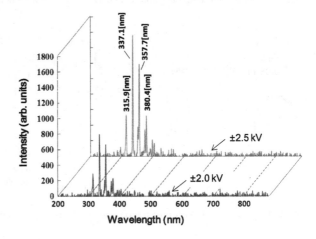

Fig. 5 Results of colony-counting analysis of *E. coli* treated by cylindrical DBD plasmas at ±2.0 and ±2.5 kV

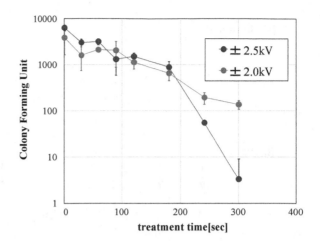

$5\text{–}7 \times 10^3$. In both the cases, the survival bacteria numbers decreased with the plasma irradiation time. Especially, the 3-digit reduction of *E. coli* population number was achieved at 300 s plasma discharge at ±2.5 kV.

It is noted here that the present experiment was performed at unrealistically high bacteria concentration, that is, there are $10^3\text{–}10^4$ *E. coli* bacteria buried in 5 mg green tea powder. However, after 5 min plasma treatment, 3-order reduction of *E. coli* was confirmed in the present experiment.

On the other hand, it is also important to study the effect of plasma treatment on the flavor components of green tea powder. Figure 6 shows the results of SPME-GC/MS analysis; (a) untreated green tea powder, (b) plasma treated green tea powder under air, and (c) plasma treated green tea powder under nitrogen gas circumstance, respectively. These peaks are not identified at present, because of difficulty to determine respective chemical component one by one. However, it is clearly found that several component peaks (indicated by arrows in Fig. 6) are newly appeared in the plasma treated sample under air, compared with those in the untreated sample. These additional components might be produced by plasma treatment under air, possibly due to oxidation of tea powder by the ozone generated by DBD plasma under air. For comparison, we also present the results of plasma treated sample under nitrogen gas background, as shown in Fig. 6c, which is analogous to the GC/MS result of untreated sample, as shown in Fig. 6a. The present results clearly indicate that it is essential to inactivate the bacteria in the tea powder without oxidation. These issues should be figured out in the future study.

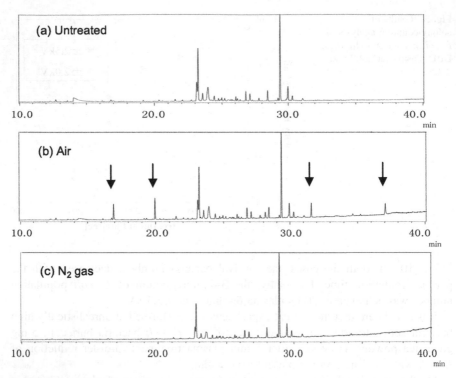

Fig. 6 Results of SPME-GCMS analysis of **a** untreated sample, **b** plasma treated sample under air, and **c** plasma treated sample under nitrogen gas circumstance, respectively. The *arrows* in (**b**) indicate additional components produced by DBD plasma treatment in air

4 Conclusion

In this study, we presented the preliminary experimental results of inactivation of *E. coli* attached to the tea powder by using a cylindrical dielectric-barrier discharge (DBD) at atmospheric pressure. We also presented the results of SPME-GC/MS measurement on flavor components of the tea powder after plasma treatment. The preliminary experimental results showed a successful disinfection of green tea powder by 3-order reduction of *E. coli* population by using the cylindrical DBD plasmas. The flavor components of the green tea powder treated by DBD plasmas under air and nitrogen gas circumstances were investigated by using SPME-GC/MS analysis. Results showed the appearance of additional components in the samples treated by DBD plasma under air, while there are no significant changes in GC/MS results in the sample treated under nitrogen gas. Therefore, it is considered that observed additional components were possibly generated by the oxidation of tea powder with ozone.

Acknowledgments This work was supported by Kumeta Manufacturing Co. Ltd., as collaborative research. The authors would like to thank Dr. Xiaoli Yang and Dr. Anchu Viswan for their help in the present *E. coli* inactivation experiment.

References

1. Ahmed, J., Ramaswamy, H.S.: Microwave pasteurization and sterilization of foods. In: Handbook of Food Preservation, 2nd edn, pp. 691–711. Taylor & Francis Group, LLC (2007)
2. Vadivambal, R., Jayas, D.S., White, N.D.G.: Wheat disinfestation using microwave energy. J. Stored Prod. Res. **43**(4), 508–514 (2007)
3. Izumi, H.: Electrolyzed water as a disinfectant for fresh-cut vegetables. J. Food Sci. **64**(3), 536–539 (1999)
4. Kim, C., Hung, Y.-C., Brackett, R.E.: Roles of oxidation–reduction potential in electrolyzed oxidizing and chemically modified water for the inactivation of food-related pathogens. J Food Prod. **1**, 3–140 (2000)
5. Eto, H., Ono, Y., Ogino, A., Nagatsu, M.: Low-temperature internal sterilization of medical plastic tubes using a linear dielectric barrier discharge. Plasma Proc. Polym. **5**(3), 269–274 (2008)
6. Eto, H., Ono, Y., Ogino, A., Nagatsu, M.: Low-temperature sterilization of wrapped materials using flexible sheet-type dielectric barrier discharge. Appl. Phys. Lett. **93**, 221502 (3 pp.) (2008)
7. Zambonin, C.G., Quinto, M., Vietro, N.D., Palmisano, F.: Solid-phase microextraction—gas chromatography mass spectrometry: a fast and simple screening method for the assessment of organophosphorus pesticides residues in wine and fruit juices. Food Chem. **86**(2), 269–274 (2004)
8. http://www.dupont.com/products-and-services/membranes-films/polyimide-films/brands/kapton-polyimide-film.html

Part V
Measurement, Signal Processing, Identification, Control

E-vehicle Predictive Control for Range Extension

Pavel Steinbauer, Florent Pasteur, Jan Macek, Zbyněk Šika
and Josef Husák

Abstract The range is currently one of main drawbacks of e-mobility, as energy storage capacity is limited. On the other hand, various information and computational resources in the cloud can be used. The control scheme uses model based predictive controllers with hierarchy of prediction horizons with various lengths. A detailed range estimation model of a Doblo e-vehicle is basis with the main subsystems: vehicle 1D model, e-motor, battery pack, air-conditioning/heating and EVCU. Due to the system substantial nonlinearity, a broad grid of linearized model is selected to rebuild a piecewise linear model. Trajectory velocity profile, designed by cloud control layer serves as input. Resulting controllers are merged using gain scheduling approach.

Keywords E-mobility · Range extension · Ecorouting · Model predictive control

1 Introduction

Current vehicles, especially electric ones are being equipped with more and more complex controllable structures and subsystems. It means that control strategy for complex, highly non-linear dynamic systems must be designed. In addition, a lot of information about planned route is known in advance. Especially slope of the route can be obtained from open resources as well as geographic coordinates of the route. These pieces of information are linked with legal velocity limits, traffic density etc. On the other hand, the amount of energy stored on-board is quite limited in case of e-vehicle. In this paper, control strategies based on detailed simulation model of the vehicle are evaluated.

The mechatronic nature of current vehicles together with widely available high capacity data connection and various information resources attract a lot of research

P. Steinbauer (✉) · J. Macek · Z. Šika · J. Husák
Faculty of Mechanical Engineering, CTU in Prague, Praha 6, Czech Republic

F. Pasteur
Siemens Industry Software S.A.S. DF PL STS CAE 1D, Saint Priest, France

© Springer International Publishing AG 2017   279
R. Jabłoński and R. Szewczyk (eds.), *Recent Global Research and Education:
Technological Challenges*, Advances in Intelligent Systems and Computing 519,
DOI 10.1007/978-3-319-46490-9_38

effort to increase energy efficiency of vehicle operation. In particular, algorithms to find optimal routing with respect to energy consumption (called eco-routing, e.g. [1–4]) based on 3D map data are being developed. Also, there are eco-driving initiatives which provide drivers with guidelines for efficient behaviour and eco-driving systems studying strategies for efficient driving within traffic with varying congestion. Presented approach is using prior optimization of velocity profile and complements it by non-linear model based predictive controller which adapts required velocity profile according to the current vehicle and traffic state.

2 Methodology

The task is solved by following approach. Detailed range estimation simulation model was created and calibrated. It was used to validate fast optimization models for pre-trip velocity profile design and optimization. To maintain pre-optimized velocity, the non-linear model predictive controller (PMC) was designed, it's stability checked. The approach based on multi-criteria optimization was used to select PMC parameters in the best way.

2.1 Model Design

The Vehicle Energy Management (VEM) simulator is a LMS.IMAGINE.Lab Amesim 1D virtual model of the Fiat Doblo electric vehicle. This simulation platform is well suited to carry on global vehicle energy consumption evaluation [3, 4]. It is basically range prediction model, containing main subsystems: Vehicle dynamics 1D model with front and rear axles, model of an electric machine and DC converter using real measurement data-files, which define the losses, the minimum and the maximum torques, High voltage battery, modelled as a quasi-static equivalent electric circuit model, defined with the datasheet measure information, taking into account the OCV and resistance dependence as a function of SOC and temperature and simplified aging model, HVAC system is defined as a simplified reduced model, the vehicle auxiliary consumers i.e. all the vehicle electric equipment that belongs to the low voltage on board network of the vehicle (12 V) and finally vehicle control unit (VCU), which computes the motor torque demand according to vehicle state, the pedal position, battery SOC, inverter power limitation (high frequency control dynamics such as ESP have been neglected).

The LMS model has been modified to be integrated into the MATLAB/Simulink environment. A coupling block is added (Fig. 1) to define the quantities to be exchanged between LMS-Amesim and Simulink.

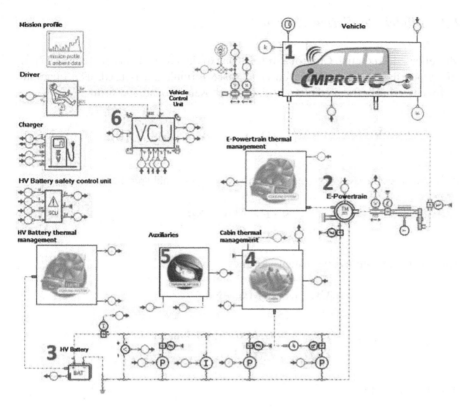

Fig. 1 LMS-Amesim vehicle energy management model

2.2 Trajectory Planning

Optimal velocity defined along the route can be optimized before the trip really starts [5]. It is based on suitable discretization of the route into sections with constant characteristic properties (maximum achievable velocity, slope, rolling resistance etc.). Velocity profile in each section is described by five parameters, which fully describe four phases of each section: Acceleration phase followed by constant velocity, then coasting and braking. Any piecewise velocity profile can be achieved, as the length of the phase can be of zero length.

The velocity profile is optimized using dynamic optimization [5, 6] and provides optimal strategy to achieve savings trading within the given—available travel time. However, the real drive always differs due to the surrounding traffic and vehicle model differences. That's why the feedback controller must be used to adapt the actual torque to achieve optimal behaviour.

2.3 Model Predictive Control

Model-based Predictive Control (MPC) is one of the most advanced methods [7]. In this paper, the approach based on state space formulation is used. The quadratic optimality criteria for prediction horizon is chosen in quadratic form

$$J_k = \sum_{j=1}^{N} \left(\left(\mathbf{y}_{k+j}^T - \mathbf{w}_{k+j}^T \right) \mathbf{Q}_y \left(\mathbf{y}_{k+j} - \mathbf{w}_{k+j} \right) + \mathbf{u}_{k+j-1}^T \mathbf{R}_y \mathbf{u}_{k+j-1} \right) \qquad (1)$$

If the linear controlled system is described in usual state-space form

$$\begin{aligned} \mathbf{x}_{k+1} &= \mathbf{M}\mathbf{x}_k + \mathbf{N}\mathbf{u}_k \\ \mathbf{y}_k &= \mathbf{C}\mathbf{x}_k \end{aligned} \qquad (2)$$

The outputs in future time instants follows

$$\mathbf{y}_{k+1} = \mathbf{C}\mathbf{x}_{k+1} = \mathbf{C}(\mathbf{M}\mathbf{x}_k + \mathbf{N}\mathbf{u}_k)$$

$$\cdots$$

$$\begin{aligned} \mathbf{y}_{k+N} = \mathbf{C}\mathbf{x}_{k+N} &= \mathbf{C}(\mathbf{M}\mathbf{x}_{k+N-1} + \mathbf{N}\mathbf{u}_{k+N-1}) \\ &= \mathbf{C}\mathbf{M}^N \mathbf{x}_k + \mathbf{C}\mathbf{M}^{N-1}\mathbf{N}\mathbf{u}_k + \cdots + \mathbf{C}\mathbf{N}\mathbf{u}_{k+N-1} \end{aligned} \qquad (3)$$

Introducing matrices

$$\mathbf{f} = \begin{bmatrix} \mathbf{C}\mathbf{M} \\ \mathbf{C}\mathbf{M}^2 \\ \cdots \\ \mathbf{C}\mathbf{M}^N \end{bmatrix} \mathbf{x}_k = \mathbf{f}_1 \mathbf{x}_k \quad \mathbf{G} = \begin{bmatrix} \mathbf{C}\mathbf{N} & \cdots & 0 \\ \vdots & \ddots & \vdots \\ \mathbf{C}\mathbf{M}^{N-1}\mathbf{N} & \cdots & \mathbf{C}\mathbf{N} \end{bmatrix} \quad \mathbf{U} = \begin{bmatrix} \mathbf{u}_k \\ \mathbf{u}_{k+1} \\ \cdots \\ \mathbf{u}_{k+N-1} \end{bmatrix} \qquad (4)$$

Writing outputs $\mathbf{y}_{k+i}$ and required outputs $\mathbf{w}_{k+i}$ over whole prediction horizon N in the matrix form

$$\mathbf{Y} = \begin{bmatrix} \mathbf{y}_{k+1} \\ \mathbf{y}_{k+2} \\ \cdots \\ \mathbf{y}_{k+N} \end{bmatrix} = \mathbf{f} + \mathbf{G}\mathbf{U}, \quad \mathbf{W} = \begin{bmatrix} \mathbf{w}_{k+1} \\ \mathbf{w}_{k+2} \\ \cdots \\ \mathbf{w}_{k+N} \end{bmatrix} \qquad (5)$$

The criteria J_k can be written in the form

$$J_k = (\mathbf{Y} - \mathbf{W})^T \mathbf{Q}(\mathbf{Y} - \mathbf{W}) + \mathbf{U}^T \mathbf{R}\mathbf{U} \qquad (6)$$

Then for minimum of J_k following holds as long as input u_k is unlimited:

$$\frac{\partial \mathbf{J}_k}{\partial \mathbf{U}^T} = 0 \tag{7}$$

And substituting from (4) and (5) and several treatments leads into control matrix for whole prediction horizon

$$\mathbf{U} = \left(\mathbf{G}^T\mathbf{Q}\mathbf{G} + \mathbf{R}\right)^{-1}\mathbf{G}^T\mathbf{Q}(\mathbf{W} - \mathbf{f}) \tag{8}$$

To choose only first control action at time instant k, we can, using zero and unit square matrices with size of number of inputs, form selection matrix

$$\begin{aligned}\mathbf{e_I} &= [\,\mathbf{I_i} \quad \mathbf{0_i} \quad \cdots \quad \mathbf{0_i}\,] \\ \mathbf{u_k} &= \mathbf{e_I}\mathbf{U}\end{aligned} \tag{9}$$

Considering (4), the control action can be written as

$$\mathbf{u_k} = \mathbf{e_I}\left(\mathbf{G}^T\mathbf{Q}\mathbf{G} + \mathbf{R}\right)^{-1}\mathbf{G}^T\mathbf{Q}(-\mathbf{f_1}\mathbf{x_k}) + \mathbf{e_I}\left(\mathbf{G}^T\mathbf{Q}\mathbf{G} + \mathbf{R}\right)^{-1}\mathbf{G}^T\mathbf{Q}\mathbf{W} \tag{10}$$

Introduction of closed loop gain matrix $\mathbf{KK} = \mathbf{e_I}\left(\mathbf{G}^T\mathbf{Q}\mathbf{G} + \mathbf{R}\right)^{-1}\mathbf{G}^T\mathbf{Q}\mathbf{f_1}$ and feedforward part $\mathbf{K_w} = \mathbf{e_I}\left(\mathbf{G}^T\mathbf{Q}\mathbf{G} + \mathbf{R}\right)^{-1}\mathbf{G}^T\mathbf{Q}\mathbf{W}$ yields control law, consisting of state feedback and feedthrough components

$$\mathbf{u_k} = -\mathbf{K}\mathbf{x_k} + \mathbf{K_w} \tag{11}$$

So stability of the controlled system can be easily checked by condition for closed loop pole position

$$|eig(\mathbf{M} - \mathbf{N}\mathbf{K})| < 1 \tag{12}$$

Unfortunately, Eq. (12) does not hold for MPC control with constraints, e.g. introduced in linear form

$$\mathbf{H}\mathbf{U} + \mathbf{L}\mathbf{Y} < b \tag{13}$$

In this case, (12) is only necessary stability condition, but not sufficient. The really used control action must be calculated by quadratic programming methods [7], which search for solution which minimize criteria (1) and conforms to constraints (13). There is no straightforward stability condition for such case. Approach based on numerically evaluated Lyapunov function (based on optimality criteria J_k) and its numerical time derivative can be used [8], if stability of the control is important issue. Suitably large subspace of the state space must be selected.

As the behaviour of the vehicle is quite non-linear, especially with respect to state of charge (SOC) and actual velocity, set of MPC controllers has been designed in the space of these two variables (Fig. 2) and combined using gain-scheduling approach.

Fig. 2 Control grid

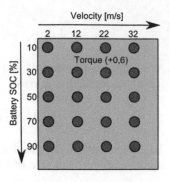

2.4 Multi-objective Optimization

MPC design algorithm finds optimal controller with respect to quadratic criteria (1). However, there are additional design parameters, like sampling period, length of prediction horizon, maximum controlled variable rate etc.

The ultimate objective of this control design is to decrease energy consumption. Pre-optimization of the vehicle profile [5] has shown that maximum time of travel affects maximum energy consumption. To determine best controller settings, multi-objective criteria optimization based on genetic algorithms was used with two conflicting criteria

$$J_1 = \int_0^t 1 dt \tag{14}$$

Fig. 3 Pareto set for conflicting criteria

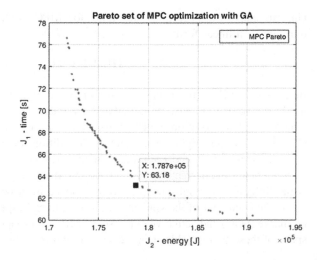

$$J_2 = E_{MPC} = \int_0^t P_{mot}dt = \int_0^t \omega_{mot}.M_{mot}dt \qquad (15)$$

The individual (controller) was tested on shortened route (500 m) for acceptable optimization time. The genetic algorithms used 150 individuals in 5 generations (Fig. 3).

The result of the optimization is organized into Pareto set, which shows possible combinations of J_1 and J_2 criteria.

3   Results

The resulting selected non-linear MPC controller was tested on reference route of the length 18 km. Reference control was a discrete PID controller, trying to follow pre-optimized velocity profile. The P, I and D components of the controller were tuned using the Ziegler-Nichols method (Fig. 4).

Fig. 4 Control behaviour for reference PID controller and MPC controller

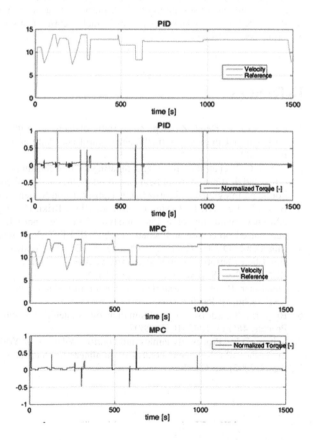

Fig. 5 Energy consumption
comparison

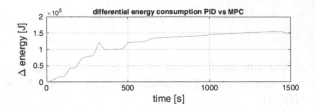

The PID controller exhibits high fluctuation of controlled variable (torque), even switching between acceleration and deceleration (braking) in subsequent control inputs. MPC controller in contrary exhibits quite smooth control trajectory without braking and the energy consumption is also improved (Fig. 5).

The Pareto set (Fig. 3) shows the influence of travel time on the energy savings. About 9 % of energy savings can be achieved if the travel time is extended for about 25 %. So the potential of the described control technology was demonstrated on realistic vehicle model.

Acknowledgments This research has been realized using the support of EU FP 7 Project No. 608756, Integration and Management of Performance and Road Efficiency of Electric Vehicle Electronics and using the support of The Ministry of Education, Youth and Sports program NPU I (LO), project # LO1311 Development of Vehicle Centre of Sustainable Mobility. This support is gratefully acknowledged.

References

1. Fu, M., Li, J., Deng, Z.: A practical route planning algorithm for vehicle navigation system. Fifth World Congress on Intelligent Control and Automation, 2004. WCICA 2004, vol. 6. IEEE (2004)
2. Minett, C.F., et al.: Eco-routing: comparing the fuel consumption of different routes between an origin and destination using field test speed profiles and synthetic speed profiles. In: 2011 IEEE Forum on Integrated and Sustainable Transportation System (FISTS), pp. 32–39. IEEE (2011)
3. Badin, F., Le Berr, F., Castel, G., Dabadie, J.C., Briki, H., Degeilh, P., Pasquier, M.: Energy efficiency evaluation of a Plug-in Hybrid Vehicle under European procedure, Worldwide harmonized procedure and actual use, EVS28 KINTEX, Korea, 3–6 May 2015
4. Maroteaux, D., Le Guen, D., Chauvelier, E.: Development of a fuel economy and CO_2 simulation platform for hybrid electric vehicles—Application to Renault EOLAB prototype. RENAULT SAS; SAE International (2015)
5. Steinbauer, P., et al.: Dynamic optimization of the e-vehicle route profile. No. 2016-01-0156. SAE Technical Paper (2016)
6. Biegler, L.T.: An overview of simultaneous strategies for dynamic optimization. Chem. Eng. Process. **46**(11), 1043–1053 (2007)
7. Seborg, D.E.: Process dynamics and control. Wiley, New York (2010)
8. Steinbauer, P.: Nonlinear control of nonlinear mechanical systems, CTU in Prague. Ph.D. thesis (2002) (in Czech)

Tilt Measurements in BMW Motorcycles

Sergiusz Łuczak

Abstract Motorcycles are subjected to various orientation changes, both while riding as well as while parked (including pitch as well as roll angles). Equipping them with a tilt sensor, based e.g. on a micromachined accelerometer or gyroscope, provides many interesting possibilities. On one hand, the driver can be informed of results of measurements of various important parameters, which are disturbed by a tilt of the motorcycle. On the other hand, operation of various motorcycle control systems can be considerably improved and enhanced owing to tilt measurements. The existing applications of the measurements (anti-theft alarms, new generation of ABS systems) as well as possible options are surveyed, discussed and briefly analyzed.

Keywords Tilt · MEMS · Accelerometer · Gyroscope · Motorcycle · BMW

Abbreviations

ABS	Anti-lock braking system
ABS Pro	Enhanced ABS [1]
ASC	Automatic stability control [2]
DDC	Dynamic damping control [1]
DTC	Dynamic traction control [1]
ESA	Electronic suspension adjustment [2]
ESC	Electronic stability control
IMU	Inertial measurement unit
MEMS	Microelectromechanical systems
MSC	Motorcycle stability control [3]
TCS	Traction control system
VDS	Vertical down sensor [1]

S. Łuczak (✉)
Faculty of Mechatronics, Warsaw University of Technology, Warsaw, Poland
e-mail: s.luczak@mchtr.pw.edu.pl

© Springer International Publishing AG 2017
R. Jabłoński and R. Szewczyk (eds.), *Recent Global Research and Education: Technological Challenges*, Advances in Intelligent Systems and Computing 519,
DOI 10.1007/978-3-319-46490-9_39

1 Introduction

Application of micromachined (MEMS) sensors, serving various kinds of purposes, has significantly intensified over the last years, with an ever-increasing tendency, what can be proved e.g. by a growing number of emerging applications [4]. One of them is detection of tilt—a typical type of measurement realized by means of accelerometers [5] and gyroscopes, minutely addressed in [6]. Some common applications of these MEMS sensors (photo cameras, cellular phones, autonomous robots and vehicles—presented e.g. in [7, 8]) are well known. However, new unexpected ideas for their use are constantly introduced, e.g. in safety systems of orthotic robots [9] or diving computers [10].

Another surprise is a strategic application of MEMS inertial sensors in motorcycles. As a leader in motorcycle production and development, the brand of BMW is a very good example of the considered innovations. Ever since the company of Robert Bosch GmbH introduced their MSC system (enhanced version of standard ABS and TCS) in late 2014 [3], BMW along with two other European motorcycle manufacturers: KTM and Ducati [3], have started to implement it in new models of their vehicles.

This innovation not only dramatically increased safety of the motorcycle drivers [3], but made it also possible to inform them about such important training data as roll of the motorcycle. As an example, a multifunction display of a BMW sport motorcycle is presented in Fig. 1. In a racing mode, it indicates the current value and direction of roll: (1), (2), as well as maximal values of left- and right-side rolls (3) [1].

Fig. 1 Multifunction display of BMW S 1000 RR motorcycle [1]

2 Tilt of a Motorcycle

Tilt of a motorcycle takes place when:

- the motorcycle is propped on the side stand—roll angle,
- the motorcycle is parked on a gradient (uphill or downhill)—pitch angle,
- the motorcycle rides along a bend and is leaned in order to maintain its balance —roll angle,
- the motorcycle rides forward and a "wheelie" or "stoppie" stunt is preformed (either front or rear wheel is elevated)—pitch angle,
- reasonable combination of the above—pitch and roll angles,
- motorcycle is dropped—roll angle.

Even though some of the aforementioned cases are exceptional and rare, the other are common. A general case of a motorcycle tilt is presented in Fig. 2; the x-y-x coordinate system is fixed to the motorcycle and the x_0-y_0-x_0 system is the reference.

The most accurate way to determine the component tilt angles (pitch and roll) by means of an accelerometer is to use the following formulas [6]:

$$\alpha = \arctan \frac{g_x}{\sqrt{g_y^2 + g_z^2}} \tag{1}$$

$$\gamma = \arctan \frac{g_y}{g_z} \tag{2}$$

where: α—pitch angle; γ—roll angle; $g_{x...z}$—components of the gravitational acceleration g in x-, y- and z-axis.

Fig. 2 Tilt of a motorcycle

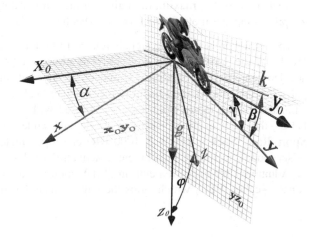

Having analyzed Eqs. (1) and (2) it can be noted that pitch is independent on roll (just as in the case of angle β being independent on pitch), whereas roll is influenced by pitch (just as angle φ is). The result is a variable sensitivity of determining the roll angle [10].

3 New Control Systems in Motorcycles

As aforementioned, a real breakthrough in the development of control systems for motorcycles took place when the Robert Bosch GmbH Corp. introduced to the market their MSC system, using an array of sensors to monitor the riding dynamics with frequency over 100 Hz—among them an IMU [3] (including MEMS accelerometer and gyroscope). The first version of the MSC employed measurements of the roll angle in order to reduce the braking force while riding along a bend.

The latest version of the MSC (9th generation), implemented in some of the new BMW motorcycles (released in 2015 and 2016), features the following functionalities, both of the ABS as well as TCS (ABS Pro and DTC in BMW motorcycles) [3]:

- distribution of the braking force between the front and rear wheel in such a way as to eliminate an unintentional change of the trajectory while braking along a bend—ABS Pro,
- rear-wheel lift-up control—ABS Pro,
- hill hold control function preventing the motorcycle from rolling back unintentionally—ABS Pro,
- the slope dependent control function detecting a danger of the rear wheel lifting up on a downhill—ABS Pro,
- integrated braking system activating always both front and rear brakes—ABS Pro,
- regulation of the maximum engine torque while riding along a bend—DTC,
- preventing the motorcycle form a "wheelie"—ASC, DTC.

Except for processing the signals related to motorcycle tilt (pitch and roll) from the IMU, the MSC system constantly monitors rotational speed of the wheels, acceleration, braking pressure [3] and other physical quantities (e.g. rate of yaw [1]). This data allows the system to recognize critical situations and intervene then. Character of the intervention can be adjusted with respect both to ABS Pro and DTC—the user can choose between pre-defined modes of their operation (e.g. Rain, Sport, Race, Slick in BMW S 1000 RR) or set parameters of the systems himself (User mode), including a complete deactivation of the systems.

Additional tilt measurements in BMW motorcycles are realized by a separate sensor—called VDS, which stops the engine once the motorcycle dropped [1].

Tilt sensors are also used for triggering anti-theft motorcycle alarms; since this idea has been employed for a long time, even before inexpensive MEMS accelerometers were commercially available, other kinds of sensors were also used for tilt detection (e.g. electrolytic, inductive).

4 Tilt-Compensated Measurements

Except for providing the driver with a direct indication of roll angle (see Fig. 2), there are other important data influenced by tilt. In order to indicate a reliable information on such data, a tilt measurement with accuracy of $1°–2°$ seems to be satisfactory. MEMS accelerometers perfectly fit such requirement (should a higher accuracy be ensured, an electrolytic tilt sensor can be additionally applied). Application of MEMS gyroscopes significantly improves tilt measurements under dynamic conditions. So, a whole integrated IMU (including accelerometer, gyroscope, magnetometer, pressure sensor and embedded data fusion algorithms) could be applied. Pressure sensor might be additionally used for analyzing the altitude of the motorcycle (especially of enduro-type, i.e. BMW GS series) in order to intensify supply of air at a low pressure.

Either employing an additional sensor for tilt measurements or using the existing ones (e.g. IMU of MSC or accelerometer of anti-theft alarm), the following new features could be easily obtained:

- enabling the on-board computer to display a correct information on fuel level and engine-oil level (see Fig. 3) even if the bike is tilted (pitch and roll); then such information on the fluid levels could be correct at any attitude of the motorcycle,
- warning the driver against an excessive roll or pitch while propping the motorcycle on the side stand (which may result in a drop of the motorcycle),
- warning (e.g. by an acoustic signal) the driver of a drop of the motorcycle while parked, reducing thus a risk of a fire due to inflammation of the spilled gasoline.

However, it must not be neglected that it is necessary to either align the sensitive axes of the sensors or compensate for the existing misalignments, otherwise large errors of tilt measurement can be the result, as proved in [11].

Fig. 3 Multifunction display of the on-board computer in BMW K 1300 R motorcycle indicating fuel level and checking engine oil level

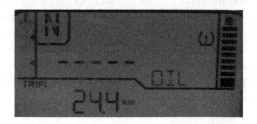

Another well-known interesting application of roll measurements in motorcycles is to display/record its value on a camera attached to the motorcycle frame (option often applied in the case of racing motorcycles).

5 New Functionalities

Application of a tilt sensor (choosing between the aforementioned options), would also make it possible to implement some new functionalities of the motorcycle, such as the following:

- in case of exceeding a critical value of roll, an actuator of a special side-stand could be activated in order to prevent the motorcycle from dropping (both while riding and while stopped),
- when the motorcycle is under a considerable pitch ("wheelie" stunt or attacking a steep incline—GS series motorcycles) the engine lubrication could be intensified.

It seems that such functionalities are very useful as they protect the motorcycle from a damage of the bodywork or the engine, which result in costly repairs.

6 Summary

Application of MEMS inertial sensors in automotive industry has become strictly connected with the safety of human life (activation of airbags, ABS, TCS, ESC). Even though range of applications of these sensors constantly widens in this field, new original ideas have been proposed.

Since other kinds of sensors (position sensors, pressure or force sensors) are also used for tilt measurements, e.g. in ASC system that is capable of detecting a case when the front wheel losses contact with the ground [2], a data fusion of the output signals from various sensors is a very important issue. It seems that enhancing this fusion with recognition of driver's intentions will improve operation of such crucial systems as ABS Pro and DTC.

Having employed such recognition, a new quality standard could be achieved, taking into consideration a rider's riding habits learned at the initial phase of using the motorcycle. For instance, certain driver actions (e.g. a quick double activation of clutch lever sensor) could activate/deactivate DTC or ABS Pro.

It is also worthwhile mentioning that new generation of motorcycles (especially manufactured by BMW) feature not only enhanced ABS and TCS, but other electronically controlled systems such as DDC (dynamic change of damping properties of the suspensions) or ESA (pre-set spring rate of the suspensions).

Acknowledgments The author would like to thank graduates of the Faculty of Mechatronics, Warsaw University of Technology, for creating the high-quality 3D graphics presented in Fig. 2: Mr. Wojciech Załuski, M.Sc. Eng. (employee of the Faculty) and Mr. Grzegorz Ekwiński, M.Sc. Eng.

References

1. Rider's Manual S 1000 RR, BMW Motorrad, Germany (2015)
2. Rider's Manual K 1300 R, BMW Motorrad, Germany (2012)
3. Active Safety Systems. Motorcycle Stability Control (MSC) Enhanced. Robert Bosch GmbH, Germany (2015)
4. Kaajakari, V.: Practical MEMS. Small Gear Publishing, Las Vegas (2009)
5. Wilson, J.S.: Sensor Technology Handbook. Newnes, Burlington (2005)
6. Łuczak, S.: Guidelines for tilt measurements realized by MEMS accelerometers. Int. J. Precis. Eng. Manuf. **15**, 2011–2019 (2014)
7. Bodnicki, M., Kamiński, D.: In-pipe microrobot driven by SMA elements. In: Březina, T., Jabloński, R. (eds.) Mechatronics 2013. Recent Technological and Scientific Advances, pp. 527–533. Springer International Publishing (2014)
8. Osiński, D., Szykiedans, K.: Small remotely operated screw-propelled vehicle. In: Szewczyk, R., Zieliński, C., Kaliczyńska, M. (eds.) Progress in Automation, Robotics and Measuring Techniques, vol. 2: Robotics, pp. 191–200. Springer International Publishing (2015)
9. Bagiński, K., Jasińska-Choromańska, D., Wierciak, J.: Modelling and simulation of a system for verticalization and aiding the motion of individuals suffering from paresis of the lower limbs. Bull. Pol. Ac.: Tech. **61**, 919–928 (2013)
10. Łuczak S., Grepl R., Bodnicki M.: Selection of MEMS Accelerometers for Tilt Measurements, SpringerPlus 6 (to be published)
11. Łuczak, S.: Effects of misalignments of MEMS accelerometers in tilt measurements. In: Březina, T., Jabloński, R. (eds.) Mechatronics 2013. Recent Technological and Scientific Advances, pp. 393–400. Springer International Publishing (2014)

Novel Measurement Method of Longitudinal Wave Velocity of Liquid Using a Surface Acoustic Wave Device

Jun Kondoh and Michiyuki Yamada

Abstract A longitudinal wave is radiated from a surface acoustic wave (SAW) at the solid/liquid interface and non-linear phenomena are observed, such as acoustic streaming. When a short burst signal was applied to the SAW device, the reflected signal from a droplet on it. Experimental results show that the observed time depends on a droplet volume. This means that the wave does not reflect at the air/droplet interface. Propagation model of the longitudinal wave in the droplet was assumed and the equation was derived using it. The experimental results of the reflex time agree with the calculated results using the equation. As the longitudinal wave velocity is involved in the equation, the velocity is obtained by measuring the arrival time.

Keywords SAW · Longitudinal wave radiation · SAW streaming · Nonlinear phenomena · Longitudinal wave velocity measurement · Droplet

1 Introduction

A surface acoustic wave (SAW) device is one of the important electronic devices in a mobile communication system, such as smartphones [1]. The SAW device is fabricated on a piezoelectric substrate surface and is generated and received by an interdigital transducer (IDT). When a liquid is loaded onto the SAW propagation surface, the SAW becomes a leaky-SAW and the longitudinal wave is radiated into the liquid (see Fig. 1). Nonlinear phenomena, such as droplet manipulation, are caused by the longitudinal wave radiated [2]. In the liquid, an acoustic streaming is

J. Kondoh (✉)
Graduate School of Science and Technology, 3-5-1 Johoku, Naka-ku,
Hamamatsu, Shizuoka 432-8561, Japan
e-mail: kondoh.jun@shizuoka.ac.jp

J. Kondoh · M. Yamada
Faculty of Engineering, Shizuoka University, 3-5-1 Johoku, Naka-ku,
Hamamatsu, Shizuoka 432-8561, Japan

© Springer International Publishing AG 2017
R. Jabłoński and R. Szewczyk (eds.), *Recent Global Research and Education:
Technological Challenges*, Advances in Intelligent Systems and Computing 519,
DOI 10.1007/978-3-319-46490-9_40

observed due to the longitudinal wave radiation from the SAW, the phenomena are called the SAW streaming. Many applications based on the SAW streaming are proposed [3–5]. Also, we proposed a digital microfluidic system by integrating a sensor on to the droplet manipulation surface [6]. It is important to know the droplet position. Several papers are published to determine the droplet position by the SAW [7, 8]. For example, Renaudin et al. proposed to use reflected signals from the droplet [8]. However, they do not clarify the mechanism of the reason why the reflected wave is observed. In this paper, we assumed the propagation path of the longitudinal wave in the droplet and then the equation was derived from the model. The experimental results were compared with the equation derived and good agreement was obtained.

2 SAW Streaming

The SAW is generated from an interdigital transducer (IDT), which is fabricated on a piezoelectric substrate. When a liquid is placed onto the SAW propagating surface, the longitudinal wave is radiated into liquid. Figure 1 schematically shows the radiation of the longitudinal wave. Radiation angle, θ, is called the Rayleigh angle and defined as follows.

$$\theta = \sin^{-1}\left(\frac{V_L}{V_{SAW}}\right) \tag{1}$$

where V_L and V_{SAW} are the longitudinal wave velocity of the liquid and the phase velocity of the SAW, respectively. Visualization result of the longitudinal wave from the SAW is shown in Fig. 2. The SAW device was placed on the bottom of the water vessel. As the SAW propagates to both the left and right sides from the IDT, two longitudinal waves are observed. In Fig. 3, droplet manipulation, droplet ejection, and jetting phenomena are summarized. The amplitude of the SAW depends on the applied electrical power. The phenomena in the figure can be selected by tuning the electrical power applied. When a thin water layer is formed on the surface, mists are generated from it. This is called an atomization by the SAW [4].

Fig. 1 Schematic illustration of the longitudinal wave radiation from the SAW

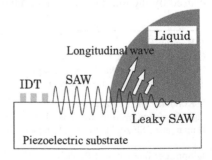

3 Measurement Setup

The SAW device was fabricated on the 128° rotated Y-cut X-propagating LiNbO$_3$. The center frequency of the SAW device used was 50 MHz. The experimental setup is shown in Fig. 4. The short burst wave was applied to the IDT on the SAW device and the reflected signal was observed by using a conventional oscilloscope.

Fig. 2 Observation result of the longitudinal wave from the SAW device

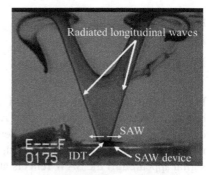

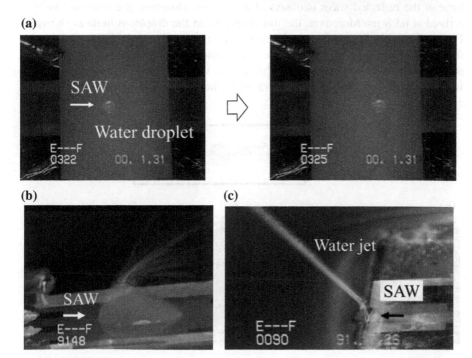

Fig. 3 Nonlinear phenomena, SAW streaming, caused by the longitudinal wave radiated from the SAW. **a** Droplet manipulation, **b** droplet ejection, and **c** pump

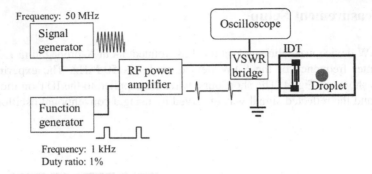

Fig. 4 Measurement setup in this study

4   Results and Discussions

Figure 5 shows device structure used and the results observed without or with water droplet on the device surface. The reflected waves from the edges of the SAW device are observed in Fig. 5b. When the droplet was loaded on the surface, the reflected waves from the edges of the device were vanished and a new reflected signal appeared. If the SAW reflects at the air/water inter face, the estimated arrival time of the reflected wave is observed at 3.71 μs. However, the reflected wave is arrived at 6.98 μs. Moreover, the time depends on the droplet volume as shown in Fig. 6. The dots in the figure represents the measured results. In the figure, the distances between the IDT and droplet were varied. The difference of the distance is the SAW propagation time. From the results, we concluded that the longitudinal wave radiated from the leaky-SAW travels in the droplet and return to the IDT. The

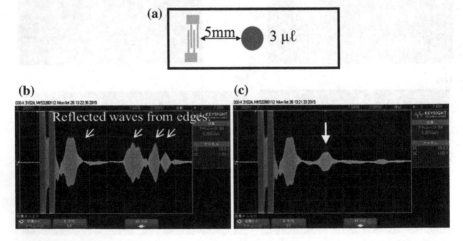

Fig. 5 a Illustration of the SAW device with water droplet, b, c observation results by the oscilloscope without or with water droplet

Fig. 6 Experimental and
calculated results of the
arrival time of the reflected
wave as a function of droplet
volume

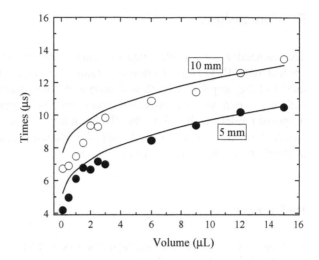

model of the propagation path in the droplet was assumed as shown in Fig. 7. This
model was based on the observation results in the droplet [9]. Based on the model,
the simple equation was derived as follows.

$$t = \frac{2L}{V_{SAW}} + \frac{\left(1 + \frac{\pi}{2}\right)d}{V_L} \tag{2}$$

Here, t is arrival time, V_{SAW} and V_L are the phase velocity of the SAW and the
longitudinal wave velocity of the liquid, respectively, and L and d are shown in
Fig. 7. The calculated results are also plotted in Fig. 6. The calculated results, solid
lines, agree with the experimental results. We also measured the glycerol-water
binary mixture and obtained reasonable results. Therefore, the results indicate that
the droplet position, L, is estimated by measuring the arrival time of the reflected
wave. When the distance, L, and the droplet diameter, d, are known, a longitudinal
wave velocity, V_L, will be estimated. Therefore, this method can be applied for a
novel measurement method of a liquid longitudinal wave using the SAW device.

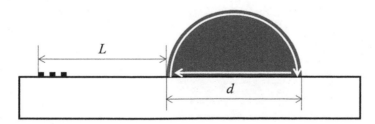

Fig. 7 Propagation path model of the longitudinal wave in the droplet

5 Conclusions

The reflected wave was observed from the droplet when the short burst wave was applied to the IDT. The detection mechanism was experimentally discussed and the model of the propagation path was proposed. The equation derived was compared with the experimental results and good agreement was obtained. However, the proposed model was based on the 2D. Therefore, it is necessary to consider the 3D model. Also, we must visualize the actual propagation path of the longitudinal wave in the droplet. These are our future works.

References

1. Campbell, C.K.: Surface Acoustic Wave Devices for Mobile and Wireless Communications. Academic Press, London (1998)
2. Shiokawa, S., Matsui, Y., Moriizumi, T.: Experimental study on liquid streaming by SAW. Jpn. J. Appl. Phys. **28**(Suppl. 28-1), 126–128 (1989)
3. Shiokawa, S., Matsui, Y.: The dynamics of SAW streaming and its application to fluid devices. Mater. Res. Soc. Symp. Proc. **360**, 53–64 (1995)
4. Chono, K., Shimizu, N., Matsui, Y., Kondoh, J., Shiokawa, S.: Development of novel atomization system based on SAW streaming. Jpn. J. Appl. Phys. **43**(5B), 2987–2991 (2004)
5. Kondoh, J., Shimizu, N., Matsui, Y., Sugimoto, M., Shiokawa, S.: Development of temperature-control system for liquid droplet using surface acoustic wave device. Sens. Actuators A **149**, 292–297 (2009)
6. Yasuda, N., Sugimoto, M., Kondoh, J.: Novel micro-laboratory on piezoelectric crystal. Jpn. J. Appl. Phys. **48**, 07GG14 (2009)
7. Bennes, J., Alzuaga, S., Cherioux, F., Ballandras, S., Vairac, P., Manceau, J.F., Bastien, F.: Detection and high-precision positioning of liquid droplets using SAW systems. IEEE Trans. UFFC **54**, 2146–2151 (2007)
8. Renaudin, A., Sozanski, J.P., Verbeke, B., Zhang, V., Tabourier, P., Druon, C.: Monitoring SAW-actuated microdroplets in view of biological applications. Sens. Actuators B **138**, 374–382 (2009)
9. Kondoh, J., Sato, M., Miyata, J., Sugimoto, M., Matsui, Y.: Experimental study on SAW streaming phenomenon. In: Proceedings of Piezoelectric Materials & Devices Symposium, pp. 5–8 (2008) (in Japanese)

The Effective Method to Search the Optimal Experimental Conditions in a Micro Flow Reactor

M. Abe and K. Takeda

Abstract A batch reactor has been used for organic synthesis and pharmaceutical synthesis. But reaction intermediates and impurities have to be analyzed and this analysis requires a lot of time and costs. The micro flow reactor is expected to replace a batch reactor in organic synthesis. The micro flow reactor has the advantages such as significant reduction in the reaction time and energy consumption, Improving the yield and on-line measurement because it is capable of quickly and uniformly heating by uses of micro wave. However, we need to conduct a lot of experiments and find optimal conditions to obtain the high yield in this reactor. Therefore, the aim of this study is to develop the effective method to search the optimal experimental conditions in the micro flow reactor. The approximated curved surface for the experimental data was generated by experimental data such as flow rate, temperature and yield. The approximated curved surface was optimized by least squares method. This procedure was repeated until the difference between the measurement data and the analyzed data were within one percent. The data which was larger than the average was used in order to improve the approximation precision within the range of the high yield. The proposed method was able to search the optimal experimental conditions at one-third fewer runs than ordinary methods.

Keywords Micro flow reactor · Optimal experimental condition · Search method

1 Introduction

A Batch reactor is used for many organic synthesis but that requires a lot of time and costs. Because products have to be analyzed reaction. Micro flow reactor is focused on to take the place of a batch reactor since that can be measured on-line. Figure 1 shows schematic of a micro flow reactor.

M. Abe (✉) · K. Takeda
Graduate School of Engineering, Shizuoka University, Shizuoka, Japan
e-mail: abechan312@yahoo.co.jp

© Springer International Publishing AG 2017
R. Jabłoński and R. Szewczyk (eds.), *Recent Global Research and Education: Technological Challenges*, Advances in Intelligent Systems and Computing 519,
DOI 10.1007/978-3-319-46490-9_41

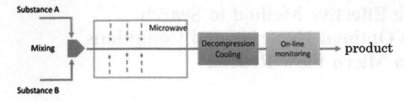

Fig. 1 Schematic of micro flow reactor

Substance A and B are mixed at small channel in this reactor. The Micro flow reactor enables to rapidly heat solvents and generate more uniform heating profile, which significantly reduces reaction times and, in many cases, increases yields and saves energy. Junichi Yosihda proposed that a micro flow reactor is able to synthesis safely and effectively [1]. Minjing Shanga proposed that a micro flow reactor is able to get high yield product when experimental condition is set at optimal value [2]. However, the micro flow reactor obtain high yield, many experimental conditions should be strictly set at optimal value and many experiments is necessary to search for experimental conditions. And optimal experimental condition is done in a non-systematic way based upon previous laboratory experiences. Therefore our objective is to develop the effective method to search the optimal experimental conditions.

2 Method

This study used experimental data of flow rate, temperature and yield. These data were used to search the optimal condition. We show below a procedure.

1. Set the first experimental condition.
   Initial experimental conditions have to prevent miss search when search the next experimental condition. Therefore initial experimental conditions cover wide space.
2. Execute at set experimental condition.
   First experiments were executed at initial experimental conditions. The other case, next experiments set at step 4 were executed.
3. Select data.
   For better approximation accuracy at high yield region, above the average data were selected.
4. Search the optimal experimental condition.
   The approximated curved surface was made from high yield data and the maximum yield data was searched from the approximated curved surface. The maximum yield condition was next experimental condition.
5. Terminate if the convergence condition was satisfied.

If the convergence condition was satisfied, search method is terminated, else return to step 2.

3 Results and Discussion

Figure 2 shows an object reaction and Table 1 shows used data.

An Approximated curved surface is made from initial experimental data. Figure 3 shows an initial approximated curved surface.

In the axis of the approximated curved surface of Fig. 3 are flow rate, temperature and yield. Experimental data is blue dot. The maximum yield data is red dot which is next experimental condition. Next experimental condition was flow rate 0.1 mL/min temperature 250 °C. The experimental data above the average are selected because approximation accuracy improved of high yield. Figure 4 shows an approximated curved surface which was made after selected data.

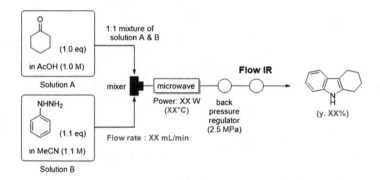

Fig. 2 Object reaction

Table 1 First experimental data

Residence time (s)	GC yield (%) and exit temperature (°C) 上段: 実験データ, 下段: 論文データ				
	Irradiation power (W)				
	70	90	110	130	150
60	97, 243 (*n.d.*)	–	–	–	–
40	54,209 (*n.d.*)	89,235 (*n.d.*)	–	–	–
30	18,175 (*n.d.*)	42,206 (*n.d.*)	70,229 (24, 179)	83,240 (52, 191)	–
24	44,148 (*n.d.*)	63,176 (*n.d.*)	35, 207 (27,188)	89,222 (54,205)	90, 240 (83,232)

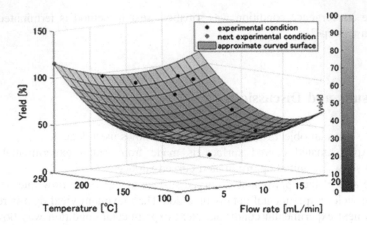

Fig. 3 Initial approximated curved surface

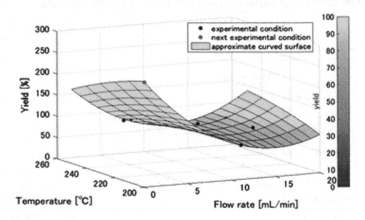

Fig. 4 Approximated curved surface of high yield

The next experimental condition of Fig. 4 is flow rate 0.1 mL/min temperature 200 °C. But this condition flow rate is very slow. So even if micro wave is very weak, temperature exceeds 200 °C. Thus next experimental condition is searched from all experimental data. Next experimental condition was flow rate 0.1 mL/min and temperature 250 °C.

Approximate curved surface was made after next experiment data was added. Table 2 shows an experimental date after step 4. The bottom data of Table 2 was searched from Fig. 3. Second approximate curved surface was made from Table 2. Figure 5 shows a second approximate curved surface.

The next experimental condition of Fig. 5 is flow rate 20 mL/min temperature 250 °C. As a result of having repeated the search method. Figure 6 shows a final approximate curved surface.

Table 2 Experimental date after step 4

Flow rate (mL/min)	Temperature (°C)	Yield (%)
12.8	175	18
16	207	35
12.8	206	42
16	148	44
9.6	209	54
16	176	63
12.8	229	70
12.8	240	83
9.6	235	89
16	222	89
16	240	90
6.4	247	97
0.1	250	96

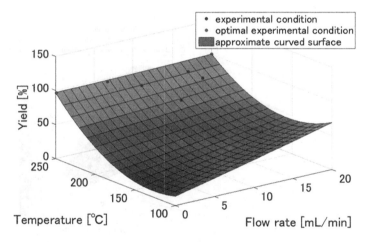

Fig. 5 Second approximate curved surface

The blue arrow is the optimal experimental condition and the red arrow is searched optimal experimental condition from final approximate curved surface. The blue arrow is at flow rate 6.0 mL/min temperature 250 °C. The red arrow is at flow rate 6.4 mL/min temperature 248 °C. These two experimental conditions almost accorded. Thus search method was finished. The number of experiment were made a comparison between article [3] and search method. Figure 7 shows an approximate curved surface of article. The next experimental condition of Fig. 7 is flow rate 1.0 mL/min temperature 162 °C predictive yield 76.4 %. Figure 8 shows an approximate curved surface of searched method.

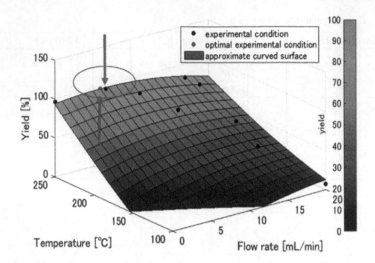

Fig. 6 Final approximate curved surface

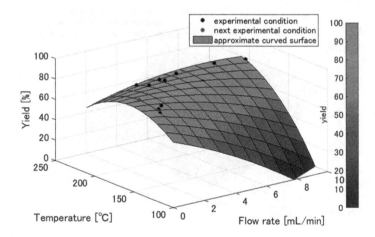

Fig. 7 Approximate curved surface of article

The next experimental condition of Fig. 8 is flow rate 1.0 mL/min temperature 155 °C predictive yield 74.9 %. Figure 7 of optimal experimental condition almost accorded with Fig. 8 of optimal experimental condition. Figure 7 is made from 24 experiment data. Figure 8 is made from 16 experiment data. There for search method be able to search the optimal experimental conditions at fewer runs than article data.

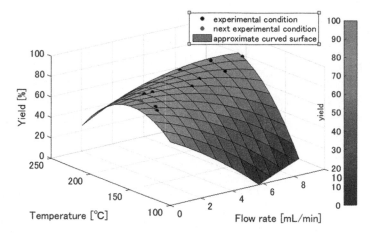

Fig. 8 Approximate curved surface of searched method

4 Conclusion

To search the optimal conditions, an approximated curved surface was generated by the experimental data such as flow rate, temperature and yield. The approximated curved surface was generated by quadratic approximation to decrease necessary data. This approximated curved surface couldn't search within the range of the high yield because using data include in low yield data. Therefore, the data which were larger than the average were used in order to improve the approximation precision within the range of the high yield. Until the convergence condition is not satisfied, we add the next experimental data and searched the optimal conditions again. If convergence condition is met, we finished the searching. In addition, we compared the proposed method with the treatise for each the necessary number of experiments. Consequently the proposed method was able to search the optimal experimental conditions at one-third fewer runs than reference methods.

References

1. Yosihida, J.-I., Nagaki, A., Yamada, D.: Continuous Flow Synthesis. Drug Discovery Today: Technologies Spring, vol. 10, pp. 53–59 (2013)
2. Shang, M., Noël, T., Wang, Q., Su, Y., Miyabayashi, K., Hessel, V., Hasebe, S.: 2- and 3-Stage temperature ramping for the direct synthesis of adipic acid in micro-flow packed-bed reactors. Chem. Eng. J. **260**(15), 454–462 (2015)
3. Satori, Y., Noriyuki, O., Tadashi, O., Hiromichi, O.: Development of a high efficient single-mode microwave applicator with a resonant cavity and its application to continuous flow syntheses. RSC Adv. **5**, 10204–10210 (2015)

Displacement Field Estimation for Echocardiography Strain Imaging Using B-Spline Based Elastic Image Registration—Synthetic Data Study

Aleksandra Wilczewska, Szymon Cygan and Jakub Żmigrodzki

Abstract Strain imaging in echocardiography allows identification of myocardium dysfunctions. This paper describes the use of own implementation of elastic image registration, to calculate displacement field in two-dimensional echocardiographic data. Performance of the algorithm was examined on synthetic ultrasonic data. A series of tests examining the influence of algorithm parameters on the outcome was conducted. Displacement fields were compared with reference data from the finite element model used for generation of the synthetic data. Quality of image registration was evaluated using two error measures: mean absolute error and median absolute error. The displacement field errors obtained in the direction transverse to the ultrasound wave had an order of magnitude 10^{-5} m and errors in the direction along ultrasound wave: 10^{-6} m, which is close to the accuracy of two state-of-the-art methods for displacements estimation tested on the same input data.

Keywords Strain imaging · Elastic image registration

1 Introduction

Cardiological problems are one of the most common death causes and their proper diagnosis is an important challenge faced by modern medicine [1]. When it comes to assessment of myocardial mechanical function there is a number of commonly used global parameters, such as stroke volume and ejection fraction that provide useful information [2]. However, there are diseases, that cause only local changes in

A. Wilczewska (✉) · S. Cygan · J. Żmigrodzki
Institute of Metrology and Biomedical Engineering, Warsaw University
of Technology, Warsaw, Poland
e-mail: a.wilczewska1@gmail.com

S. Cygan
e-mail: s.cygan@mchtr.pw.edu.pl

J. Żmigrodzki
e-mail: j.zmigrodzki@mchtr.pw.edu.pl

© Springer International Publishing AG 2017
R. Jabłoński and R. Szewczyk (eds.), *Recent Global Research and Education:*
Technological Challenges, Advances in Intelligent Systems and Computing 519,
DOI 10.1007/978-3-319-46490-9_42

309

myocardial function, which makes diagnosis based on global parameters much harder [3, 4]. Regional heart muscle function evaluation enables better analysis of acute myocardial infarction consequences [3], and estimation of postinfarction scar size [5]. Such evaluation can be achieved by calculating displacement field and strain in the tissue, based on images of the heart, acquired using inter alia ultrasonography or MR imaging. This paper describes an implementation of elastic image registration algorithm and its application to calculate displacement field in echocardiographic data.

2 Materials and Methods

2.1 Data

In this study a synthetic ultrasonic data set was used. The data was generated as described in the work of Żmigrodzki et al. [6].

A phantom of the left ventricle was modeled as a solid, homogenous ellipsoid with a wall thickness of 15 mm in relaxed state, corresponding to the end-systolic phase of cardiac cycle. Numerical model of the geometry, created using the Autodesk Inventor 2012 software (Autodesk Inc., USA) was exported in CAD format to the Finite Element software—Abaqus 6.13-3 (Simulia, USA). The phantom was meshed into linear hexahedral elements (C3D8H). Phantom material was modeled as hyperelastic with 3-rd order Yeoh definition [7]. Deformation was forced by applying pressure load to the inner surface of the phantom. Shape of the pressure curve resembled the physiological volume change of the left ventricle with magnitude of 36 kPa. Sets of deformed meshes, exported to Matlab environment (Mathworks USA) were used to generate synthetic ultrasonic data in the long axis view LAX (Fig. 1) using the FIELD II package [8]. A sequence of simulated echocardiographic data covering the whole deformation cycle was created. For further analysis a pair of images corresponding to maximum deformation rate was used. Underlying meshes provided a reference data for evaluation of the proposed algorithm.

2.2 Algorithm

The implemented method is an iterative algorithm, matching two images by manipulating a grid of control points superimposed on one of them. It consists of four parts: image transformation, using B-splines; image interpolation; calculating cost function and optimization strategy [9].

Image transformation. Image transformation used in the implemented algorithm is based on cubic B-splines, commonly used in elastic image registration, which are

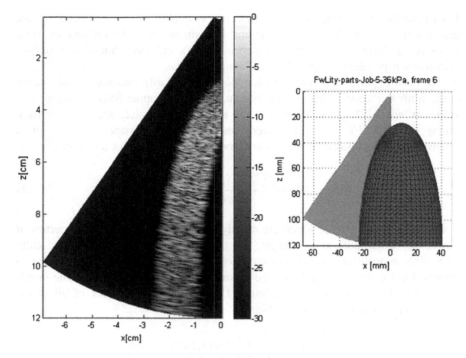

Fig. 1 Synthetic data in long axis view of the phantom model (*left*) and the 3D mesh with the scanning sector marked in *red* (*right*)

a basis of all continuous (n − 1)—differentiable polynomial functions—all such functions can be described as a proper sum of B-splines [10]. One of their main assets is their limited span, which assures local impact of control points displacement, leading to shorter computation time and easier application of optimization strategy [9].

Image transformation requires a grid of n × m evenly spaced control points. Image pixels displacement caused by movement of grid knots can be calculated using B-spline tensor product [11], given by Eq. (1).

$$T(x, y) = \sum_{m=0}^{3} \sum_{n=0}^{3} B_m(u) B_n(v) \emptyset_{i+m,j+n} \tag{1}$$

where $T(x, y)$—pixel location after transformation; $\emptyset_{i+m,j+n}$—grid control point location indexed $i + m, j + n$; $B_m(u)$—mth basis cubic B-spline function [11].

Interpolation. Image interpolation is essential for proper presentation of a transformed image and thus for calculating cost function value. In the work, due to its accuracy, bicubic interpolation was used [12].

Cost function. Cost function is a way to assess quality of image matching. Cost function used in the described algorithm, is a simple sum of similarity measure Q (Sum of Squared Differences criterion—SSD) and smoothness and surface conservation penalties Ps and Pv, with weights ω_s, ω_v.

The most important part of a cost function is a similarity measure. As all images come from the same modality, the simple, easy to compute SSD between pixel values criterion [9], given by the Eq. (2) is a suitable choice, where i, j—image points indices, $p1(i, j)$—pixels intensities in reference image, $p2(i, j)$—pixel intensities in transformed image, $ixmax, iymax$—image size in x and y directions.

$$SSD = \frac{\sum_{i=1}^{ixmax} \sum_{j=1}^{iymax} (p1(i,j) - p2(i,j))^2}{ixmax \cdot iymax} \qquad (2)$$

Penalties are functions, which are strictly connected with physical properties of the tissue. Their goal is to restrain the transformation and prevent physically impossible movement. Surface conservation penalty, given by Eq. (3) is used to restrain local extension and shrinkage of small tissue segments, impossible in relatively small deformation of the tissue [13]. Where: N_D—number of all image pixels, $J_T(p)$—transformation jacobian.

$$P_v = \frac{1}{N_D} \sum_{x \in D} |\log(J_T(p))| \qquad (3)$$

Smoothness penalty assures smoothness of the transformation and is based on thin metal sheet bending energy definition [13], as given by the Eq. (4).

$$P_s = \int\limits_x \int\limits_y \left(\frac{\partial^2 T}{\partial x^2}\right)^2 + \left(\frac{\partial^2 T}{\partial y^2}\right)^2 + 2 \cdot \left(\frac{\partial^2 T}{\partial x \partial y}\right)^2 dx \qquad (4)$$

Optimization strategy. Due to relatively short computing time, accuracy and low memory use, a quasi-Newton L-BFGS (Low Memory Broyden-Fletcher-Goldfarb-Shanno) method was used. It is an iterative method using cost function value and gradient to find its minimum and thus the best image match [14]. This optimization strategy is particularly useful while working on multivariable functions [14].

2.3 Competitive Methods

Accuracy of obtained displacements estimation results was compared against two methods. A block matching speckle tracking method [6] and another spline based non-rigid image registration method [11] were used.

3 Results

Tests of the algorithm were conducted on two neighboring images from the data set described in previous point. The main goal of the test was to evaluate the impact of four factors: grid size, weights of penalties and displacement used for calculating cost function gradient, on achieved results. Obtained displacement fields were compared with reference displacement fields both visually and using quantitative criterions: mean absolute error and median absolute error. Table 1 shows, how changes of parameters affected mean absolute error, and median absolute error in x (transverse to ultrasound beam) and y (along ultrasound beam) directions (while one of the parameters was tested, others maintained constant values).

4 Discussion

In all tests, regardless of parameters' values, errors in the direction transverse to the ultrasound beam had the order of magnitude of 10^{-5} m and in direction along the ultrasound beam—10^{-6} m.

Test results showed, that grid size has an impact on the accuracy in both directions. Among grid sizes under assessment, 5×20 grid offers best results, both in x and y directions. Although increasing control points number leads to regularization of transformation it does not improve image match quality.

Control point movement used to calculate the cost function gradient has significant impact on registration results. Step of 0.01 (in grid coordinates) offered the best results. Larger step made the transformation too irregular, creating small areas,

Table 1 Obtained error values

Factor	Factor value	Mean absolute error y (m)	Median absolute error y (m)	Mean absolute error x (m)	Median absolute error x (m)
Displacement	0.001	4.65E-06	3.69E-06	1.62E-05	1.64E-05
Displacement	0.04	4.08E-06	2.95E-06	1.63E-05	1.69E-05
Displacement	0.1	5.33E-06	4.57E-06	2.06E-05	1.63E-05
Smoothness	0.005	4.24E-06	3.40E-06	1.54E-05	1.27E-05
Smoothness	0.05	4.54E-06	3.53E-06	1.74E-05	1.57E-05
Smoothness	0.3	5.68E-06	4.59E-06	1.66E-05	1.67E-05
Incompressibility	0.00005	4.50E-06	3.84E-06	1.73E-05	1.46E-05
Incompressibility	0.0005	4.04E-06	3.13E-06	1.69E-05	1.47E-05
Incompressibility	0.005	5.22E-06	4.53E-06	2.57E-05	2.22E-05
Grid size	5×20	4.23E-06	3.31E-06	1.48E-05	1.50E-05
Grid size	8×32	4.56E-06	3.94E-06	1.38E-05	1.51E-05
Grid size	10×10	3.52E-06	2.31E-06	2.15E-05	2.20E-05

where error values were particularly high. On the other hand decreasing the step caused general drop in match quality, caused by excessive displacement of large image parts.

Smoothness penalty, as expected, affects smoothness of transformation. Its increase results in improved algorithm behavior on image edges. It also causes decrease of local displacement differences, which may have negative consequences, in attempts of detecting tissue regions with changed compressibility. Compressibility constraint weight value has an impact on displacement scale in the entire image. If the value is too high, it results in excessive increase in transformation stiffness and makes effective detection of displacement impossible. In this particular case the most suitable weight value was 0.0005.

Accuracy in the best achieved case reached 15 μm in axial and 30 μm in lateral direction while for the two reference methods, the block matching accuracy reached 6/23.3 μm and spline based registration accuracy and 9/23.6 μm respectively.

5 Conclusions

The implemented algorithm was used to calculate displacement fields in synthetic echocardiography data of a left ventricle phantom. Obtained results were compared with reference displacement fields, which helped assess method's effectiveness.

Registration errors were caused inter alia by the fact that speckle patterns in echocardiography, despite being directly caused by interaction of ultrasounds with tissue, do not reflect its structure precisely. Speckles "slide" over the material, and therefore their movement is not unequivocal with movement of the tissue itself. Additionally deformations of the myocardium are three dimensional, which makes their full, accurate analysis using 2D data impossible. Further development of the algorithm should include computation time optimization, tests on heterogeneous data and clinical data analysis.

References

1. Gillum, R.F.: Epidemiology of heart failure in the United States. Am. Heart J. **126**(4), 1042–1047 (1993)
2. Folland, E.D., Parisi, A.F., Moynihan, P.F., Jones, D.R., Feldman, C.L., Tow, D.E.: Assessment of left ventricular ejection fraction and volumes by real-time, two-dimensional echocardiography. A comparison of cineangiographic and radionuclide techniques. Circulation **60**(4), 760–766 (1979)
3. Møller, J.E., Hillis, G.S., Oh, J.K., Reeder, G.S., Gersh, B.J., Pellikka, P.A.: Wall motion score index and ejection fraction for risk stratification after acute myocardial infarction. Am. Heart J. **151**(2), 419–425 (2006)
4. Heyde, B., Cygan, S., Choi, Hon Fai, Lesniak-Plewinska, B., Barbosa, D., Elen, A., Claus, P., Loeckx, D., Kaluzynski, K., D'hooge, J.: Regional cardiac motion and strain estimation in

three-dimensional echocardiography: a validation study in thick-walled univentricular phantoms. IEEE Trans. Ultrason. Ferroelectr. Freq. Control **59**(4), 668–682 (2012)

5. Gjesdal, O., Helle-Valle, T., Hopp, E., Lunde, K., Vartdal, T., Aakhus, S., Smith, H.J., Ihlen, H., Edvardsen, T.: Noninvasive Separation of Large, Medium, and Small Myocardial Infarcts in Survivors of Reperfused ST-Elevation Myocardial Infarction. Circ. Cardiovasc. Imaging **1** (3), 189–196 (2008)

6. Żmigrodzki, J., Cygan, S., Leśniak-Plewińska, B., Kałużyński, K.: Identification of subendocardial infarction—a feasibility study using synthetic ultrasonic image data of a left ventricular model. In: Computational Vision and Medical Image Processing V: Proceedings of the 5th Eccomas Thematic Conference on Computational Vision and Medical Image Processing, pp. 137–142 (2016)

7. Cygan, S., Żmigrodzki, J., Leśniak-Plewińska, B., Karny, M., Pakieła, Z., Kałużyński, K.: Influence of polivinylalcohol cryogel material model in FEM simulations on deformation of LV phantom. In: van Assen, H., Bovendeerd, P., Delhaas, T. (eds.) Functional Imaging and Modeling of the Heart, pp. 313–320. Springer International Publishing (2015)

8. Jensen, J.A.: FIELD: a program for simulating ultrasound systems. Presented at the 10th nordicbaltic conference on biomedical imaging, vol. 4, pp. 351–353 (1996)

9. Kybic, J., Unser, M.: Fast parametric elastic image registration. IEEE Trans. Image Process. **12**(11), 1427–1442 (2003)

10. Unser, M., Aldroubi, A., Eden, M.: Fast B-Spline transforms for continuous image representation and interpolation. IEEE Trans. Pattern Mach. Intell. **13**(3), 227–285 (1991)

11. Rueckert, D., Sonoda, L.I., Hayes, C., Hill, D.L.G., Leach, M.O., Hawkes, D.J.: Nonrigid registration using free-form deformations: application to breast MR images. IEEE Trans. Med. Imaging **18**(8), 712–721 (1999)

12. Keys R.: Cubic convolution interpolation for digital image processing. IEEE Trans. Acoust. Speech Signal Proc. **29**(6) (1981)

13. Rohlfing, T., Maurer, C.R. Jr., Bluemke, D.A., Jacobs, M.A.: Volume-preserving nonrigid registration of MR breast images using free-form deformation with an incompressibility constraint. IEEE Trans. Med. Imaging **22**(6), 730–741 (2003) (str.)

14. Liu, Dong C., Nocedal, J.: On the limited memory BFGS method for large scale optimization. Math. Prog. **45**(1), 503–528 (1989)

Numerical Simulation of the Self-oscillating Vocal Folds in Interaction with Vocal Tract Shaped for Particular Czech Vowels

Petr Hájek, Pavel Švancara, Jaromír Horáček and Jan G. Švec

Abstract The study presents a two-dimensional (2D) finite element (FE) model which consists of the vocal folds (VF), the trachea and idealized vocal tract (VT) shaped for Czech vowels [a:], [i:] and [u:] created from magnetic resonance images (MRI). Such configuration enables solving fluid-structure-acoustic interaction, flow-induced self-oscillations of the VF and acoustic wave propagation in the VT by explicit coupling scheme with two separate solvers for structure and fluid domain. The FE model of the VF includes the VF pretension and setting to phonatory position, large deformation of tissues and VF contact. Unsteady viscous compressible airflow through the FE model of the trachea, glottis and the VT is modelled by using Navier-Stokes (NS) equations. Moving boundary of the fluid domain (according to the VF motion) is solved by the Arbitrary Lagrangian-Eulerian approach. The solution obtained for the FE models is analyzed and the effect of the VT shape on the spectra of the generated acoustic pressure at the lips is discussed and the results are compared with measured data published in literature.

Keywords Simulation of phonation · Fluid-structure-acoustic interaction · Finite element method · Biomechanics of voice · Speech recognition

---

P. Hájek (✉) · P. Švancara
Institute of Solid Mechanics, Mechatronics and Biomechanics, Brno University
of Technology, Technická 2896/2, 616 69 Brno, Czech Republic
e-mail: y126528@stud.fme.vutbr.cz

P. Švancara
e-mail: svancara@fme.vutbr.cz

P. Švancara · J. Horáček
Institute of Thermomechanics, Academy of Sciences of the Czech Republic,
Dolejškova 1402/5, 182 00 Prague, Czech Republic
e-mail: jaromirh@it.cas.cz

J.G. Švec
Department of Biophysics, Palacky University Olomouc, 17. listopadu 12,
771 46 Olomouc, Czech Republic
e-mail: jan.svec@upol.cz

© Springer International Publishing AG 2017   317
R. Jabłoński and R. Szewczyk (eds.), *Recent Global Research and Education:*
Technological Challenges, Advances in Intelligent Systems and Computing 519,
DOI 10.1007/978-3-319-46490-9_43

1   Introduction

At a basic level the vowel production is described by the source-filter theory [1]. For deeper insight one has to take account of the myoelastic-aerodynamic theory of phonation [2, 3] and vowel production turns into a problem involving fluid-structure-acoustic interaction. To deal with such a complicated problem the low degrees of freedom (DOF) lumped-mass models are not exact enough [4–7] and high DOF computational models are used. One of the first FE models was a 2D model of the VF, which employed small deformations of the three layer VF tissue with transversely isotropic material properties [8]. This model was extended to the 3D FE model with aerodynamics modelled by the NS equations and acoustics using wave-reflection method in [9]. Zhao et al. [10] developed 2D axisymmetric FE model with prescribed movement of the VF and fluid-acoustic coupling based on Lighthill's acoustic analogy. Similar approach used Pořízková et al. [11] with 2D finite volume model of VT aerodynamics modelled as unsteady flow of a viscous fluid, which caused acoustic perturbations. Realistic results were obtained by Alipour and Scherer [12] from 2D FE model of the self-oscillating VF in laryngeal tract with false VF, where aerodynamics and structure was coupled by two solvers for the NS equations and for the solid mechanics equations.

In earlier studies of the authors [13–16] both the FE model of VF self-oscillations and the FE model of the trachea with the VT shaped for [a:] were created. In the present work new 2D FE models for the Czech vowels [a:], [u:] and

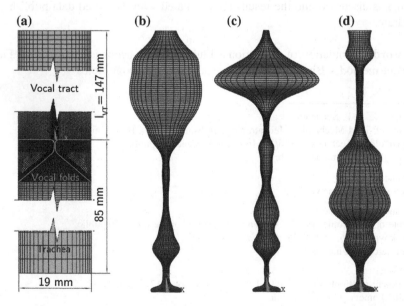

Fig. 1 Schema of the complete 2D FE model with parts of the trachea and simplified VT (*left*) and the VT models for vowels [a:], [u:] and [i:] (from *left* to *right*)

[i:] are created (see Fig. 1) and the spectra of the computed acoustic pressure at the lips are compared with data published in literature.

2 Methods

The geometry of the VF was adapted according to Scherer's M5 geometry [17], the geometry of the VT was obtained from MRI [18] and the trachea was idealized. All FE models were created using program system ANSYS 15.0. The homogeneous isotropic linear-elastic material of the VF was considered according to [15] with the Young's moduli: 2000 Pa for the superficial lamina propria, 25,000 Pa for the epithelium, 8000 Pa for the ligament and 65,000 Pa for thyroarytenoid muscle. The following Poisson's ratios were considered: 0.49 for all layers except 0.40 for the muscle. Density was 1040 kg m^{-3} for all layers. The compressible air for 36 °C with the following characteristics was considered: sound speed 353 m s^{-1}, density 1.205 kg m^{-3}, fluid viscosity 1.81351 × 10^{-5} kg m^{-1} s^{-1}.

At the entrance to the trachea the constant lung pressure of 250 Pa was prescribed for the VF excitation, which matches clinical results [3]. Zero pressure boundary condition was applied to the oral orifice, absolute reflectivity and zero flow velocity were prescribed for the walls of fluid FE model.

Detailed characteristics of the FE model can be found in previous papers of the authors [15, 16] including overall dimensions of the four-layered VF with layers thicknesses, represented boundary conditions and computational algorithm.

3 Results and Discussion

Tube-shaped VT. At first the simplified tube-shaped VT model for the vowel [a:] was created, analysed and compared with the analytical solution. Assume that the first natural frequency for [a:] shaped VT is f_{N1} = 600 Hz. Now we create the

Table 1 Natural frequencies computed from modal analysis for all VT models

	f_{N1} (Hz)	f_{N2} (Hz)	f_{N3} (Hz)	f_{N4} (Hz)
Tube-shaped [a:]	600	1800	3000	4200
[a:]	600	1126	2829	3283
[i:]	316	1959	2721	3238
[u:]	430	889	1931	2525

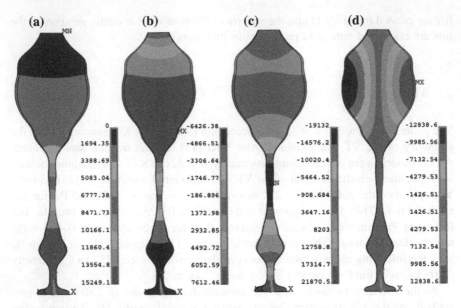

Fig. 2 Pressure eigenmode shapes for the first three natural frequencies of **a** 600 Hz, **b** 1126 Hz and **c** 2829 Hz, and **d** for the first transversal mode (sixth natural frequency) at 4336 Hz

tube-shaped VT (see Fig. 1) closed at the VF level and open at the lips level. Using the simple analytical solution for acoustic wave propagation in a straight pipe we obtain the $l_{VT} \approx 147$ mm for the f_{N1}. Dependence between the second, third and fourth natural frequency and f_{N1} for such VT is $f_{N2} = 3 \cdot f_{N1}$, $f_{N3} = 5 \cdot f_{N1}$ and $f_{N4} = 7 \cdot f_{N1}$ (see Table 1).

Real-shaped VT. Different results were obtained for the VT models shaped for the vowels [a:], [u:] and [i:]. The first four computed natural frequencies for the particular VT models are summarized in Table 1 and the first three longitudinal and the first transversal pressure mode shapes are shown in Fig. 2. Time course of the acoustic pressures and power spectral densities (PSD) of the acoustic pressure computed at proximity of the lips for all vowels are shown in Figs. 3 and 4. One can observe the relationship between the natural frequencies of the VT and the formant frequencies (peaks) in the sound spectra in accordance with the source-filter theory (especially for [u:] and [i:]). The computed results for [u:] and [i:] match the formant frequencies known from literature, see Table 2.

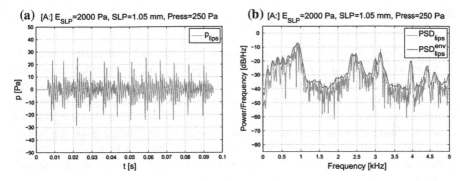

Fig. 3 a Numerically simulated acoustic pressure in time domain computed near the lips for the VT model of vowel [a:] and **b** PSD of this acoustic pressure signal

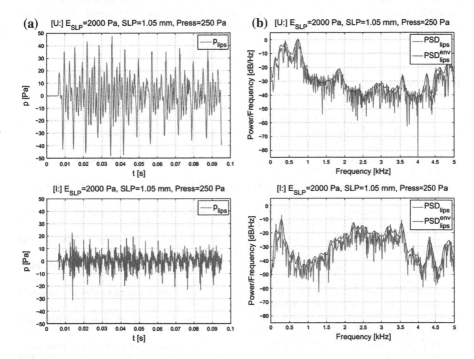

Fig. 4 a Numerically simulated acoustic pressure computed near the lips for the VT model of vowels [u:] and [i:] and **b** PSD of the acoustic pressure signals

Table 2 Formant frequencies of the Czech vowels measured by Merhaut [19]		F_1 (Hz)	F_2 (Hz)
	[a:]	420–800	1100–1800
	[i:]	200–450	1800–3800
	[u:]	240–400	600–1100

4 Conclusion

Vowels are one of the main parts of a spoken language and their analysis is a core of many computer programs for speech recognition. In this work the spectral characteristics of three Czech corner vowels [a:], [u:] and [i:] are discussed and mainly first and second formant are studied, because they are playing key role in vowel differentiation. Modal characteristics of appropriate VT models are analysed using the finite element models. Obtained results are in relative good agreement with experimental data published in literature. The results can be used in design of intelligent artificial voices or for improvement of methods in computer speech analysis.

Acknowledgments This research is supported by the Brno University of Technology by the project *Fond Vědy* FV 16-09 and by the Grant Agency of the Czech Republic project No 16-01246S *Computational and experimental modelling of self-induced vibrations of vocal folds and influence of their impairments on human voice.*

References

1. Fant, G.: Acoustic theory of speech production. Mouton, The Netherlands, The Hague (1960)
2. van den Berg, J.: Myoelastic-aerodynamic theory of voice production. J. Speech Lang. Hearing Res. **1**(3), 227–244 (1958)
3. Titze, I.R.: The myoelastic aerodynamic theory of phonation. National Centre for Voice and Speech, Denver and Iowa City (2006)
4. Flanagan, J.L.: Source-system interaction in the vocal tract. Ann. N. Y. Acad. Sci. **155**(1), 9–17 (1968)
5. Ishizaka, K., Flanagan, J.L.: Synthesis of voiced sounds from a two-mass model of the vocal cords. Bell Syst. Tech. J. **51**(6), 1233–1268 (1972)
6. Hirano, M.: Morphological structure of the vocal cord as a vibrator and its variations. Folia Phoniatrica et Logopaedica **26**(2), 89–94 (1974)
7. Story, B.H., Titze, I.R.: Voice simulation with a body-cover model of the vocal folds. J. Acoust. Soc. Am. **97**, 1249 (1995)
8. Alipour, F., Berry, D.A., Titze, I.R., Introduction, I.: A finite-element model of vocal-fold vibration. J. Acoust. Soc. Am. **108**(6), 3003–3012 (2000)
9. Alipour, F., Scherer, R.C.: Vocal fold bulging effects on phonation using a biophysical computer model. J. Voice **14**(4), 470–483 (2000)
10. Zhao, W., Zhang, C., Frankel, S.H., Mongeau, L.: Computational aeroacoustics of phonation, Part I: Computational methods and sound generation mechanisms. J. Acoust. Soc. Am. **112** (5), 2134–2146 (2002)
11. Pořízková, P., Kozel, K., Horáček, J.: Numerical solution of compressible and incompressible unsteady flows in channel inspired by vocal tract. J. Comput. Appl. Math. **270**, 323–329 (2014)
12. Alipour, F., Scherer, R.C.: Time-dependent pressure and flow behavior of a self-oscillating laryngeal model with ventricular folds. J. Voice Official J. Voice Foundation **29**, 649–659 (2015)
13. Švancara, P., Horáček, J., Hrůza, V.: FE modelling of the fluid-structure-acoustic interaction for the vocal folds self-oscillation. In: Náprstek, J., Horáček, J., Okrouhlík, M., Marvalová,

B., Verhulst, F., Sawicki, J.T. (eds.) Vibration Problems ICOVP 2011 SE—108. Springer Proceedings in Physics, vol. 139, pp. 801–807. Springer, Netherlands (2011)

14. Švancara, P., Horáček, J., Martínek, T., Švec, J.G.: Numerical simulation of videokymographic images from the results of the finite element model. Eng. Mech. 640–643 (2014)

15. Hájek, P., Švancara, P., Horáček, J., Švec, J.G.: Finite element modelling of the effect of stiffness and damping of vocal fold layers on their vibrations and produced sound. Appl. Mech. Mater. **821**, 657–664 (2016)

16. Hájek, P., Švancara, P., Horáček, J., Švec, J.G.: Numerical simulation of the effect of stiffness of lamina propria on the self-sustained oscillation of the vocal folds. Eng. Mech. 182–185 (2016) (Svratka)

17. Scherer, R.C., Shinwari, D., De Witt, K.J., Zhang, C., Kucinschi, B.R., Afjeh, A.A.: Intraglottal pressure profiles for a symmetric and oblique glottis with a divergence angle of 10 degrees. J. Acoust. Soc. Am. **109**, 1616–1630 (2001)

18. Radolf, V.: Direct and inverse task in acoustics of the human vocal tract. Ph.d. thesis, Czech Technical University in Prague, 2010

19. Merhaut, J.: Základy fyziologické akustiky a teorie přirozených akustických signálů, ÈVUT, Praha, 1972

Distance Metric for Speech Commands of Dysarthric Users in Smart Home Systems

Gabriella Simon-Nagy and Annamária R. Várkonyi-Kóczy

Abstract Chronic neuromuscular diseases often cause dysarthria (speech distortions, impaired articulation, etc.), that becomes more severe over time. This aspect of the disease represents a serious problem in voice-controlled smart home systems. Medical research suggests that some speech features are impaired considerably, while others remain relatively unharmed. Therefore, it is possible to create a distance metric based on medical data that measures difference between two speech commands in a dysarthria-specific way: the contribution of various features to the distance is based on the extent of dysarthric impairment. Specifying a minimal distance between speech commands contributes to a more effective recognition during later stages of the disease.

Keywords Speech recognition · Smart home · Dysarthria · Distance metric

1 Introduction

Chronic neuromuscular disorders such as amyotrophic lateral sclerosis (ALS) or multiple sclerosis (MS) are also known to cause dysarthria; dysarthric symptoms include speech distortions, impaired articulation, prosody, respiration and phonation. These diseases tend to be progressive, causing the speech symptoms as well as the motor impairment to become more and more severe over time. The distortion of

---

G. Simon-Nagy (✉)
Doctoral School of Applied Informatics and Applied Mathematics,
Óbuda University, Budapest, Hungary
e-mail: nagy.gabriella@nik.uni-obuda.hu

G. Simon-Nagy
Integrated Intelligent Systems Japanese-Hungarian Laboratory, Budapest, Hungary

A.R. Várkonyi-Kóczy
Department of Mathematics and Informatics, J. Selye University, Komarno, Slovakia
e-mail: varkonyi-koczy@uni-obuda.hu

© Springer International Publishing AG 2017
R. Jabłoński and R. Szewczyk (eds.), *Recent Global Research and Education: Technological Challenges*, Advances in Intelligent Systems and Computing 519,
DOI 10.1007/978-3-319-46490-9_44

speech presents a challenge to patients in several aspects of their daily lives; in later stages it can even make social participation nearly impossible.

Although motion disabled persons could make extensive use of voice-controlling their household items, this aspect of the disease represents a serious problem for the creators of speech recognition in smart home systems: not only the user's pronunciation is changing, but it also becomes less and less intelligible. Additionally, smart home systems for the severely motion disabled users require the minimization (possibly elimination) of false command recognition. For example, in the case of house heating control, a healthy user could walk to a wall thermostat and press a button easily to correct a faulty setting followed from a miscategorized speech command, whereas a severely motion disabled person may be unable to do the same. Therefore, speech command recognition should be made as unambiguous as possible, and one way to accomplish this is a suitable set-up of command set regarding typical dysarthric symptoms of the patients.

Our goal is to specify a distance metric based on medical data that measures the difference between two speech command phrases in a dysarthria-specialized way. The contribution of various command parts and features to the distance is calculated according to the extent of dysarthric impairment on that feature, so that more difference in the more stable features results in a larger distance, while severely impaired features matter less. Specifying a minimal distance between speech commands of a smart home system may contribute to a more effective recognition during later stages of the disease.

2   Related Work

2.1   Distance Metrics for Strings

There are several metrics for the calculation of differences between character strings [1–3]. The simplest one is Hamming distance, which is the number of mismatches between corresponding positions in the strings. It is only defined for sequences of equal length; therefore it is rarely used for general strings of characters. Hamming distance could be generalized to strings of different lengths by padding the shorter string with empty characters, but a method like this would be very similar to the Levenshtein distance (also known as edit distance).

The Levenshtein distance of any two strings is defined as the number of single-character modifications (substitution, insertion or deletion) required to transform one string to the other. This metric is often used in applications where short strings are matched to short or longer texts and a few small differences are to be expected (e.g. dictionary queries, spell checkers), but it is impractical for the calculation of distances between two long strings because of its computational cost.

Damerau–Levenshtein distance extends the original Levenshtein distance by allowing the transposition of two adjacent characters. In fact, the term "edit distance" is often used to reference to a general string metric where the allowed modifications with their associated costs are considered to be parameters of the metric.

There are methods that regard N-grams of the two strings instead of single characters during matching. For example, Q-gram distance gives the number of absolute differences between N-grams of the strings. Jaccard distance for strings can be determined as: 1 minus the quotient of shared N-grams and all occurring N-grams. (Originally Jaccard distance was defined for finite sets.)

There are other popular methods to calculate distances that are not a metric in a mathematical sense such as Jaro-Winkler distance developed for duplicate detection. This method gives a measure of similarity based on the number of matching characters and the number of required transpositions between them. It also has a similarity bias toward strings that have their first few (up to 4) characters in common. This bias can be scaled by a factor between 0 and 0.25.

All of these methods share the same property: they cannot differentiate between characters by "importance", i.e. the matching of some characters cannot produce a larger similarity measure than the matching of other characters.

2.2 Medical Research on Dysarthria

Medical research [4–8] has been revealing considerable amount of data about the changing aspects of speech in the progress of neuromuscular diseases. An overview of the literature suggests that some speech features are impaired considerably, while others remain relatively unharmed. Although every disease affects speech differently and every patient has somewhat different set of symptoms, later changes of speech characteristics can be predicted according to the initial manifestation of symptoms. For example, in multiple sclerosis it is observed that those patients, who already have some dysarthric symptoms early on, will have more severe dysarthria in later stages. Also the types of symptoms can be predicted based on the area of damage in the brain of the patient.

Speech impairments causing considerable difficulties in speech recognition are the following:

- Articulation problems: the lack of quick, precise movements of lips and tongue results in slurred speech and less intelligible consonants
- Impaired volume control: bursts of loud speech or constantly low volume or changing (decreasing) volume within a phrase
- Impaired timing: lengthening of word segments or abnormally fast speech in short rushes because of problems of breath control
- Repetition: repetition of speech segments or short words in spontaneous speech (contrary to overlearned material like childhood poems or singing that tend to remain clear)

- Scanning speech: words are broken into syllables with pauses in between, the natural rhythm of speech is impaired
- Dysphonia: coarseness, nasal resonance, breathy voice, impaired pitch control.

Gender differences in dysarthric symptoms are not apparent, there are however some socio-cultural aspects that can affect speech production, for example in societies with heavy substance abuse, the drug of choice may cause a gender-difference in speech quality (e.g. in Hungary it is observed by medical staff, that men prefer drinking, whereas women prefer smoking).

3 Description of the Dysarthria-Specific Metric

First of all, a metric must meet the criteria of non-negativity, identity, symmetry, and the triangle-equality, therefore we had to take these properties into consideration. We have based our metric on a parametrized edit distance with the edit operations insertion, deletion and substitution, and non-negative costs assigned to the operations. This kind of distance is known to be a metric if for every operation there is an inverse operation with equal cost. Distances comparing N-grams do not make sense in this context because of the frequent distortion or loss of phonemes in the speech.

Dysarthric symptoms of the user add some special requirements to the above criteria:

- If breath control is impaired then the length of syllables (vowels) should not make any difference (in some languages it does normally), i.e. the substitution cost between them should be zero.
- If scanning speech is involved then word boundaries should not make any difference, e.g. the phrases "a part" and "apart" should have zero distance.
- If the patient has articulation problems then the editing of a vowel should have a larger weight than editing a consonant, e.g. the words "sun" and "sum" should be more similar than "sun" and "sin".
- If speech volume is decreasing during speaking then a prefix-bias similar to that of Jaro-Winkler distance can be utilized (although it may violate the "inverse operation with equal cost" criterion of metrics).

Involuntary repetitions in speech do not have to be regarded in this case because they mostly occur in spontaneous speech, whereas speech commands of a smart home fall under the category of overlearned text, especially after longer use. According to the above, the following metric can be specified.

For the pre-processed command phrases $P = p_1 p_2 \ldots p_m$ and $R = r_1 r_2 \ldots r_n$, the distance $D(m, n)$ can be calculated based on the following recurrent rules:

$$D(0,j) = \sum_{k=1}^{j} b_k c_{insert}(r_k) \tag{1}$$

$$D(i,0) = \sum_{k=1}^{i} b_k c_{delete}(p_k) \tag{2}$$

$$D(i,j) \begin{cases} D(i-1,j-1) & \text{if } p_i = r_j, \\ \min \begin{cases} D(i,j-1) + b_j c_{insert}(r_j) \\ D(i-1,j) + b_i c_{delete}(p_i) & \text{if } p_i \neq r_j \\ D(i-1,j-1) + b_j c_{substitute}(p_i, r_j) \end{cases} \end{cases} \tag{3}$$

where c_{insert}, c_{delete}, and $c_{substitute}$ are the costs of the appropriate edit operations according to the medical evaluation of the symptoms of the patient, b_i corresponds to the prefix bias for the ith position in the string, and $D(i, j)$ stands for the distance between the first i letters of P and the first j letters of R (of course $D(0, j)$ means the distance between the empty string and the first j letters of R). In case of consonant articulation problems, substitution operation can be restricted to the same class of letters, i.e. vowel can be substituted for another vowel and consonant for another consonant.

If the dysarthric symptoms make it necessary, pre-processing can include the removal of spaces between words and in case of some languages, the substitution of short vowels for their long counterparts.

4 Usage Examples

In this section, two examples are introduced to show the difference between our dysarthria-specific metric versus the Levenshtein metric.

In Hungarian, frequently used smart home commands contain the following words: "vészhívás" (emergency call) and "megnyitás" (open, e.g. an application). The Levenshtein distance, if calculated with unit costs for every operation, is 6, because 6 substitutions are required to transform one to another. (In Hungarian, the letter combinations "sz" and "ny" count as single letters). According to this calculation, these two words seem to be very different.

Let us assume that the user of the smart home system has articulation problems and tends to lengthen syllables. Therefore a pre-processing is necessary to substitute short vowels for long ones. After pre-processing, the two words are "veszhivas" and "megnyitas". Moreover, let us assume that the medical evaluation of the speech determined that operation cost for vowels is $c(vowel) = 1$ and the cost for consonants is only $c(consonant) = 0.4$. In this case, only 4 consonant substitutions are required that's costs add up to 1.6. This number reflects much more precisely, how similar these words sound in the speech of the user.

Another example can be the problem of the commands with similar beginnings in case of respiratory symptoms that cause a volume decrease during speech. If a user has these symptoms, the application of a prefix-bias is in order: operations in the beginning of phrases should have more impact on the difference. For example, according to medical evaluation $b_i = 1$ for $i < 8$, then linearly decreasing until $i = 17$. Using the commands "Indítsd a tévét" (start TV controller) and "Indítsd a rádiót" (start radio controller), the effect of prefix-bias can be demonstrated. Levenshtein distance is 5 (1 insertion and 4 substitutions), while the dysarthria-specific distance is only 3.

In these examples, a smaller distance is more appropriate from the viewpoint of a speech recognition system because the example phrase-pairs can sound much more similar in dysarthric speech than in healthy speech.

5   Conclusion

The dysarthria-specific distance metric could be used to create speech controlled smart home systems for patients with severe symptoms of dysarthria. The appropriately chosen command phrases can be recognized with more certainty, even in later stages of the disease. A minimum distance can be determined based on the medical predictions about the expected severity of the dysarthric symptoms.

Acknowledgment This work has been partially sponsored by the Hungarian National Scientific Fund under contract OTKA 105846 and the Research and Development Operational Program for the project "Modernization and Improvement of Technical Infrastructure for Research and Development of J. Selye University in the Fields of Nanotechnology and Intelligent Space", ITMS 26210120042, co-funded by the European Regional Development Fund.

References

1. Bilenko, M., et al.: Adaptive name matching in information integration. IEEE Intell. Syst. **18** (5), 16–23 (2003)
2. Cohen, W.W., Ravikumar, P., Fienberg, S.E.: A comparison of string distance metrics for name-matching tasks. In: IJCAI-03 Workshop on Information Integration, pp. 73–78 (2003)
3. Ukkonen, E.: Approximate string-matching with q-grams and maximal matches. Theor. Comput. Sci. **92**(1), 191–211 (1992)
4. Hartelius, L., Runmarker, B., Andersen, O.: Prevalence and characteristics of dysarthria in a multiple-sclerosis incidence cohort: relation to neurological data. Folia Phoniatr. Logop. **52**, 160–177 (2000)
5. Tomik, B., Guiloff, R.J.: Dysarthria in amyotrophic lateral sclerosis: a review. Amyotroph. Lateral Scler. **11**, 4–15 (2010)
6. Rosen, K.M., Goozée, J.V., Murdoch, B.E.: Examining the effects of Multiple Sclerosis on speech production: Does phonetic structure matter? J. Commun. Disord. **41**(1), 49–69 (2008)
7. Feijó, A.V., et al.: Acoustic analysis of voice in multiple sclerosis patients. J. Voice **18**(3), 341–347 (2004)
8. Kuo, C., Tjaden, K.: Acoustic variation during passage reading for speakers with dysarthria and healthy controls. J. Commun. Disord. **62**, 30–44 (2016)

E-learning Environment for Control of Form Measuring Machines

Rafał Kłoda, Kacper Kurzejamski, Jan Piwiński and Konrad Parol

Abstract This paper presents the results of the adaptation of E2LP project to the needs of a remote learning laboratory focused on a PIK-2 form measuring machine (FMM). The E2LP (Embedded Engineering Learning Platform) was an FP7 finalized project which focused on Embedded Systems Education at University level. This work covers the integration between the hardware measurement technologies and high-level programming for access and administration of the FMM Remote Lab. The innovative approach incorporates the use of LabVIEW services as a bridge between the physical signals from FMM and an accessible user interface. The use of Moodle open-source learning platform opens the way for translation of the low-level web services into the easily administrated and accessible form. It is possible due to the development of custom software blocks and modules that may be easily adapted to communicate with existing web services. The result of the FMM Remote Lab is an e-learning portal which provides remote real-time control of the measurement processes. Moreover, this project demonstrates the possibility of modernisation of obsolete systems and adaptation to up-to-date standards without the need of interfering in their base structure.

Keywords Remote laboratory · E-learning · Form measuring machine · PIK-2

R. Kłoda (✉) · K. Kurzejamski · K. Parol
Institute of Metrology and Biomedical Engineering, Warsaw
University of Technology, Św. Andrzeja Boboli 8, 02-525 Warsaw, Poland
e-mail: kloda@mchtr.pw.edu.pl

K. Kurzejamski
e-mail: k.kurzejamski@gmail.com

K. Parol
e-mail: konrad.parol@gmail.com

J. Piwiński
Industrial Research Institute for Automation and Measurements PIAP, Al. Jerozolimskie 202,
02-486 Warsaw, Poland
e-mail: jpiwinski@piap.pl

© Springer International Publishing AG 2017
R. Jabłoński and R. Szewczyk (eds.), *Recent Global Research and Education:
Technological Challenges*, Advances in Intelligent Systems and Computing 519,
DOI 10.1007/978-3-319-46490-9_45

1   Introduction

Growing tendencies for online teaching have been observed for many years [1], but remote laboratories for measurement of surface macrogeometry are very rare. This is not surprising as the form measurement machines (FMM) used in this field are sophisticated and expensive devices. Moreover, proper setting-up of e-learning environment is costly, which poses a problem and a challenge for e-learning systems developers, who are responsible for providing secure access to measuring machines for many students. The works undertaken by the authors concern integration between the hardware measurement technologies and high-level programming to develop a flexible multiuser system which would enable remote access to and administration of the FMM. This innovative approach incorporates the use of LabVIEW web services (WS) as a bridge between the physical signals from FMM and the accessible user interface. In previous work [2], the architecture of such system was based on Internet Toolkit for LabVIEW.

This paper is organized as follows: Sect. 2 describes an overview of PIK-2 form measuring machine and its technical specification. Section 3 presents an e-learning platform which was used for adaptation and illustrates its main modules. This section also discusses some security issues connected with remote control of PIK-2 system. Section 4 presents the user interface implemented in order to provide remote access measuring machine. Finally, Sect. 5 concludes the paper.

2   PIK-2 Form Measuring Machine

The laboratory of surface macrogeometry measurements at Faculty of Mechatronics, Warsaw University of Technology has become a venue for an innovative and challenging project—an opportunity to refurbish obsolete interfaces of the measuring systems and to transform a standard, on-site laboratory into a fully interactive, online platform. The laboratory consists of many form measuring devices, one of which is the heart of the project—a PIK-2 machine, dedicated to roundness and cylindricity measurements [3]. A diagram of the system with a description of its components is presented in Fig. 1.

A measured element is placed on a rotating table (1). The contact sensor is an inductive probe (2), mounted on a moveable carriage allowing for height control. Direct control of the system is possible with a keyboard (6), and the results of the operations are displayed on a monitor (4). In the original system the data was transmitted to a printer (5) through RS232 protocol. The control task and functions related to the data processing are performed by a modular specialized computer (3). The system is placed on a granite block (8) for stability.

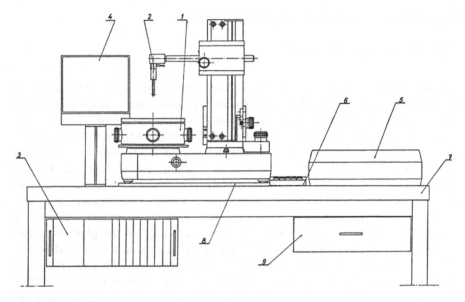

Fig. 1 Diagram of PIK-2 form measuring machine

Although it was designed in the early 1990s at the Institute of Metrology and Biomedical Engineering, its parameters are still comparable to those achieved by modern measurement systems of similar purpose [4, 5]. Crucial parameters of PIK-2 FMM, taken from its technical datasheet [6], were presented in Table 1.

As the measuring system provides relatively low uncertainty with a wide range of applications in metrology, it is still used in the research laboratories [7] and for teaching purposes [3]. However, its interface is visibly outdated and unintuitive. Compatibility is also an issue, as the main control device, a ten button keypad, is not standardised in any way and was designed specifically for PIK-2, making it impossible to find a substitute. The idea behind the project is to create an online platform that would not only allow for remote work, but also provide a universal interface that may be applied in various similar systems and laboratories.

Table 1 PIK-2 FMM parameters

Type of sensor	Inductive
Measuring range in typical configuration	8, 25, 80, 250, 800, 2500 μm
Rotational speed	7 rpm
Roundness measurement uncertainty (depending on the distance H between the measurement plane and rotating table surface)	$(0.3 + 0.0005H)$ μm
Parallelism measurement uncertainty	1 μm/100 mm

3 Adaptation of an E-learning Platform

3.1 Moodle Framework

The main framework used in both E2LP [8] and PIK-2 FMM remote laboratories is Moodle (Modular Object-Oriented Dynamic Learning Environment). It is an open-source e-learning platform that perfectly meets the needs of a highly inter-active environment for students of technical universities [9]. From the teacher's point of view, it provides a variety of administrative tools, allowing for easy course management and creation, review of students' grades and solutions, etc. For the students, it is an accessible, intuitive platform with interactive exercises, forums and a messaging system [10]. From the programming vantage point, Moodle is a great choice for developing an interactive remote laboratory that requires a high level of software and hardware integration. The main advantage of the platform is the possibility of including custom blocks and modules, created or modified by the developer, that exactly meet the needs of the system. The idea and process of creation of such blocks was widely described in [11]. In both E2LP and PIK-2 projects this feature allowed for the incorporation of external services into the Moodle environment, FPGA board programming for the former and control of PIK-2 FMM for the latter. Some of the modules concerning remote laboratory administration were transferred almost directly from E2LP to PIK-2 FMM, as the main idea of the system remains the same [12, 13]. Among the aforementioned modules are:

- Student's calendar—with the possibility of booking and managing time slots reservations for PIK-2 access, as shown in Fig. 2
- Check solution—each exercise is a new challenge for a student and it has to be completed correctly. This module allows for an automatic verification of the solution
- Additional teaching material layout—in the text form, introducing the topic of the exercises to a student with appropriate instructions
- Database administration—defines automatic connections with the database, retrieving necessary information and saving data generated during users' activity on the platform.

The modules directly responsible for the control of PIK-2 FMM had to be created specifically for the system, though their functionality was still based on the E2LP project.

3.2 Secure Control of the PIK-2 System

The main control of the PIK-2 system is provided by incorporating the LabVIEW web services into the Moodle platform. An additional LabVIEW validation module

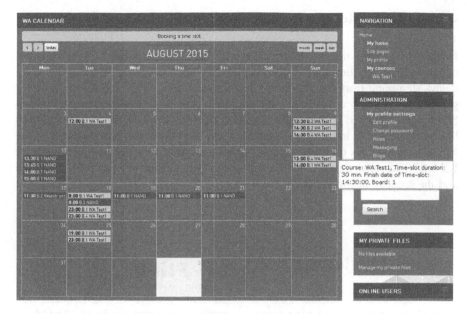

Fig. 2 Timeslot reservation module

serves as a security bridge between Moodle and the control modules of the FMM. The structure of the access to the system is presented in Fig. 3.

The LabVIEW security validation is based on three parameters—current time, Moodle user ID and current module ID (exercise). The module checks in the database whether a specific user has a booking for such module in a given time period. If authentication succeeds, a connection to other LabVIEW web services is open and control over PIK-2 FMM is allowed. Otherwise, any unauthorised user's action is blocked at the LabVIEW Security level. In this way, obtaining some other student's credentials for Moodle platform is not enough to gain access to the laboratory. Moreover, even possessing both public and secret parts of API key for the web services does not grant authorisation as they only enable connection to the LabVIEW Security Module according to LabVIEW Security Key algorithm [14].

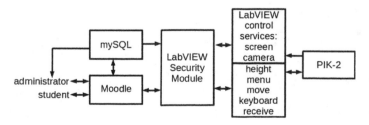

Fig. 3 Structure of the PIK-2 remote access from Moodle environment

4   The User Interface for Remote Control of PIK-2 System

In the E2LP project the user interface was implemented as a direct reference to the
active web service through the *<iframe>* HTML tag. This led to some unnecessary
delays stemming from the additional data transmission, defining the web service's
graphics. Furthermore, obtaining the address of the service would create a possibility of unauthorised control from outside of the Moodle platform. A new approach
has been introduced in the PIK-2 system, where the interface is based strictly on the
Moodle components (buttons, inputs, graphic, etc.) that are virtually linked with the
LabVIEW Security module. As a result, the system's efficiency has been improved
and the security has increased.

Fig. 4 Screenshot of PIK-2 remote control interface

From the user's point of view the interface is divided into three sections. The first one is the feedback section (Fig. 4), where images from the screen and the camera preview are located. In order to improve the system's speed, instead of streaming, the pictures are refreshed after any input occurrence (via keyboard) or manually. Below is the virtual keyboard, enabling menu navigation. The last section is the solution module. It requires from the student to input his calculated parameter (according to the topic of the current exercise), for example roughness of the surface R_a. Meanwhile, the program calculates the correct value of the parameter in the background on the basis of the raw data from the measurement. On submission, the student's value is compared with the calculated one and an appropriate mark is displayed.

5 Conclusion

The use of a Moodle-based platform provides a great variety of tools for efficient creation of technical laboratories for e-learning, as was shown on the example of E2LP (FPGA programming) and PIK-2 (form-measuring machine) projects. The possibility of creation of custom blocks and modules allows for an incorporation of external resources, such as LabVIEW web services, hence making high level software and hardware integration possible. In addition, Moodle's flexibility makes the use of multiple programming techniques possible (PHP/HTML web development, mySQL database administration, LabVIEW components), allowing the developer to choose an optimal tool for a certain purpose.

Furthermore, these successful examples of the two projects demonstrate that this approach allows for modernization of systems with obsolete interfaces without a need to modify their core structure. It may not only be very convenient, but also, in some circumstances, crucial while modifying systems with precise and fragile components, such as micrometric displacement sensors. The universal platform is easily reconfigurable so that it may be adapted for use in a variety of systems and laboratories with different hardware, thus enabling further development.

References

1. Jennifer, O., Codde, J., deMaagd, K., Tarkleson, E., Sinclair, J., Yook, S., Egidio, R.: An analysis of e-learning impacts & best practices in developing countries. East Lansing, Michigan, Michigan State University (2011)
2. Żebrowska-Łucyk, S., Kłoda, R.: System for remote control of form measuring machines via the Internet. Adv. Coordinate Metrol. (2010)
3. Żebrowska-Łucyk, S.: Maszyny do pomiaru odchyłek kształtu, położenia i kierunku. Materiały pomocnicze do ćwiczeń laboratoryjnych. Warszawa, 2014
4. Roundness & Form Measurement. Taylor Hobson (accessed 01 June 2016). http://www.taylor-hobson.com/products/roundness-form.html

5. Roundness & Form Measurement Product Information, Mitutoyo (accessed 01 June 2016), URL http://www.mitutoyo.co.jp/eng/products/shinen/shinen.html
6. System pomiarowy PIK-2 do badania odchyłek kształtu i położenia. Opis zastosowań i obsługi, Warszawa (1992)
7. Pawlowski, M., Gapinski, B., Rucki, M.: Experimental check of the simulated cylinder's geometrical characteristics obtained from the expert program. XIX IMEKO World Congress Fundamental and Applied Metrology, Lisbon, Portugal, 2009
8. E2LP project website. http://www.e2lp.org
9. Kotzer, S., Yossi, E.: Learning and teaching with moodle-based e-learning environments, combining learning skills and content in the fields of math and science and technology. In: Proceedings of 1st Moodle Research Conference, Heraklion, Greece (2012)
10. William, Rice: Moodle: e-learning course development: a complete guide to successful learning using moodle. Packt Publishing, Birmingham (2006)
11. Moore, J., Michael, C.: Moodle 1.9 Extension Development: Customize and Extend Moodle by Using Its Robust Plugin Systems. Birmingham, Packt Publishing (2010)
12. Kłoda, R., Piwiński, J.: E2LP remote laboratory: e-learning service for embedded systems education. Advances in Intelligent Systems and Computing, Embedded Engineering Education, vol. 421. Springer (2015). ISSN 2194-5357
13. E2LP Remote Laboratory, e-learning portal for E2LP. http://e2lp.piap.pl
14. LabVIEW web services security, national instruments (accessed 01 June 2016). http://www.ni.com/white-paper/7749/en/

Cathodoluminescent Properties and Particle Morphology of Eu-Doped Silicate Phosphors Synthesized in Microwave Furnace

Igor A. Turkin, Mariia V. Keskinova, Maxim M. Sychov,
Konstantin A. Ogurtsov, Kazuhiko Hara, Yoichiro Nakanishi
and Olga A. Shilova

Abstract Eu-doped silicate phosphors have been synthesized in microwave furnace using different methods to prepare charge mixture (solid-state mixing, sol-gel method) and various experimental setups. Influence of synthesis condition on cathodoluminescent (CL) properties of phosphors (CL spectra, color coordinates and dependence of CL brightness from voltage) and particle size and shape was shown in this article. Sol-gel process is found to be inappropriate for charge mixture preparation at cathodoluminescent phosphor synthesis due to an amorphous phase formation in large quantities. The sample synthesized under optimal conditions is featured with high brightness, white color of luminescence, linear brightness-voltage dependence and can be used in cathodoluminescent light sources.

Keywords Silicate phosphors · Microwave synthesis · Cathodoluminescence · Europium

1 Introduction

Nowadays, manufacturing of modern light sources including CL light sources [1] is an important task. A lot of new phosphors such as $LaPO_4:Pr_{3+}$, $ZnAl_2O_4$ are being developed [2].

I.A. Turkin · M.V. Keskinova (✉) · M.M. Sychov · K.A. Ogurtsov
Saint-Petersburg Institute of Technology (Technical University), Saint Petersburg, Russia
e-mail: Keskinova88@gmail.com

M.V. Keskinova · K. Hara · Y. Nakanishi
Research Institute of Electronics, Shizuoka University, Shizuoka, Japan

O.A. Shilova
Institute of Silicate Chemistry of RAS, Saint-Petersburg, Russia

© Springer International Publishing AG 2017
R. Jabłoński and R. Szewczyk (eds.), *Recent Global Research and Education:
Technological Challenges*, Advances in Intelligent Systems and Computing 519,
DOI 10.1007/978-3-319-46490-9_46

339

Silicate phosphors doped with Eu ions are of interest because of high chemical and thermal stability, durability and ability to form wide range of solid solutions. Moreover, they have high quantum yield and intensity of emission [3].

Description of the synthesis of silicate phosphors doped with Eu^{2+} in microwave furnace was shown in our works [4, 5]. Time of microwave synthesis (10 min) was much lower comparing to 150 min needed to prepare efficient phosphor using muffle oven. Energy consumed by microwave furnace is also much lower (300 W while muffle consumes 1000 W). Finally, temperature during microwave synthesis (700 °C) was also lower compared to conventional synthesis (900–1200 °C). Nevertheless, phosphor synthesis using microwave furnace provided much higher brightness compared to conventional method. This result may be explained by the effect of pondermotoric forces (microwave vibration) produced by currents emerged from electromagnetic wave and electrodiffusive mass transfer [6].

2 Experimental

The charge mixtures for synthesis were prepared using two methods: mechanical mixing of starting powders; sol-gel process. Analytical purity grade reagents were used for synthesis of the samples. In the case of mechanical mixing SiO_2, $SrCl_2.6H_2O$, $Ca(OH)_2$, Eu_2O_3 powders were preliminary ground in a mortar, sieved and homogenized in a drum mixer for 3 h.

To perform sol-gel synthesis $Ca(OH)_2$ and Eu_2O_3 were dissolved in concentrated nitric acid, $SrCl_2.6H_2O$ was dissolved in water. Then salt solutions and TEOS (tetraethoxysilane $(C_2H_5O)_4Si$) were sequentially added to water-alcohol mixture containing nitric acid as a catalyst. The gelation was carried out at 25 °C for 24 h. The obtained gels were dried in an oven at 150 °C during 5 h, then dried gels were ground in the agate mortar and used for the synthesis of phosphors.

The phosphor synthesis was performed in the custom made microwave furnace with the frequency of 2.45 GHz using 3 different experiment setups described in details in our work [5].

Setup 1 represents a fibrous corundum container. Quarts tube filled with charge mixture is placed in it. Containers of setups 2 and 3 additionally consist of 4 quarts tubes abutted to the middle one with charge mixture. These 4 tubes are filled with susceptor for increasing temperature during the synthesis. SiC featuring with a very strong intensity of interaction with 2.45 GHz microwaves was used as a susceptor [7].

In order to transform Eu^{3+} into Eu^{2+} reducing conditions were provided in the system by the introduction carbon powder via either mixing with the initial charge mixture (2 setup) or as separate layers between the charge mixture layers (3 setup).

In the case of the syntheses without preliminary annealing, the charge mixture contains crystallization water, which intensively interact with microwave radiation. Powder was introduced into the system without a susceptor and heated by separate microwave pulses (each pulse is 1 min heating time, 1 min interval between pulses) to avoid intense outgassing from the reactor.

Before the synthesis in 2 and 3 setups, charge mixture was preliminary heated during 10 min in microwave furnace in setup 1 in order to carry out the further synthesis in continuous mode.

CL spectra were measured by Hamamatsu Photonics C7473 multichannel analyser. CL brightness and CL color coordinates were measured by Topcon BM-5A luminance meter. Electron micrographs of the samples were obtained using a VE-8800 Scanning Electron Microscope. The phosphor particle sizes and shape factors were calculated using ImageJ and Excel software.

3 Results and Discussion

The synthesized silicate phosphors have a multiphase structure comprising Eu^{3+}- and Eu^{2+}-doped strontium orthosilicate as well Eu^{2+}-doped mixed calcium-strontium chlorsilicate phases.

Synthesis conditions and performances of the prepared phosphors are shown in the Table 1.

SEM microphotographs of the samples and particle size distribution are shown in the Figs. 1 and 2, correspondingly.

From all used synthesis methods of silicate phosphors, layer-by-layer process (setup3) let us to obtain particles with the smallest average size (350 nm) due to the highest temperature gradients during synthesis because of additional layers of carbon powder which interact intensively with microwave energy.

The phosphor synthesized from charge mixture prepared by sol-gel method is featured with smaller average particle size compared with the phosphor obtained by using mechanical mixing for charge mixture preparation. This effect is caused by the formation of higher number of nucleation centers during synthesis by sol-gel process.

Shape factor was calculated by formula $f = 4\pi * area/perimeter^2$. The sample synthesized from the charge mixture prepared by sol-gel method is featured with the smallest shape factor (extended form of particles) while the sample obtained from charge mixture prepared by mechanical mixing has the biggest shape factor (roundish form of particles).

In general, microwave synthesis let us to obtain more round particles with less average size due to the effects as microwave vibrations and electrodiffusive mass transfer. In complex system, like charge mixture, consisting of components with different dielectric permittivity and tangent of dielectric losses, non-uniform polarization field develops as a result of interaction with microwave energy. This process intensifies mass transfer proportionally to squared electric field intensity [6].

CL spectra (Fig. 3) of the samples prepared in microwave furnace consist of 2 bands of $Sr_2SiO_4:Eu^{2+}$, ~ 490 and ~ 560 nm, which are attributed to 5d–4f

Table 1 Synthesis conditions and performances of the prepared phosphors

Sample ID	Setup	Method of charge mixture preparation	CL brightness (a.u.)	FWHM of CL spectrum (nm)	Color coordinates	Correlated color temperature (K)	Average particle size (μm)	Particle shape factor
B	2	mechanical mixing	59	177	x = 0.404 y = 0.412	3700	0.82	0.84
C	3 (top layer)	mechanical mixing	150	170	x = 0.370 y = 0.440	4500	0.35	0.81
F	2	sol-gel	56	272	x = 0.339 y = 0.337	5500	0.58	0.68
Ref	Muffle oven, 950 °C and 2.5 h	mechanical mixing	117	180	x = 0.433 y = 0.253	1800	0.86	0.76

Fig. 1 Sem microphotographs of the samples: *B*, *F*, *C*–magnification is 20,000, *Ref*—magnification is 10,000

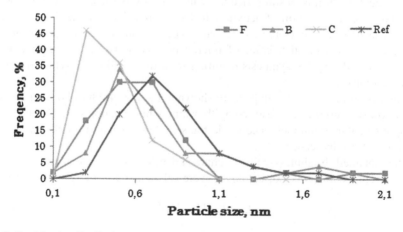

Fig. 2 Particle size distribution

Fig. 3 CL spectra

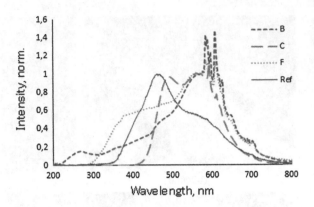

transitions of Eu^{2+} in 10 coordinated (M(I)) and 9 coordinated (M(II)) positions of Sr^{2+}, correspondingly [8]. Another band observed at ~ 580 nm relates to Eu^{2+}-doped mixed calcium-strontium chlorsilicate [9]. The bands at ~ 590, ~ 610, ~ 653 and ~ 703 nm corresponded to 5D0-7F1, 5D0-7F2, 5D0-7F3 and 5D0-7F4 transitions of Eu^{3+} in silicate matrix [10]. Two bands at ~ 268 and ~ 365 nm were observed on CL spectra of B and F samples are supposedly attributed to 5d–4f transition in Eu^{2+} in the amorphous phase of the studied phosphor. The band at ~ 365 nm is much more intense in the case F phosphor (sol-gel charge mixture preparation) compared with B sample (mechanical mixing). For F sample, the band attributed to Eu^{2+} transitions in 10-coordinated (M(I)) position in silicate matrix is also higher than in the case of mechanical mixing preparation of the charge mixture. This difference is probably determined by the fact that in the case of sol-gel process Eu introduction in this position occurred more easily because of smaller particle size and better distribution of activator inside the reactive volume. In the case of layer-by-layer synthesis (setup 3) the bands corresponding to Eu^{3+} transitions are less intensive than in other cases of microwave synthesis due to higher concentration of CO during the synthesis resulting from the additional layers of reducing agent (carbon).

Color coordinates of the samples are shown in the Fig. 4. Phosphors synthesized in microwave furnace are featured with warm white (samples B, C) and white (sample F) color of luminescence while the sample obtained in muffle furnace has red color of luminescence.

The obtained brightness-voltage characteristics data are summarized in the Table 2. The coefficients L0, U0 and n were calculated according to the formula:

$$L = L_0(U - U_0)^n \tag{1}$$

where L is brightness and U is voltage, Lo—coefficient, n—power coefficient, Uo—threshold voltage [11]. The samples B and F synthesized in setup 2 have high threshold voltages and quite nonlinear dependence of brightness from voltage due to the formation of amorphous phase. This results in low CL brightness and

Fig. 4 Color coordinates

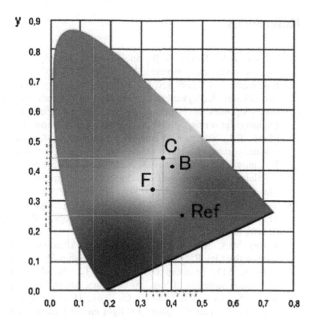

Table 2 Threshold voltages and power coefficients of the samples

Sample ID	Uo	n
B	587	2.5
C	0	1.3
F	423	2.8
Ref	0	1.4

therefore, these two samples are not suitable for cathodoluminescent light sources. On the contrary, the sample C synthesized using a layer-by-layer method (setup 3) and the sample obtained in muffle furnace have threshold voltages about zero their brightness-voltage dependence is almost linear. These advantageous features make the phosphor C and muffle furnace prepared phosphor useful as white and red cathodoluminescent light source components respectively.

4   Conclusion

The variation of microwave synthesis conditions affords a precise adjustment of the resulting Eu-doped mixed silicate phosphor particle shape and size, their cathodoluminescence spectra and color coordinates. A layer-by-layer synthesis provides phosphors featuring with white color of luminescence and linear brightness-voltage dependence promising for the application in cathodoluminescent light sources, whereas sol-gel preparation of the reaction mixture leads to the formation of amorphous phase making the resulting phosphors inefficient.

References

1. Bugaev, A.S., Kireev, V.B., Sheshin, E.P., Kolodyazhnyj, A.J.: Cathodoluminescent light sources: status and prospects. Phys. Usp. **58**(8), 853–883 (2015)
2. Bakhmetyev, V., Lebedev L., Malygin, V., Podsypanina, N., Sychov, M., Bogdanov, S., Bondarenko, A., Vyvenko, O.: Synthesis of UV cathodoluminescent phosphors and study of their properties. In: Materials 5th International Scientific Conference «Lighting and Power Engineering: history, problems and perspectives», pp. 21–22 (2015)
3. Marsh, P.J., Silver, J., Vecht, A., Newport, A.: Cathodoluminescence studies of yttrium silicate: cerium phosphors synthesised by a sol–gel process. J. Lumin. **97**, 229–236 (2002)
4. Keskinova, M.V., Ogurtsov, K.A., Sychov, M.M., Kolobkova, E.V., Turkin, I.A., Nakanishi, Y., Hara, K.: Synthesis of chlorine-silicate phosphors for white light-emitting diodes. Adv. Mater. Res. **1117**, 48–51 (2015)
5. Turkin, I.A., Keskinova, M.V., Sychov, M.M., Ogurtsov, K.A., Hara, K., Nakanishi, Y., Shilova, O.A.: Microwave synthesis of Eu-doped silicate phosphors. JJAP Conf. Proc. **4**, 011108 (2016)
6. Janney, M.A., Kimrey, H.D., Kiggins, J.O.: Microwave proceedings of ceramics: guide-lines used of the Oak Ridge Laboratory. MRS Symp. Proc. **269**, 173–185 (1992)
7. Ramesh, P.D., Brandon, D., Schächter, L.: Use of partially oxidized SiC particle bed for microwave sintering of low loss ceramics. Mater. Sci. Eng. **266**(1), 211–220 (1999)
8. Kim, J.S., Jeon, P.E., Choi, J.C., Park, H.L.: Emission color variation of M_2SiO_4:Eu^{2+} (M = Ba, Sr, Ca) phosphors for light-emitting diode. Solid State Commun. **133**, 187–190 (2005)
9. Sasaki, Y., Daicho, H., Aoyagi, S., Sawa, H.: US Patent 0256222 A1 (2012)
10. Yanmin, Qiao, Xinbo, Zhang, Xiao, Ye, Yan, Chen, Hai, Guo: Photoluminescent properties of Sr_2SiO_4:Eu^{3+} and Sr_2SiO_4:Eu^{2+} phosphors prepared by solid-state reaction method. J. Rare Earth **27**(2), 323 (2009)
11. Ozawa, L.: Cathodoluminescence and photoluminescence: theories and practical applications. Taylor & Francis Group (2007)

Part VI
Robotics, Computing, Modelling, Diagnostics

Integration of Machine Learning and Optimization for Robot Learning

Amir Mosavi and Annamaria R. Varkonyi-Koczy

Abstract Learning ability in Robotics is acknowledged as one of the major challenges facing artificial intelligence. Although in the numerous areas within Robotics machine learning (ML) has long identified as a core technology, recently Robot learning, in particular, has been witnessing major challenges due to the theoretical advancement at the boundary between optimization and ML. In fact the integration of ML and optimization reported to be able to dramatically increase the decision-making quality and learning ability in decision systems. Here the novel integration of ML and optimization which can be applied to the complex and dynamic contexts of Robot learning is described. Furthermore with the aid of an educational Robotics kit the proposed methodology is evaluated.

Keywords Machine learning · Optimization · Robotics

1 Introduction

Today learning has become a major part of the research in Robotics [1]. Machine learning (ML) algorithms in robotics in particular, within autonomous control and sensing, are being used to tackle difficult problems where large quantities of datasets are available which enable Robots to effectively teach themselves [2]. ML as a sub-field of computer science has evolved from the study of pattern recognition and computational learning theory [3]. Furthermore ML is considered as a field of study in artificial intelligence that gives computers the ability to learn from data [4]. To do so ML explores the development of models that can predict and learn from an

A. Mosavi · A.R. Varkonyi-Koczy (✉)
Institute of Mechatronics and Vehicle Engineering, Obuda University,
Budapest, Hungary
e-mail: varkonyi-koczy@uni-obuda.hu

A.R. Varkonyi-Koczy
Department of Mathematics and Informatics, J. Selye University,
Komarno, Slovakia

© Springer International Publishing AG 2017   349
R. Jabłoński and R. Szewczyk (eds.), *Recent Global Research and Education:*
Technological Challenges, Advances in Intelligent Systems and Computing 519,
DOI 10.1007/978-3-319-46490-9_47

available dataset [5]. Such models operate with the aid of algorithms capable of making data-driven predictions rather than following explicit codes [6]. Consequently ML is often used in a range of problems where designing precise algorithms is not practical. In this sense ML can replace the human expertise in information treatment [7]. For that matter ML provides the algorithmic tools for dealing with datasets and providing predictions. In fact ML tends to imitate human skills, which in most cases, act exceptional in identifying satisfactory solutions by theoretical or experience-based considerations [8].

Applications of ML in Robotics which highly contribute to Robot learning is vast and yet progressing in a fast pace [1]. Robot vision [9], Robot navigation [10], field Robotics [11], humanoid Robotics [12], legged locomotion [13], modeling vehicle dynamics [14], medical and surgery Robotics [15], off-road rough-terrain mobile Robot navigation [16], are few of the areas within Robotics for which utilizing ML technologies has become popular. It is therefore clearly evidenced that ML has in recent years become an essential part of Robotics. And this has been in fact a response to the frustration with the problems for which it has been proven difficult to conventional coding solutions. For instance in a variety of Robotics platforms the imitation learning techniques [17], inverse optimal control methods [18], programming-by-demonstration [19], and supervised learning techniques [7] have become norm. Further most notable ML technologies utilized in Robotics include; ML techniques for big data [20], self-supervised learning [16], reinforcement learning [21], and multi-agent learning [22].

2 Integration of Machine Learning and Optimization

The intersection research area of optimization and ML has recently engaged leading scientists [23]. ML has made benefit from optimization and on the other hand ML contributed to optimization as well. Today ML is seen as an exceptional replacement for human expertise in information manipulation [24]. In addition ML has the proven ability to simplify optimization functions [25]. Optimization on the other hand is the source of immense power for automatically improving decisions [26]. However in real-life applications, including Robotics, optimization has not had the chance to be used to its full potential [27]. This has been often due to the absence, complexity, or inefficient optimization functions of the complicated problem at hand [28]. Yet in such cases ML has shown the ability of modeling whole or part of the optimization functions on the basis of the availability of a reliable dataset [24]. A number of case studies concerning Robotics problems have been surveyed in literature, e.g. [29], where ML technologies simplify complicated optimization functions.

Nevertheless the long-term vision for Robot learning would be the development of a fully automated system with self-service usage [19]. To reach this goal the novel idea of integration of ML and optimization [20] aims at simplifying the whole learning process by automating the decision-making tasks in an effective manner without requiring a costly learning curve [1]. In this context the learning process is seen as a byproduct of an automated optimal decision.

Learning from the available dataset integrated with optimization can be applied to a wide range of complex, dynamic, and stochastic problems [23]. Such integration has been reported exceptional in increasing the automation level by putting more power at the hands of final-user [30]. Final-user should however specify dataset, desired outputs and CPU time. CPU time is to be set to put a limitation on optimization algorithms' run-time which can be referred as "learning time". The novel integration of ML and optimization has already been used in solving numerous complex cases [31]. Considering these examples, it is observed that once a combination of right ML technologies and optimization designed, suitable for the problem at hand, further algorithm selection, adaptation, and integration, are done in an automated way, and a complete solution for learning is delivered to the final-user [6].

2.1 Integration

Depending on the characteristics of the problem at hand and availability of dataset an arrangement of local optimization algorithms [32] is essential to come up with an optimal decision. Yet local searches leading to locally optimum is an essential principle for solving the discrete and continuous optimization problems. In this context designing a system that is capable of curing local optimum traps is desirable. To do so reactive search optimization (RSO) methodology [33] is used. RSO methodology implements an integration of ML techniques into local and heuristics search [25] for solving real-life optimization problems [5]. RSO includes a so-called "ML application builder" [34] employed to design a system which receives dataset, guide the research, and delivers a competitive application. In fact the "ML application builder" imitates the human skill in providing the automation to the system which is responsible for algorithm selection and parameter tuning. In fact human brain quickly learns and drives future decisions based on previous observations [8]. This is the main inspiration source for inserting ML techniques into the learning curve. This is referred as brain-computer optimization [35] which is an important building block of RSO. Building blocks of RSO include neural networks, statistics, artificial intelligence, reinforcement learning, and active or query learning [24]. Characteristics of RSO include learning on the job, rapid generation and analysis of many alternatives, efficient analysis of what-if scenarios, flexible decision support, diversity of solutions and anytime solutions [25].

3 Case Study

The proposed case study aims at evaluating the RSO methodology with the aid of the educational Robotics kit of LEGO Mindstorms. The objective is to evaluate the ability of learning of a mobile Robot in locating the darkest spot of a paper sheet

(Fig. 1a). The number of the light intensity inspections is limited to a total of nine sample points within the Cartesian coordinate. Today numerous universities around the world teach artificial intelligence classes with the aid of LEGO Mindstorms platform and many literatures describe the educational benefits of this practice [36].

3.1 Implementation

In order to move in a controlled manner within the Cartesian coordinate system in the identified territory the Robot has been upgraded to a new arrangement (Fig. 1b). There are around 200 LEGO parts coming as the standard LEGO Mindstorms kit to build a Robot [36]. In the presented case study a simple arrangement provides the limited straight movements of the mobile Robot. With adding a Matrix kit to a conventional matrix building system a x-y table is created. In addition the mobile Robot is equipped with a color sensor which measures the light intensity (Fig. 1b).

The prospector here is a color sensor which measures the light intensity and reports to a simple code via a USB cable configuration. According to the simple code as the Robot moves along the Cartesian coordinate over the black and white paper nine samples of light intensity are taken. Then to connect this external code to the Robot learning system the measured points are connected to a design of experiment module to further import into a function generator. In this stage we can plot the results on a 3D graph as it is presented in the Fig. 2a. A second degree

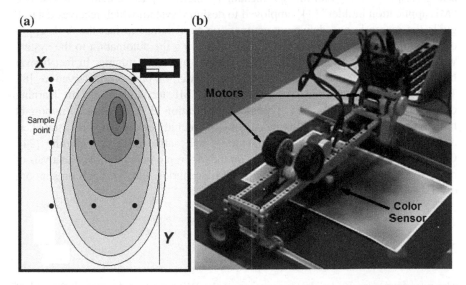

Fig. 1 Robot arrangement; **a** *Black* and *white* paper sheet presenting a random intensity of light over a Cartesian coordinate system with the coordinates of nine sample points **b** Robot in action; arrangement of LEGO Mindstorms Robot equipped with a color sensor and a Matrix kit

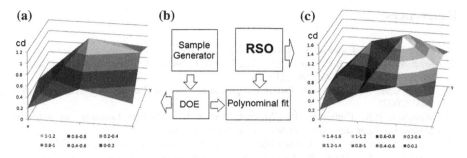

Fig. 2 **a** 3D graph of the points primary been measured **b** Building blocks of RSO methodology of learning **c** 3D graph of the newly generated points

Polynomial fit estimates the distribution of the nine sample points. And RSO runs a continues optimizer in order to predict the optimal points and generate the optimum (Fig. 2b). Robot then is directed to the newly generated optimum accordingly. Matching the predicted optimum with the darkest spot of the sheet proves the accuracy of the model (Fig. 2c).

4 Conclusions

The paper considers the novel integration of ML and optimization for the complex and dynamic context of Robot learning. RSO is introduced as a methodology to implement an integration of ML techniques into local and heuristics optimization for Robot learning. In the proposed case study RSO presents an effective framework based on solving continuous optimization problem with an efficient use of memory and self-adaptive local optimization with self-improvement capabilities in identifying the global optimum. In the case study the ability of learning of a mobile Robot in locating the darkest spot of a paper sheet is evaluated. Matching the predicted optimum with the darkest spot of the sheet proves the accuracy of the model. RSO shown to be able to well imitate the human skills in providing the automation to the system which is responsible for algorithm selection and parameter tuning.

Acknowledgment This work is sponsored by Hungarian National Scientific Fund under contract OTKA 105846 and Research and Development Operational Program for the project "Modernization and Improvement of Technical Infrastructure for Research and Development of J. Selye University in the Fields of Nanotechnology and Intelligent Space", ITMS 26210120042, co-funded by the European Regional Development Fund.

References

1. Connell, J.H., Mahadevan, S.: Robot Learning. Springer Science & Business Media (2012)
2. Knox, W.B., Glass, B.D., Love, B.C., Maddox, W.T., Stone, P.: How humans teach agents. Int. J. Social Robot. **4**(4), 409–421 (2012)
3. Bishop, CM., Nasrabadi, NM.: Pattern recognition and machine learning. J. Electron. Imaging **16**(4) (2007)
4. Michalski, R.S., Carbonell, JG., Mitchell, TM. (eds.): Machine Learning: an Artificial Intelligence Approach. Springer Science & Business Media (2013)
5. Han, J., Kamber, M., Pei, J.: Data Mining: Concepts and Techniques. Elsevier (2011)
6. Mosavi, A., Vaezipour, A.: Reactive search optimization; application to multiobjective optimization problems. Appl. Math. 1572–1582 (2012)
7. Battiti. R., Brunato. M.: Reactive search optimization: learning while optimizing. In: Handbook of Metaheuristics, pp. 543–571. Springer, US (2010)
8. Murphy, R.R.: Human-robot interaction in rescue robotics. Syst. Man Cybernetics Appl. Rev. IEEE Trans. **34**(2), 138–153 (2004)
9. Rosten, E., Drummond, T.: Machine learning for high-speed corner detection. In: Computer Vision–ECCV, pp. 430–443. Springer, Berlin (2006)
10. Sofman, B., et al.: Improving robot navigation through self-supervised online learning. J. Field Robotics. **23**, 59–75 (2006)
11. Yang, S.Y., Jin, S.M., Kwon, S.K.: Remote control system of industrial field robot. In: 6th IEEE International Conference on Industrial Informatics, pp. 442–447 (2008)
12. Peters, J., Vijayakumar, S., Schaal, S.: Reinforcement learning for humanoid robotics. In: Proceedings of the Third IEEE-RAS International Conference on Humanoid Robots (2003)
13. Kohl, N., Stone, P.: Machine learning for fast quadrupedal locomotion. In: AAAI, pp. 611–616 (2004)
14. Popp, K., Schiehlen, W.: Ground Vehicle Dynamics. Springer, Berlin (2010)
15. Taylor, R.H., Menciassi, A., Fichtinger, G., Dario, P.: Medical robotics and computer-integrated surgery. In: Handbook of Robotics, pp. 1199–1222. Springer, Berlin (2008)
16. Stavens, D., Thrun, S. A.: self-supervised terrain roughness estimator for off-road autonomous driving. arXiv preprint arXiv:1206.6872 (2012)
17. Nehaniv, C.L., Dautenhahn, K.: Imitation and Social Learning in Robots, Humans and Animals: Behavioural, Social and Communication. Cambridge University Press (2007)
18. Mombaur, K., Truong, A., Laumond, J.P.: From human to humanoid locomotion—an inverse optimal control approach. Auton. robots. **28**(3), 369–383 (2010)
19. Argall, B.D., Chernova, S., Veloso, M., Browning, B.: A survey of robot learning from demonstration. Robot. Auton. Syst. **57**(5), 469–483 (2009)
20. Mosavi, A., Varkonyi-Koczy, A., Fullsack, M.: Combination of machine learning and optimization for automated decision-making. In: Conference on Multiple Criteria Decision Making MCDM, Hamburg, Germany (2015)
21. Kober, J., Bagnell, J.A., Peters, J.: Reinforcement learning in robotics: a survey. Int. J. Robot. Res. **32**(11), 1238–1274 (2005)
22. Panait, L., Luke, S.: Cooperative multi-agent learning: The state of the art. Auton. Agent. Multi-Agent Syst. **11**(3), 387–434 (2005)
23. Sra, S., Nowozin, S., Wright, S.J.: Optimization for Machine Learning. Mit Press (2012)
24. Battiti, R., Brunato, M., Mascia, F.: Reactive Search and Intelligent Optimization. Springer Science & Business Media (2008)
25. Battiti, R.: Reactive search: Toward self-tuning heuristics. Mod. Heuristic Search Methods 61–83 (1996)
26. Battiti, R., Brunato, M.: Reactive Business Intelligence. From Data to Models to Insight. Reactive Search Srl, Italy (2011)

27. Toussaint, M., Ritter, H., Brock, O.: The optimization route to robotics—and alternatives. KI-Künstliche Intelligenz **29**(4), 379–388 (2015)
28. Battiti, R., Campigotto, P.: Reactive search optimization: Learning while optimizing. an experiment in interactive multi-objective optimization. In: Proceedings of MIC (2009)
29. Stone, P., Veloso, M.: Multiagent systems: a survey from a machine learning perspective. Auton. Robot. **8**(3), 345–383 (2000)
30. Battiti, R., Brunato, M.: The LION way. Machine learning plus intelligent optimization. Appl. Simul. Model. (2013)
31. Mosavi, A.: Decision-Making in Complicated Geometrical Problems. Int. J. Comput. Appl. **87**(19) (2014)
32. Horst, R., Pardalos, P.M.: editors. Handbook of Global Optimization. Springer Science & Business Media (2013)
33. Brunato, M., Battiti, R.: Learning and intelligent optimization (LION): one ring to rule them all. Proc. VLDB Endowment **6**(11), 1176–1177 (2013)
34. Battiti, R., Brunato, M.: The LION Way: Machine Learning Plus Intelligent Optimization. Trento University, LIONlab (2014)
35. Battiti, R., Brunato, M., Delai, A.: Optimal Wireless Access Point Placement for Location-Dependent Services. Technical Report # DIT-03-052, University of Trento, Italy (2010)
36. Parsons, S., Sklar, E.: Teaching AI using LEGO mindstorms. In: AAAI Spring Symposium (2004)

Application of Model Reference Control for MIMO System

Jerzy E. Kurek

Abstract The Model Reference Control algorithm for continuous-time MIMO system without dead-time is presented. The proposed approach is illustrated by numerical example.

Keywords Model reference control · Continuous-time systems

1 Introduction

Model reference control is well known control algorithm for SISO control systems. Its main advantages are simplicity, easy way for controller parameters tuning and easy way for design of adaptive model reference control system [1–3]. The control algorithm can be also easily implemented for instance with neural networks [4] or genetic algorithms. This makes that the controller is very popular.

The main purpose of this paper is to present an algorithm for design of Model Reference Controller for MIMO system. In Sect. 2 there is presented the considered problem. Then, in Sect. 3 an algorithm for controller design is given. There is also attached numerical example of MIMO model reference control system. Finally, in Sect. 4, concluding remarks are given.

2 Problem Formulation

Consider time-invariant, linear continuous-time square MIMO system

J.E. Kurek (✉)
Institute of Automatic Control and Robotics, Warsaw University
of Technology, ul. Boboli 8, 02-525 Warsaw, Poland
e-mail: jkurek@mchtr.pw.edu.pl

© Springer International Publishing AG 2017
R. Jabłoński and R. Szewczyk (eds.), *Recent Global Research and Education:
Technological Challenges*, Advances in Intelligent Systems and Computing 519,
DOI 10.1007/978-3-319-46490-9_48

357

Fig. 1 Block diagram of control system

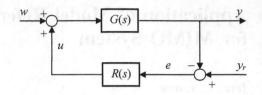

$$G(s) = \begin{bmatrix} G_{11}(s) & \cdots & G_{1p}(s) \\ \vdots & & \vdots \\ G_{p1}(s) & \cdots & G_{pp}(s) \end{bmatrix} \qquad (1)$$

where $\det G(s) \neq 0$ for $s = 0$. It is assumed that the system is asymptotically stable.

The control system for the plant is presented in Fig. 1, where and $u \in R^p$ is an input vector, $y \in R^p$ is an output vector, $y_r \in R^p$ is a reference output vector, $w \in R^p$ is a disturbance vector, and R is the controller

$$R(s) = \begin{bmatrix} R_{11}(s) & \cdots & R_{1p}(s) \\ \vdots & & \vdots \\ R_{p1}(s) & \cdots & R_{pp}(s) \end{bmatrix}$$

Then, there is also given reference model of the control system

$$F(s) = \begin{bmatrix} F_{11}(s) & \cdots & F_{1p}(s) \\ \vdots & & \vdots \\ F_{p1}(s) & \cdots & F_{pp}(s) \end{bmatrix}, \qquad F(s)|_{s=0} = I$$

i.e. $F_{ii}(s)|_{s=0} = 1$ and $F_{ij}(s) = sF_{ij1}(s)$ for $i \neq j$.

The problem can be now formulate das follows: find a controller $R(s)$ such that the closed loop control system has required transfer function $F(s)$

$$y(s) = F(s)y_r(s)$$

Such control system is known as model reference control system.

3 Controller Design

For the control system one easily finds

$$y(s)|_{w=0} = G_{yr}(s)y_r(s)$$

where

$$G_{yr}(s) = [1 + G(s)R(s)]^{-1}G(s)R(s) \tag{2}$$

and

$$y(s)|_{y_r=0} = G_{yw}(s)w(s)$$

where

$$G_{yw}(s) = [1 + G(s)R(s)]^{-1}G(s) \tag{3}$$

Then calculating MRF controller one finds

$$[1 + G(s)R(s)]^{-1}G(s)R(s) = F(s)$$

Next one obtains

$$G(s)R(s) = [1 + G(s)R(s)]F(s)$$

and

$$G(s)R(s)[I - F(s)] = F(s)$$

Thus, one gets

$$R(s) = G^{-1}(s)F(s)[I - F(s)]^{-1} = G^{-1}(s)[I - F(s)]^{-1}F(s) \tag{4}$$

It is important to note that for the controller we have abbreviation of $G(s)$ and $G^{-1}(s)$ in the closed-loop control system transfer function

$$\begin{aligned}
G_{yr}(s) &= [1 + G(s)R(s)]^{-1}G(s)R(s) \\
&= \{1 + G(s)G^{-1}(s)F(s)[I - F(s)]^{-1}\}^{-1}G(s)G^{-1}(s)F(s)[I - F(s)]^{-1} \quad (5) \\
&= \{1 + F(s)[I - F(s)]^{-1}\}^{-1}F(s)[I - F(s)]^{-1} = F(s)
\end{aligned}$$

and

$$\begin{aligned}
G_{yw}(s) &= [1 + G(s)R(s)]^{-1} = \{1 + G(s)G^{-1}(s)F(s)[I - F(s)]^{-1}\}^{-1}G(s) \\
&= \{1 + F(s)[I - F(s)]^{-1}\}^{-1}G(s) = [I - F(s)]G(s)
\end{aligned} \tag{6}$$

In Fig. 2 there is presented block diagram of the controller.
In the case that control system is designed based on uncertain model of plant

Fig. 2 Block diagram of
MIMO MRC controller with
reference control system
model $F(s)$

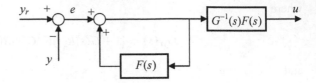

$$G_m(s) = \frac{1}{M_m(s)} L_m(s) \approx G(s)$$

where $M_m(s)$ is a polynomial, and $L_m(s)$ jest a polynomial matrix, $M_m(s) \in R[s]$ and $L_m(s) \in R[s]^{p \times p}$, one obtains

$$R(s) = G_m^{-1}(s)F(s)[I - F(s)]^{-1} = G_m^{-1}(s)[I - F(s)]^{-1}F(s) \qquad (7)$$

Then, we have

$$
\begin{aligned}
G_{yr}(s) &= [1 + G(s)R(s)]^{-1}G(s)R(s) \\
&= \{1 + G(s)G_m^{-1}(s)F(s)[I - F(s)]^{-1}\}^{-1}G(s)G_m^{-1}(s)F(s)[I - F(s)]^{-1}
\end{aligned} \qquad (8)
$$

and

$$G_{yw}(s) = [1 + G(s)R(s)]^{-1} = \{1 + G(s)G_m^{-1}(s)F(s)[I - F(s)]^{-1}\}^{-1}G(s) \qquad (9)$$

In this case there is no cancelation of $G(s)$ and $G^{-1}(s)$ and it is easy to see that the closed-loop control system is asymptotically stable only if plant $G(s)$ is asymptotically stable and all zeros s_i of control plant model $G_m(s)$, rank $G_m(s_i) < p$, are 'asymptotically stable', i.e. Re $s_i < 0$

$$G_m^{-1}(s) = \left(\frac{1}{M_m(s)} L_m(s)\right)^{-1} = M_m(s)L_m^{-1}(s) = M_m(s)\frac{1}{\det L_m(s)} \operatorname{adj} L_m(s)$$

We illustrate design of the controller in numerical example.

Example 1 Consider system (1) with the following transfer function $G(s)$:

$$G(s) = \begin{bmatrix} \dfrac{1.2}{(10s+1)(5s+1)(4s+1)} & \dfrac{1.8}{(5s+1)(4s+1)} \\ \dfrac{1.1}{(6s+1)(4s+1)(2s+1)} & \dfrac{1.5}{(8s+1)(4s+1)(2s+1)} \end{bmatrix}$$

For the system one has calculated the following model for controller design

$$G_m(s) = \begin{bmatrix} \dfrac{1.2}{(14.5s+1)(4.5s+1)} & \dfrac{1.8}{(7s+1)(2s+1)} \\ \dfrac{1.1}{(9s+1)(3s+1)} & \dfrac{1.5}{(11s+1)(3s+1)} \end{bmatrix}$$

It can be easily calculated that all zeros of the model are 'asymptotically stable'

$$s_1 = -0.3322, \quad s_2 = -0.2042, \quad s_3 = -0.0995, \quad s_4 = -0.0074$$

Then, the following reference model for the control system has been designed

$$F(s) = \begin{bmatrix} \dfrac{1}{(4.35s+1)(3.25s+1)} & \dfrac{s}{(4.35s+1)^2(3.25s+1)} \\[3mm] \dfrac{s}{(3.3s+1)^2(3s+1)} & \dfrac{1}{(3.2s+1)(3s+1)} \end{bmatrix}$$

It is rather easy to check that the model satisfies required conditions.

In Fig. 3 there is presented step response of the system, system model and model of the control system. Next in Fig. 4 there are presented response of control system for reference output and disturbances. It is easy to see that system works according to the requirements.

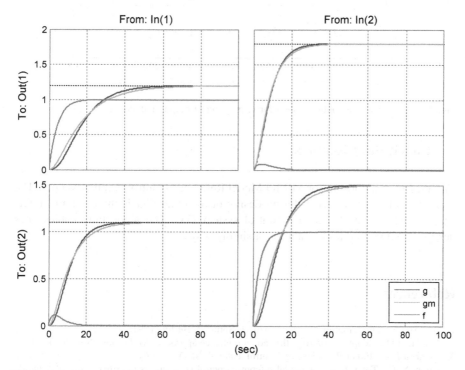

Fig. 3 Step response of the system $G(s)$, system model $G_m(s)$ and model of the control system $F(s)$

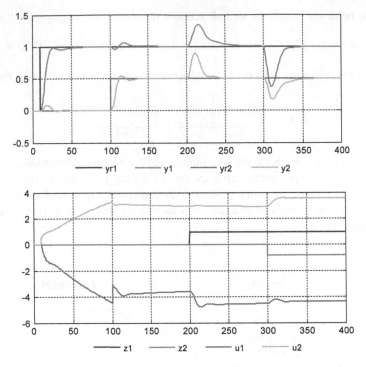

Fig. 4 Response of the control system for step reference output and step disturbance

4 Concluding Remarks

The presented algorithm for MIMO Model Reference Control is rather simply. The main problem which occurs during design of the controller is the design of reference model of the closed-loop control system $F(s)$. The algorithm enables ones design of MIMO control system in a simply way.

References

1. Iserman, R.: Digital control systems, vol. 2. Springer, Berlin (1991)
2. Landau, I.D.: Adaptive control—the model reference approach. M. Dekker, New York (1979)
3. Ioannou P.A., Fidan B.: Adaptive Control Tutorial. SIAM (2006)
4. Chen, Y.-C., Teng, C.-C.: A model reference control structure using a fuzzy neural network. Fuzzy Sets Syst. **73**, 291–312 (1995)

IT System Supporting the Security System in Plants Posing a Risk of a Major Industrial Accident

Michał Syfert, Bartłomiej Fajdek and Jan Maciej Kościelny

Abstract The article presents the tasks and structure of an IAPS IT system supporting the security system in plants posing a risk of a major industrial accident. The operation of the system consists in collecting digital documentation, as well as monitoring and supervising tasks related to security in a plant and supporting HAZOP risk analyses with the use of qualitative modelling. After a short introduction and presentation of motives for developing a system, the paper describes the basic modules constituting the system and functions realized with their use. The summary presents prospective benefits resulting from the implementation of the system and expected directions of further development.

Keywords Security system · Risk analysis · HAZOP · Qualitative modelling · Process graph

1 Introduction

Preventing major industrial accident is nowadays one of the main tasks of security systems in industrial plants. A crucial element of preventing them is the issue of early detection and elimination of the sources of prospective risks. It seems necessary to constantly monitor all industrial installations, particularly increased risk installations (IRI) and major risk installations (MRI). There is a need of developing IT systems supporting security systems, interpreted as a set of procedures and operation programs connected with security in industrial plants [1]. The necessity of

M. Syfert (✉) · B. Fajdek · J.M. Kościelny
Institute of Automatic Control and Robotics, Warsaw University
of Technology, Warsaw, Poland
e-mail: m.syfert@mchtr.pw.edu.pl

B. Fajdek
e-mail: b.fajdek@mchtr.pw.edu.pl

J.M. Kościelny
e-mail: jmk@mchtr.pw.edu.pl

© Springer International Publishing AG 2017
R. Jabłoński and R. Szewczyk (eds.), *Recent Global Research and Education:
Technological Challenges*, Advances in Intelligent Systems and Computing 519,
DOI 10.1007/978-3-319-46490-9_49

realizing such systems comes from legal regulations such as Environmental Protection Law [2] and Directive 2012/18/EU of European Parliament and Council on the control of major-accident hazards involving dangerous substances [3–7].

The Institute of Automatic Control and Robotics of the Warsaw University of Technology, together with Central Institute for Labour Protection—National Research Institute and in cooperation with the specialists from State Fire Service, Chief Environmental Protection Inspectorate and Office of Technical Inspection undertook within 3rd stage of the long-term program "Improvement of security and working conditions" the realization of a system called "Intelligent Accident Prevention System" (IAPS). This is an IT system included in the group of systems mentioned at the beginning. It is intended to support the introduction and monitoring of procedures and security system documentation in the IRI and MRI installations. The use of the system in a given plant means introducing a new generation of tools supporting security systems. The following article presents the results of the development works on the first version of IAPS system.

2 Mode of Operation and Tasks of IAPS System

IAPS system is intended to realize the following sets of tasks:

- Collecting of digital documentation related to plant security. The procedures of the security systems introduced in the IRI and MRI are related to a specific circulation of documents, which have to be implemented in the plant. Such documents are used by the plant staff and presented to external bodies. The IAPS system is a central information repository related to the implementation and functioning of security systems in the plant. The structure of information kept in the IAPS system is imposed by the safety system and based on relevant legal norms and regulations. In order to adjust the operation of the system to the specificity of different plants, users have the ability to flexibly modify the structure of the stored documents, as well as to define their content in a free way.

- Supervising of security system tasks. Introducing security system in the plant is also connected with implementing adequate procedures and monitoring changes (documentation, trainings, duties performed by the staff). In this field the IAPS system supports: (a) monitoring procedures related to the program of accident prevention, including monitoring duties performed by the workers and trainings connected with the security systems; (b) circulation of the documents required by law, e.g. monitoring of the validity periods; (c) introducing changes in security systems (storing historical data, comparing different versions of the documents).

- Supporting the performance of risk analyses. An additional task of an IAPS system is supporting risk analysis performance with particular consideration of HAZOP method. The system allows for: (a) making a qualitative model of the process reflecting the cause-and-effect relationships between process variables, (b) the analysis of the relations described by the qualitative model conducted in order to support the creation and verification of the completeness of conducted risk analysis, (c) gathering in a central database documentation related to the risk analyses.

- Keeping a journal of accidents. The IAPS system also allows for introducing and storing information on the existing emergency situations. Each emergency situation is accompanied by the details on the cause of the accident, its consequences and actions undertaken to counteract propagation and the consequences of the event.

IAPS system may operate in two basic modes:

- As a central IT system of a regional range, e.g. national (version (a) in Fig. 1). In this version, the system is managed by an independent organization or company and provides access to services in the area of support for the concerned plants. Data concerning particular plants are stored in the central system.
- As an IT platform of the company or plant (version (b) in Fig. 1). In this version the system is installed on the hardware structure of the company and is administered by it, data on the particular plants are stored on independent servers.

In both versions, the access to the system for users is realized in the form of web pages. Proper mechanism of defining particular types of users and authorization allows for full control of the access to the data on plant level, including access control to particular company data for the specified external users, such as inspectors of State Fire Service.

In order to allow the realization of basic tasks in the IAPS system, it was necessary to introduce a series of mechanisms of general and administrative nature, including:

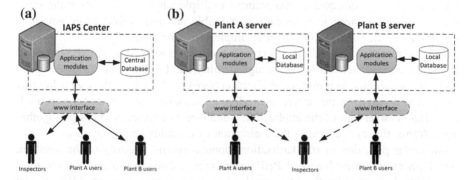

Fig. 1 Two modes of implementing IAPS system: **a** as a central system of a regional range, **b** as a dedicated platform for company or plant

- defining users and groups of users and authorizing them to preview or edit particular data and performing particular operations,
- defining in the system a description of a hierarchical structure of the process acknowledging the division into plants, departments, sections and particular technological devices,
- acquisition and storage of documentation describing technological process and installation components,
- the possibility to perform automatic evaluation of the plant as an IRI or MRI on the basis of the introduced information on the hazardous substances occurring in the plant.

A detailed list of the groups of data stored in the system contains: a description of the technological process and the general principles of the plant's functioning; a list of installation's components posing a risk together with its documentation; a list of possible faults and accidents; a list of hazardous substances in the plant; the results of risk analyses; instructions for safe operation of an installation in a normal mode, maintenance and temporary pauses; the analyses of emergency situations and internal and external operation and rescue plans; documentation maintained by the Program of Accident Prevention; documentation prepared by the Report on Security and reports on the launch, changes and stopping of the production submitted to the offices; information on the existing failures; information about persons supervising security: qualifications, powers and functions performed, required and pursued trainings etc.

3 Description of the Framework of the System

The designed system is, from a technical perspective, a web multiple access system with a relatively small load and a complicated database. Its fundamental features are: operation on a server platform, without the necessity to install any dedicated client software; modular structure, facilitating its development/adjustment to varying external requirements; maintaining multiple language versions at the same time; provided security level as high as possible and reasonable; using solutions based on free licenses in order to limit implementation costs.

The system uses a Unix/Linux server in connection with Apache web server and MySQL database. In order to obtain high stability and full control over the designed code, it was decided to introduce application performed partly in the user's Internet browser and partly on the server, according to the so-called web-desktop technology. The browser part of the application is realized in Javascript language together with Webix library facilitating the realization of complex screen layouts.

In server part, due to standardization, popularity and maturity of the solution, developer environment based on PHP programming language—Zend Framework 2 was applied. It provides the framework of routing system, supports MVC model and modular construction of the application. Doctrine, the most popular ORM library for PHP, was applied as a mechanism for mapping data and objects from the central database.

3.1 Central Database

As a central database for IAPS system, both, in regional and company's version, relational database and MySQL server were used.

Logical model of a database has a complex structure representing the structure of a business model. It was designed in such a way that it enables: flexible definition of the groups of data and the required contents allowing for quickening the implementation and easier introduction of modifications in the future; using national (language) versions; advanced tracking of the changes introduced to the documentation, including storing and providing historical versions.

3.2 Modules of Application

Internal structure of the application was divided into modules responsible for particular groups of functions (such as log-in, authentication, user management), as well as the access to specific groups of data (e.g. description of the plant, risk analyses data, journal of accidents). In each module, according to MVC model, a suitable model of data, controller and interface view was designed. User screen has been developed with the use of different access devices (computer, tablet, mobile phone) (Fig. 2).

(a)   **(b)**

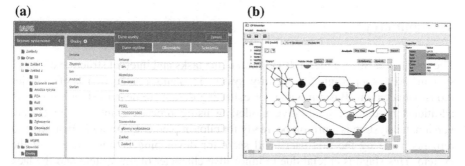

Fig. 2 Interfaces of IAPS system: **a** webpage, **a** QMod module

3.3   Module of Qualitative Modelling and HAZOP Analysis

Risk analyses are usually made by external institutions, within a dedicated group of experts. Such groups include plant's workers, specialists from the institution making the analysis at the request, or sometimes representatives of the contractor or insurance company. As a result, relevant documentation is made, which becomes a part of the description of a security system in the plant.

One of the tasks of IAPS system, as a part of collecting the documentation connected with security system, is supporting the conduction of risk analyses with particular consideration of HAZOP method. The functions of the system in this field may be divided into two task areas:

- collecting data on the performed analyses. Documentation from the performed analysis is included into data related to the description of security system in the plant. This task is realized by one of the modules of the application;
- supporting the performance of the analysis by using qualitative modelling allowing for reconstructing certain cause-and-effect relationships occurring in the process.

For designing a qualitative model of an installation and its analysis (the analysis of cause-and-effect relationships), an independent QMod module was introduced. This module is used while performing the analysis by a group of experts. It was developed as a module of the advanced monitoring and diagnostics system DiaSter [8]. In this system, tools are available for making a description of the process as regards division into component objects, defining process variables and faults, as well as determining cause-and-effect relationships in the form of process graph GP [9]. QMod module extends the functionality of the DiaSter system with the ability to easily make qualitative models and their analysis.

4   Summary

The article presents a general description of the tasks and internal structure of the IT system IAPS supporting the security system in plants posing a risk of a major industrial accident. Basic modules of collecting data were described, as well as an additional module of qualitative modelling using process graph. In the current version, the system is a prototype, ready for pilot implementation. Observations and remarks made by the first users will serve for developing another version of the system. This stage is crucial due to innovative character of the system and the lack of the settled standards and directions for using this type of application. The modular structure and the use of flexible description of the structure of data processing will make modification and development of the functionality of the designed platform easier.

Acknowledgments Publication was elaborated on the basis of the results of a 3rd stage of the long-term program "Improvement of security and working conditions" funded by the Ministry of Science and Higher Education/the National Centre for Research and Development in the years 2014–2016 in the field of scientific research and development works. Coordinator of the program: Central Institute for Labor Protection—National Research Institute.

References

1. Michalik, J. S., Domański W.: Content and objectives of the Accident Prevention System and Safety System in plants of increased and high risk of major industrial accident. Bezpieczeństwo Pracy: nauka i praktyka, pp. 22–25 (2004) (in Polish)
2. Górski, M., Kierzkowska, J. S. (eds.): Environmental Protection Law. Wolters Kluwer Polska (2009) (in Polish)
3. Seveso III Directive, Council Directive 2003/105/EC of the European Parliament amending Council Directive 96/82/EC on the control of major accident hazards involving dangerous substances. Official Journal No. L 345, 2003
4. Guidelines on a major accident prevention policy and safety management system, as required by Council Directive 96/82/EC (SEVESO II). European Commission, 1998
5. Baudisova, B., Rehacek, J., Dlabka, J., Danihelka, P.: Evaluation of risk communication in regard of SEVESO II Directive implementation in Czech. In: Steenbergen, R.D.J.M., et al. (ed.) Safety, Reliability and Risk Analysis: Beyond the Horizon, pp. 303–310 (2013)
6. Holla, K.: Complex model for risk assessment and treatment in industrial processes. IDRiM J. 4(2), 93–102 (2014)
7. Van Buren, A.: Principles and management of information process for integrated management of fire safety at SEVESO sites. Doctoral dissertation, TU Delft, Delft University of Technology, 2014
8. Syfert, M., Wnuk, P., Kościelny, J.M.: DiaSter—intelligent system for diagnostics and automatic control support of industrial processes. JAMRIS J. Autom. Mob. Robot. Intel. Syst. 5(4), 41–46 (2011)
9. Sztyber, A., Ostasz, A., Kościelny, J.M.: Graph of a process—a new tool for finding model's structures in model based diagnosis. IEEE Trans. Syst. Man Cybern. Syst. 45(7), 1004–1017 (2015)

Natural Frequencies and Multi-objective Optimization of the Model of Medical Robot with Serial Kinematical Chain

Grzegorz Ilewicz

Abstract Medical robots are used in minimally invasive surgery of soft tissue. Presently there are applied: rigid, flexible and multibody effectors in order to operate of the human body. The multibody mini-maxi structure of the medical robot that gives possibility to operate difficult accessible areas of operation is presented in the article. The finite element approximate method is used to calculate structural problems. The main goal of the article is to determine natural frequencies of the construction in order to define dangerous regions where it can occur resonance phenomenon. Then vector optimization is used to find parameters that optimized mass and frequency criterion for given boundaries. The response surface method is used to minimize the time of expensive computationally problem of optimization.

Keywords Medical robot · Natural frequencies · Finite element method · Multi-objective optimization

1 Introduction

Nowadays, medical robots with serial chain or parallel structure are used in order to perform minimally invasive surgery. Da Vinci surgical robot is generally used in cardisurgery, laparoscopy and urology and it is an example of robot with parallel structure. Other medical robot with parallel structure is Polish cardiorobot Robin Heart produced by Foundation for Cardiac Surgery Development. These robots have effectors that are similar to classical endoscopes.

The second group of medical robots are multibody structures with serial chains. The example of that structure is I-snake robot with 102° of freedom in kinematical chain or Polish construction of the medical robot ROCH-1.

G. Ilewicz (✉)
Department of Mechatronics and Control, Faculty of Mathematics
and Natural Sciences, University of Rzeszów, Rzeszów, Poland
e-mail: gilewicz@ur.edu.pl

© Springer International Publishing AG 2017
R. Jabłoński and R. Szewczyk (eds.), *Recent Global Research and Education:
Technological Challenges*, Advances in Intelligent Systems and Computing 519,
DOI 10.1007/978-3-319-46490-9_50

These structures can be used when it is necessity to reach to back wall of the operated organs in the patient body or servicing of artificial organs.

The most important problem in robotic structure is constructing of geometry of kinematical chain in such a way as to eliminate resonance phenomenon because of large amplitude of oscillation, deterioration of accuracy and repeatability of positioning of the effector. The resonance will appear when the natural frequencies of kinematical chain will be equal with frequencies of input loads, so it is important to create the numerical model that allows tuning of the robot from dangerous frequencies.

2 Natural Frequencies

Typical static problem of strength of the construction can be solved with use of the finite element method and is described by equation:

$$[K] \cdot \{u\} = \{F\} \tag{1}$$

where

$[K]$ structural matrix of stiffness,
$\{u\}$ vector of nodal displacement,
$\{F\}$ vector of external nodal forces.

If the model of Rayleigh damping can be ignored, vibration modal analysis can be described by mathematical object that has form:

$$[M] \cdot \{\ddot{u}\} + [K] \cdot \{u\} = \{0\} \tag{2}$$

where

$[M]$ structural matrix of the mass,
$\{\ddot{u}\}$ vector of second derivatives of the displacement.

The general solution of the Eq. (2) has form:

$$\{u\} = \{u_A\} \cdot \cos(\omega t) + \{u_B\} \cdot \sin(\omega t) \tag{3}$$

The second derivative of the vector of displacement (3) can be write as:

$$\{\ddot{u}\} = -\omega^2 \cdot \{u\} \tag{4}$$

Substituting (4) to (2) we receive:

$$-\omega^2 \cdot [M] \cdot \{u\} + [K] \cdot \{u\} = \{0\} \tag{5}$$

Transforming Eq. (4) we have equation that describes generalized eigenvalue problem:

$$([K] - \omega^2 \cdot [M]) \cdot \{u\} = \{0\} \tag{6}$$

where

ω eigenvalue that is natural frequency of the effector,
U eigenvector that is mode shape of the effector.

Natural frequency of vibration can be controlled changing the coefficients of stiffness or mass what is described by formula:

$$\omega_i = \sqrt{\frac{k}{m}} \tag{7}$$

where

m mass,
k stiffness.

The increasing of the frequency of vibrations results in increased stiffness and reduced strain amplitude of the effector.

3 Material of Numerical Simulation

The construction of the robot has mini-maxi structure of the kinematical chain. The maxi structure has three prismatic degrees of freedom in the cartesian configuration and it is shown on Fig. 1.

Fig. 1 Kinematical chain of the maxi structure

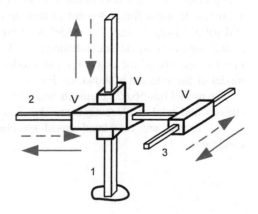

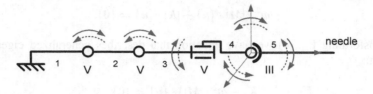
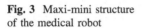

Fig. 2 Kinematical chain of the mini structure

Fig. 3 Maxi-mini structure
of the medical robot

The mini structure is the effector and has six degrees of freedom in RRRS configuration (Fig. 2). The maxi structure is intended to positioning of mini structure relatively to the trocar (access port in the human body) during minimally invasive surgery. The mini structure is dedicated to operating of soft tissue or servicing of artificial organs inside the patient. The model of the all mini-maxi structure is shown on Fig. 3.

4 Finite Element Method Model

The geometry of the model of the mini-maxi structure is discretized with use of the approximate mesh finite element method in order to calculate structural problems and natural frequencies. The model is discretized with use of the ten-node tetrahedral elements (quadratic tetrahedron). The curvature algorithm was used to improve quality of the discrete mesh model. The generated mesh of the discrete model of the effector is shown on Fig. 4.

The model has 56,625 elements and 94,962 nodes. The graphs of the equivalent stresses and total displacements are shown on the Fig. 5.

The results are calculated based on Huber hypothesis for given boundary conditions.

Fig. 4 Discrete model of the effector

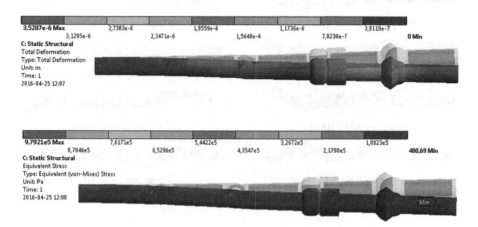

Fig. 5 The graph of stress and displacement for given boundary conditions

5 Results of Natural Frequencies

The twelve shapes of natural vibrations of the effector of medical robot is shown on Fig. 6.

The mode shapes of the effector that are shown on the Fig. 6 give a possibility to conclude about oscillation during resonant phenomenon.

6 Multi-objective Optimization of the Effector

The Pareto optimization of the model of the effector was created using multi objective genetic algorithm (MOGA) and the response surface method (Box and Wilson 1951). This algorithm uses the Pareto optimality which takes into account non dominated solutions and gives the solution in the form of global Pareto front.

The response surface is dependence between inputs (geometrical parameters of the constructions) and outputs (mass, frequency and factor of the safety).

The problem of the vector optimization for the effector of medical robot can be formulated as to find the values of decision variables $\mathbf{d}$ that optimize function:

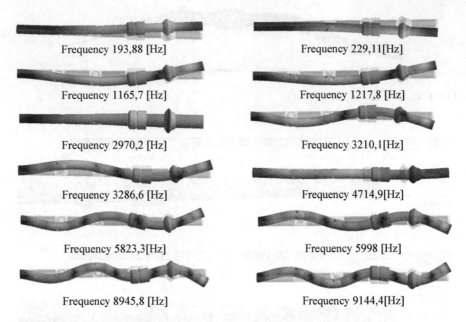

Frequency 193,88 [Hz]   Frequency 229,11[Hz]

Frequency 1165,7 [Hz]   Frequency 1217,8 [Hz]

Frequency 2970,2 [Hz]   Frequency 3210,1[Hz]

Frequency 3286,6 [Hz]   Frequency 4714,9[Hz]

Frequency 5823,3[Hz]   Frequency 5998 [Hz]

Frequency 8945,8 [Hz]   Frequency 9144,4[Hz]

Fig. 6 Mode shapes of the effector

$$f(d) = \{f_1(d), f_2(d), f_3(d)\} \tag{8}$$

with the restrictions:

$$5 \le d_1 \le 8$$
$$5 \le d_2 \le 8$$
$$f_3(d) \ge 4$$

where

$f_1(d)$ mass,
$f_2(d)$ natural frequency of the effector,
$f_3(d)$ factor of safety that is function of equivalent stress and yield strength,
d_1 the diameter of the wall of the first link,
d_2 the diameter of the wall of the second link.

The response surface of geometry relatively to first natural frequency is shown on Fig. 7. In every iteration of process of response surface creating the natural frequency, mass and factor of safety is calculating.

The optimal value of the first natural frequency calculated with use of the MOGA algorithm is equal 195.15 [Hz] for the effector, the mass of first link is equal 0.0055 kg, the mass of the second link is equal 0.007 kg and factor of the safety is better than 4.

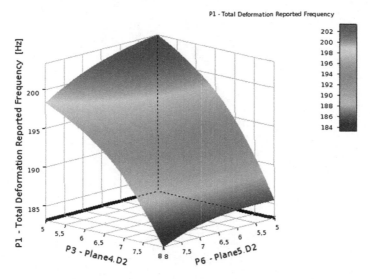

Fig. 7 The response surface for first natural frequency of the effector relatively to dimensions

7 Conclusions

The model of the mini-maxi structure of the medical robot is shown in the article. The model allows to specify strength characteristics of the construction in the field of the virtual design with use of the finite element method. This numerical method allows the elimination of expensive physical tests. Dynamical modal analysis was carried out in order to calculate natural frequencies of the effector.

The analytical metamodel (response surface) was created in order to reduce the area of the search for optimization algorithm. This approach reduces computation time. The analytical metamodel can be also used in order to inference about strength of construction when the geometry is changing without optimization process. The multi-objective optimization was performed with use of the multi objective genetic algorithm and the mass criterion was minimized and frequency criterion was maximize in order to increase the stiffness of the effector of medical robot.

References

1. Karagulle, H., Amindari, A., Akdag, M., Malgac, L., Yavuz Ş.: Kinematic-Kinetic-Rigidity evaluation of a six axis robot performing a task. Int. J. Adv. Robot. Syst. (2012)
2. Ilewicz, G., Nawrat, Z.: Optymalizacja robota medycznego z wykorzystaniem algorytmu genetycznego i metody powierzchni odpowiedzi. Medical Robots. Zabrze (2015)
3. Cook, R.: Finite Element Modeling for Stress Analysis. Wiley, USA (1994)
4. Zienkiewicz, O., Tylor, R.: The Finite Element Method, vol. 1. The Basis McGraw-Hill (2000)

Effective Testing of Precision of a Motion of the Tool Center Point of the KUKA Industrial Welding Robot in Its Various Operating Modes

Igor Košťál

Abstract The KUKA industrial robot can operate in the operating modes Manual Reduced Velocity (T1), Manual High Velocity (T2), Automatic (AUT) or Automatic External (AUT EXT). If the operator wants to test the precision of the Tool Center Point of the KUKA industrial welding robot tool motion along the programmed path in a given robot program in a serial production conditions or in conditions similar to these, he has to carry out this testing in the operating mode T2, AUT or AUT EXT. However, the operator many times changes robot operating modes during this testing and he tests the robot in T1 mode, too. We have created a .NET search application that can find required logs by the operator in the robot log file *Logbuch.txt*, which include all operating modes changes during testing of the robot program. On the basis of these found logs, the operator can choose an effective strategy for its further retesting.

Keywords Tool center point · Robot program, Log file · Testing and retesting robot · Operating mode

1 Introduction

KUKA industrial welding robots are the part of a robotic workstation which mostly, beside them, consists of manipulators and the control of whole robotic workstation. The design and creation of the robotic workstation is a relatively large-scaled and complicated process. A very important and necessary part of it is testing of particular robots in the designed and assembled robotic workstation in the developers company and their testing in the assembled workstation at a customer. This fundamentally affects the work quality of particular robots of a robotic workstation in a serial production. Part of this testing is also testing of the precision of the Tool Center Point (TCP) of each this robot tool motion along the programmed path in a

I. Košťál (✉)
Faculty of Economic Informatics, University of Economics in Bratislava, Bratislava, Slovakia
e-mail: igor.kostal@euba.sk

© Springer International Publishing AG 2017   379
R. Jabłoński and R. Szewczyk (eds.), *Recent Global Research and Education: Technological Challenges*, Advances in Intelligent Systems and Computing 519,
DOI 10.1007/978-3-319-46490-9_51

given robot program, which is carrying out in some of the operating modes T1, T2, AUT or AUT EXT, and many times in several of them. It is important, in which from these modes the operator tested this precision of the TCP of the robot tool motion, because in the mode T1 the TCP moves at reduced velocity and in the modes T2, AUT or AUT EXT the TCP moves at programmed velocity. For an effective retesting the operator needs to know, in which operating mode he tested the precision of the TCP of the robot tool motion in particular parts of a robot path in a given robot program.

We have created a .NET search application that can find this and other related information in the robot log file *Logbuch.txt* and provide their to a robot operator. In this way this application makes retesting of the precision of the TCP of the KUKA industrial welding robot tool motion along the programmed path in a given robot program more efficient and makes this process faster. This robot is the part of robotic workstation welding a cooling system radiator bracket of a passenger car. The paper deals with the functioning of our .NET search application and outputs that this application provides to the robot operator during retesting of the precision of the TCP of the welding robot tool motion.

2 Testing Precision of the TCP of a Robot Tool Motion in Various Operating Modes

Each KUKA industrial robot of a robotic workstation is equipped with some tool, for example, welding pliers. This tool is mounted on the mounting flange of a robot. During tool calibration, the robot operator assigns the working point to this tool. This point is called the TCP (Tool Center Point). All motions of the TCP are controlled by a robot program. By this is addressed through which points the TCP passes during its motion.

The TCP is a very important point of the tool robot, because its motion is programmed by a programmer during the robotic workstation design. During testing of the precision of the TCP of the KUKA industrial welding robot tool motion along the programmed path in a given robot program is tested the precision just of this point motion. This testing is carried out in various operating modes.

The KUKA industrial robot can operate in the following operating modes [1]:

- Manual Reduced Velocity (T1)—is used for a test operation, programming and teaching at programmed velocity, but maximum 250 mm/s
- Manual High Velocity (T2)—is used for a test operation at programmed velocity
- Automatic (AUT)—is used for industrial robots without higher-level controllers at programmed velocity
- Automatic External (AUT EXT)—is used for industrial robots with higher-level controllers, e.g. PLC (our case), at programmed velocity. This mode is named EX in our robot log file Logbuch.txt.

It is important, in which from presented modes the operator tested the precision of the TCP of the KUKA industrial welding robot tool motion along the programmed path in a given robot program, because in the mode T1 the TCP moves at reduced velocity and in the modes T2, AUT or AUT EXT the TCP moves at programmed velocity. If the operator wants to test this precision of the TCP of the robot tool motion along the programmed path in a given robot program in a serial production conditions or in conditions similar to these, he has to carry out this testing in the operating mode T2, AUT or AUT EXT. However, the operator many times changes robot operating modes during testing and he has to test the robot in some parts of the TCP path of the tested program in T1 mode, too. In this mode the operator tests a new or modified program, in which he founded or modified some points during its testing, because testing such programs must always be carried out first in operating mode T1.

The operator performs all additions of new points and modifications of existing points of the TCP path through the KCP (KUKA Control Panel) teach pendant in a robot program in the KUKA Robot Language (KRL) and the KUKA System Software (KSS) records them into the appropriate files to the hard drive of the robot control PC. In addition, all the operator actions on the KCP are automatically logged by KSS and he can generate the robot log file *Logbuch.txt* by KSS. All added and modified path points of the TCP and all operating modes changes during testing of the robot program are recorded into this log file, too.

For an effective retesting the operator needs to know, in which operating mode he tested the precision of the TCP of the robot tool motion in particular parts of the TCP path in a given robot program, which new points he added into this path and which points in its he modified in T1 mode. If the operator would like, after finished testing, to summarize this information, then he has to search often the large-scale robot log file *Logbuch.txt*. However, if this file has e.g. 5281 logs, which are written in 27,531 lines, it is very time consuming work which is prone to errors. Our .NET search application can help in this case because it is able to find logs of new-founded or modified points, and operating modes changes during testing of a given robot program in the robot log file *Logbuch.txt* according to the user requirements. Besides that, our .NET search application stores particular results of searches into disk files with a time stamp.

3 A .NET Application Searching for New-Founded and Modified Points and Operating Modes Changes During Testing of Programs of the KUKA Industrial Welding Robot

Our .NET search application was developed in the C# language in the development environment Microsoft Visual Studio 2013 for the Microsoft .NET Framework version 4 and for the Microsoft operating systems Windows 7 and Windows 8.1.

This .NET application searches for all new-founded and modified points in five KRL programs with names *csrb325*, *milling_caps*, *replacement_caps*, *service_- position* and *milling_cutters_test* and all operating modes changes carried out during testing of these programs in the robot log file *Logbuch.txt*. These KRL programs are part of the KUKA industrial welding robot control program. Using the *csrb325* program the robot carries out welding a given part of a cooling system radiator bracket of a passenger car. The robot uses other programs to carry out service operations. For example, if the cover caps of robot welding pliers spikes are worn, the TCP of the robot moves to the specified position, where these cover caps are milled (the *milling_caps* program). If these cover caps are worn more than the permissible value, they are exchanged for new ones (the *replacement_caps* pro- gram) in the TCP service position (the *service_position* program).

Immediately after the start-up our .NET search application attempts to connect through the Intranet to the robot control PC and searches for the directory with the *Logbuch.txt* file and the directory with DAT and SRC files of above mentioned programs on its hard drive. When the .NET search application is connected to the correct directories on the robot control PC hard drive, it will create items of both its combo boxes dynamically. The left combo box will contain items with programs names from the found directory. The .NET search application is ready to full use now.

Our .NET search application provides for each robot program a separate output to the user that includes:

- *basic statistics* about found **new-founded** and **modified points** in a given **robot program** and about found **operating modes changes** during testing of the precision of the TCP of the robot tool motion along the programmed path in a given **robot program**. The .NET search application displays these statistics in a text box.
- *all found complete logs* about **new-founded** and **modified points** in a given **robot program** and about **operating modes changes** during testing of the precision of the TCP of the robot tool motion along the programmed path in a given **robot program**. The .NET search application displays these complete logs as items in a control checked list box.
- Besides these two outputs the .NET search application creates the **disk file** with the name, for example, *NewfoundedModifPts_OperatModesChangesIn_csrb325_20160509_094723.DTX* containing the date and time when the file was created (2016-05-09 09:47:23). This file will contain the complete search results.

The output displayed in Fig. 2 provides to the robot operator, besides other information, also useful information about modifications of two points *Xp_632_src_rb_005* (Log 1662), *Xp_631_srb_rb_001* (Log 1773) and about switching robot operating modes (Log 1639, Log 1670, Log 1728, Log 1812) that relate to these modifications. From a chronology of switching operating modes the operator can see clear conjunction of these switching with carried out modifications of these two points. Before a modification of each from these points the operator switched the robot operating mode from EX to T1. After modifications of these two points the operator switched the robot operating mode from T1 to EX and he tested the precision of the TCP of the robot tool motion through the next points in EX mode, when the TCP moves at programmed velocity.

From the results of searching for new-founded points, modified points and operating modes changes during testing of the precision of the TCP of the robot tool motion along the programmed path in the robot program *csrb325* (Figs. 1 and 2),

---

The results of searching for ALL NEW-FOUNDED and MODIFIED points in the csrb325 program and operating modes changes during testing the csrb325 program
(from 2016-05-09 09:47:23Z)

The TOTAL number of ALL logs in the 'Logbuch.txt' file: 5281

The list of ALL points:
XR0, XR1, XR2, XR3, XR4, XR5, XR6, XR7, XR8, XR9, XR10, XR11, . . ., XR64,
XLZ000001, XLZ000002, Xsrb_rb_001, Xp_631_srb_rb_001, Xp_632_src_rb_001,
Xp_632_src_rb_002, Xp_632_src_rb_003, Xp_632_src_rb_004, Xp_632_src_rb_005,
XGOTO_TOV1, XTOV, Xgoto_TOV2, Xgoto_TOV

The number of new-founded points: 0
The new-founded points: 0

The number of modified points: 6
The modified points:
Xp_631_srb_rb_001, Xp_632_src_rb_001, Xp_632_src_rb_002, Xp_632_src_rb_003,
Xp_632_src_rb_004, Xp_632_src_rb_005
(The UNCHANGED points: XR0, XR1, XR2, XR3, XR4, XR5, XR6, XR7, XR8, XR9, XR10,
XR11, . . ., XR64, XLZ000001, XLZ000002, Xsrb_rb_001, XGOTO_TOV1, XTOV,
Xgoto_TOV2, Xgoto_TOV)

The points modified many times:
Xp_631_srb_rb_001 (3 times), Xp_632_src_rb_003 (2 times), Xp_632_src_rb_005
(2 times)

The operating modes used in this test: T1, EX
The number of operating modes changes: 44

Fig. 1 The part of the file *NewfoundedModifPts _OperatModesChangesIn_csrb325_ 20160509_094723.DTX* with the basic statistics of searching (the .NET search application displays the same output in its text box)

The list of ALL full logs with NEW-FOUNDED and MODIFIED points in the csrb325 program and with operating modes changes during testing the csrb325 program:

Log 1812 (User action, Warning)
2014-01-15 09:09:58'772
Kind of the mode changed from T1 to EX! Program selected. Current block: 97
Module: XEdit   MsgNo: 1

Log 1773 (User action, Error)
2014-01-15 09:05:27'373
Point modified
Point name: KRC:\R1\PROGRAM\CSRB325.DAT/Xp_631_srb_rb_001
Old coordinates: {E6POS: X -43.35685, Y -109.8231, Z 68.11096, A 103.6849,
B -1.582342, C -171.7306, S 2, T 2, E1 -38.99446, E2 0.0, E3 0.0, E4 0.0, E5 0.0, E6 0.0}
New coordinates: {E6POS: X -43.09044, Y -111.3042, Z 68.06561, A 103.6744,
B -1.765892, C -171.7293, S 2, T 2, E1 -38.99406, E2 0.0, E3 0.0, E4 0.0, E5 0.0, E6 0.0}
Module: Techhandler   MsgNo: 2

Log 1728 (User action, Warning)
2014-01-15 09:02:59'678
Kind of the mode changed from EX to T1! Program selected. Current block: 100
Module: XEdit   MsgNo: 1

Log 1670 (User action, Warning)
2014-01-15 08:53:35'423
Kind of the mode changed from T1 to EX! Program selected. Current block: 153
Module: XEdit   MsgNo: 1

Log 1662 (User action, Error)
2014-01-15 08:53:16'130
Point modified
Point name: KRC:\R1\PROGRAM\CSRB325.DAT/Xp_632_src_rb_005
Old coordinates: {E6POS: X -568.9536, Y -3.788075, Z -11.01400, A 87.51799,
B 57.78004, C 173.7463, S 6, T 18, E1 -37.88382, E2 0.0, E3 0.0, E4 0.0, E5 0.0, E6 0.0}
New coordinates: {E6POS: X -568.9494, Y -3.788010, Z -11.01087, A 87.51798,
B 57.77988, C 174.2808, S 6, T 18, E1 -37.88372, E2 0.0, E3 0.0, E4 0.0, E5 0.0, E6 0.0}
Module: Techhandler   MsgNo: 2

Log 1639 (User action, Warning)
2014-01-15 08:51:48'583
Kind of the mode changed from EX to T1! Program selected. Current block: 153
Module: XEdit   MsgNo: 1

Fig. 2 The part of the file *NewfoundedModifPts _OperatModesChangesIn_csrb325_ 20160509_094723.DTX* with found complete logs (the .NET search application displays the same output in its checked list box)

result to **the operator**, that he **will have to retest this precision of the TCP motion in the** *csrb325* **program in the points** *Xp_631_srb_rb_001*, *Xp_632_src_rb_001*, *Xp_632_src_rb_002*, *Xp_632_src_rb_003*, *Xp_632_src_rb_004*, *Xp_632_src_rb_005*, because he modified their coordinates during testing this program, some of them many times, in the operating mode T1. The TCP moves at reduced velocity (max. 250 mm/s) in this operating mode. **The operator will has to retest the precision of the TCP of the robot tool motion in these points in** operating modes **T2** or **EX**, when the TCP moves at programmed velocity. Additionally, from these results of searching the operator finds out, that **testing of this program was carried out in** operating modes **T1** and **EX**. He can also get information, that during the TCP of the robot tool motion through others points as modified, the robot operated in the operating mode EX, when the TCP moves at programmed velocity. For **the robot operator** this is an important information, because on the basis of it he knows, that he **does not have to retest** this **motion precision in these other points** *XR0, XR1, XR2, XR3, XR4, XR5, XR6, XR7, XR8, XR9, XR10, XR11, ..., XR64, XLZ000001, XLZ000002, Xsrb_rb_001, XGOTO_TOV1, XTOV, Xgoto_TOV2* and *Xgoto_TOV*. Our .NET search application will also search these points, marks their by the text *the UNCHANGED points* and displays in the output.

The .NET search application provides information of the same kind for other four robot programs *milling_caps*, *replacement_caps*, *service_position* and *milling_cutters_test*, too. Such information can make retesting of each of these programs more efficient and make this process for the operator faster.

4 Conclusion

The most valuable outputs that our .NET search application provides to the robot operator after carried out testing of the precision of the TCP of the KUKA industrial welding robot tool motion along the programmed path in a given robot program, are:

- the list of new-founded and modified points during this testing and with them related robot operating modes changes
- the list of switching robot operating modes during this testing
- the list of unchanged points during this testing.

From this information is the robot operator clear, that he has to retest this precision of the TCP of the robot tool motion in a given robot program in new-founded and modified points and in operating modes T2 or EX, when the TCP moves at programmed velocity. Additionally, from this information the operator knows, that he does not have to retest this motion precision in unchanged points, because he tested it in these points during carried out testing in EX mode, when the TCP moves at programmed velocity. This accurate information can make the retesting of the precision of the TCP of the robot tool motion in a given robot program more efficient and faster. Thus, our .NET search application can be a very effective software support for the operator during testing and retesting of a given robot program.

References

1. KUKA Roboter GmbH: KUKA System Software, KUKA System Software 5.5, Operat-ing and Programming Instructions for System Integrators. KUKA Roboter GmbH, Augsburg (2010)
2. KUKA Roboter GmbH: KUKA System Technology, KUKA.RobotSensorInterface 2.3 for KUKA System Software 5.4, 5.5, 5.6, 7.0. KUKA Roboter GmbH, Augsburg (2009)
3. KUKA Roboter GmbH: Controller, KR C4; KR C4 CK, Specification. KUKA Roboter GmbH, Augsburg (2013)
4. KUKA Roboter GmbH: KUKA System Technology KUKA.Load 5.0 KUKA.Load Pro 5.0 Valid for KSS 5.5, 5.6, 8 and VSS 8. KUKA Roboter GmbH, Augsburg (2013)
5. Microsoft Corporation 2016 "MSDN Library on-line". http://msdn.microsoft.com

Modified Flow Rate Algorithm for Leak Detection in Transient State from a Liquid Pipeline's Operating Point Change

Paweł Ostapkowicz, Mateusz Turkowski and Andrzej Bratek

Abstract This article presents the modified algorithm to improve a detection of a leakage in liquid transmission pipeline. It is focused on the leaks, which occur during changes of pipeline's operating conditions. The algorithm uses cyclical measurements of flow rate on both ends of the pipeline. In comparison to the well-known and commonly used solution, it is characterized by a new process variable structure and two options of a resulting function. The proposed algorithm was evaluated by carrying out experimental tests on the physical model of a pipeline.

Keywords Pipeline · Leak detection · Transient

1 Introduction

A common way to diagnose leakages in liquid transmission pipelines are methods based on measurements of internal flow parameters such as: flow rate, pressure and temperature. In industry practice, these methods are applied by leak diagnostic systems.

P. Ostapkowicz (✉)
Faculty of Mechanical Engineering, Bialystok University of
Technology, Bialystok, Poland
e-mail: p.ostapkowicz@pb.edu.pl

M. Turkowski
Faculty of Mechatronics, Warsaw University of
Technology, Warsaw, Poland
e-mail: m.turkowski@mchtr.pw.edu.pl

A. Bratek
Industrial Research Institute for Automation
and Measurements, Warsaw, Poland
e-mail: abratek@piap.pl

© Springer International Publishing AG 2017
R. Jabłoński and R. Szewczyk (eds.), *Recent Global Research and Education:
Technological Challenges*, Advances in Intelligent Systems and Computing 519,
DOI 10.1007/978-3-319-46490-9_52

Considering the full scope of tasks, which such systems should perform, the very first and crucial one is the ability to detect a leakage. The following steps would be: leak localization and estimation of its size. Nevertheless, in this paper we focus exclusively on the leak detection activity.

Still detecting a leak in transient states definitely causes a significant issue. In practice, it involves applying sophisticated methods, which often are based on process dynamics models [1–5]. The implementation of such solutions is very complex and expensive [2]. Some of these methods might even require performing special tests [3], which means interfering with pipeline's operational process. Due to the above mentioned reasons, less complex leak detection algorithms are being sought.

In the previous paper [6], the authors presented a few simplified techniques for a single leak detection in a transient state, which was provoked by a change in pipeline's operating point. The discussed techniques were experimentally verified.

In this paper, the authors present a modified solution of the simplified detection algorithm, aimed at identifying a single leak in the course of pipeline's operating point change. The effectiveness of the algorithm was evaluated with the data obtained from the laboratory experimental pipeline installation. It is worth mentioning that a considerable significant changes in the pipeline's operating point were taken into account.

2   Experimental Data Acquired from the Pilot Pipeline

2.1   The Pilot Pipeline Installation

The laboratory pipeline installation for water pumping is established at the Faculty of Mechanical Engineering of the Bialystok University of Technology. Its total length is close to 400 m with the main pipe section of 380 m long. The installation, assembled of polyethylene tubes (HDPE) of 34 mm internal diameter and 40 mm external diameter, is controlled with a variable flow pump. At the pipeline's inlet and outlet there are two semi-open tanks of 300 dm^3 capacity each.

On the pipeline there have been installed two electromagnetic flow meters (at the inlet and outlet at coordinates $z_{in} = -6$ and $z_{out} = 382.2$ m), six pressure transducers (at coordinates $z_{1(in)} = 1$, $z_2 = 61$, $z_3 = 141$, $z_4 = 201$, $z_5 = 281$ and $z_{6(out)} = 341$ m) and two thermometers.

2.2   Conditions of Experiments

An overview of performed experiments is shown in Fig. 1. Experiments consisted in changing nominal parameters of water pumping concurrently with simulating of a single leakage. In this way, leakages started up in the transient state. The pipe's

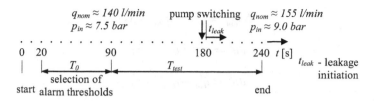

Fig. 1 Time scenario of experiments

operating point[1] was shifted by increasing the pump's rotation velocity. Prior to this setting, the pipeline was operating in steady state conditions.

The leakages were simulated in the following manner: their initiation was synchronized with the change of the pump's operating conditions to ensure a burst just on time with reaching the leak tap position by the pressure wave, provoked by the pump's handling.

The experiments were carried out in two scenarios: with and without leakage simulation.

Flow rate and pressure signals were sampled with frequency $f_p = 100$ Hz.

3 Modified Algorithm for Leak Detection and Results of Its Implementation

Besides discussing the proposed modified algorithm, the main goal of this paper is to compare the used so far algorithms. Both methods were tested in the previously described experimental conditions.

3.1 Overview of Algorithms

The structure of previous algorithm corresponds to solutions commonly applied [4, 5]. At the beginning, by using the measured flow rate signals q_{in} and q_{out}, x_1 and x_2 variables in the shape of Δq_{in} and Δq_{out} residua are calculated according to the relationships (1) and (2). However, the novelty of the proposed solution resides in applying a different approach for calculations of x_1 and x_2 variables, i.e. as differences dq and $d\bar{q}$, according to relationships (3) and (4).

$$\Delta q_{in}^k = q_{in}^k - \bar{q}_{in}^k = x_1^k \tag{1}$$

---

[1]The operating point of the pipeline is defined by the values of nominal flow rate as well as inlet and outlet pressure.

$$\Delta q_{out}^k = q_{out}^k - \bar{q}_{out}^k = x_2^k \tag{2}$$

$$dq^k = q_{in}^k - q_{out}^k = x_1^k \tag{3}$$

$$d\bar{q}^k = \bar{q}_{in}^k - \bar{q}_{out}^k = x_2^k \tag{4}$$

where $\bar{q}_{in}^k$ and $\bar{q}_{out}^k$ are the reference values at the moment k, k is the moment resulting from the applied sampling time T_P of signals q_{in} and q_{out}.

The reference variables $\bar{q}_{in}$ and $\bar{q}_{out}$, which are considered in relationships (1)–(4) might be calculated with recursive filtration of the flow rate data q_{in} and q_{out}. They may be calculated as well for example with the use of mathematical dynamic model for a pipeline without a leak [1].

The following step in both algorithms is based on calculating a cross-correlation function (5) of the variables x_1 and x_2, using N sample long data vectors (time windows). The number of N samples in the time window covers the period $\langle k - \tau, k \rangle$. Then, their average value is found according to (6). In the next step, the obtained calculations are recursively filtered (7).

$$R_{x_1 x_2}(m) = \frac{1}{N} \hat{R}_{x_1 x_2}(m - N) \tag{5}$$

where $m = 1, 2, \ldots, 2N - 1$.

$$\hat{R}_q^k = \frac{1}{2N + 1} \sum_{m=k-\tau}^{m=k+\tau} R_{x_1 x_2}(m) \tag{6}$$

$$\hat{R} f_q^k = (\beta \cdot \hat{R} f_q^{k-1}) + ((1 - \beta) \cdot \hat{R}_q^k) \tag{7}$$

where $\hat{R} f_q^{k-1}$ is the reference value in the moment $k - 1$ resulting from the applied sampling period T_P, β is the filter correction factor $0 < \beta < 1$.

Let's assume that the function (7) is denoted identically in case of the previous algorithm, i.e. as $\hat{R} f_q$. However, for the proposed algorithm the $\hat{R} f_q'$ symbol will be used.

In addition, the proposed algorithm assumes as well that for leak detection purpose, besides the $\hat{R} f_q'$ function, an additional function $\hat{R} f_q''$ might also be considered. The use of $\hat{R} f_q''$ function targets the noise avoidance. It is calculated with the means of median filtration (8)

$$\hat{Rf}_q''^k = med\left[\hat{Rf}_q'^{k-i}, \hat{Rf}_q'^{k-i+1}, \dots \hat{Rf}_q'^k, \dots \hat{Rf}_q'^{k+i-1}, \hat{Rf}_q'^{k+i}\right] \tag{8}$$

where $N_A = 2i + 1$ is the number of samples.

A leak alarm is generated, when particular functions $\hat{Rf}_q$, $\hat{Rf}_q'$, $\hat{Rf}_q''$ exceed their alarm threshold, according to conditions given in (9) and (10).

$$\hat{Rf}_q < Pal_q^- \cup \hat{Rf}_q > Pal_q^+ \tag{9}$$

$$\hat{Rf}_q' > Pal_q' \quad \hat{Rf}_q'' > Pal_q'' \tag{10}$$

3.2   The Results

Examples of the courses of several diagnostic functions $\hat{Rf}_q$, $\hat{Rf}_q'$, $\hat{Rf}_q''$ are presented in Fig. 2. Figure 2a, c consider the change in pipeline's operating point in case of the pipeline with a simulated leak, while Fig. 2b, d in case of the pipeline without a leak. The existing algorithm (function $\hat{Rf}_q$) flags indeed a change in the pipeline's operating point (in the form of a pulse), but it doesn't allow to detect a leak. The proposed algorithm (functions $\hat{Rf}_q'$, $\hat{Rf}_q''$) also signals the change of an operating point (in the shape of a pulse). However, what is crucial, it enables leak detection (when the decreasing function's values are outside the alarm threshold margin) and distinguish a state without a leak (when the decreasing function's values are again inside their alarm thresholds' margins).

The results of compared algorithms obtained for the experiments with different size of leakages at the point of coordinate 155 are presented in Table 1. The results concern two indicators. The t_{det} indicator is the response time of operational change in the pipeline (expressed in seconds), which was calculated with regards to the moment of pump's rotation velocity change. The second indicator is the leak detection, which was denoted with "+" (if it was achievable according to the rule described in the previous paragraph) or with "−" (if it wasn't possible).

While analyzing the results, we might state that the proposed algorithm enabled the detection of simulated leaks with the size of 0.5 % of nominal flow rate. However, the leak detection was not possible with the previous algorithm.

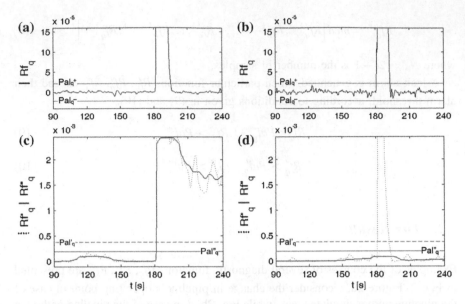

Fig. 2 Indicator functions for compared algorithms in the state with leakage (*left column*) and without leakage (*right column*): **a** and **b** earlier solution, **c** and **d** proposed solution

Table 1 Results of leak detection indicators obtained for previous and proposed algorithms

No.	z_{leak}	q_{leak} (%)	Time response of operation change t_{det} (s)			Detection of leakage		
			$\hat{R}f_q$	$\hat{R}f_q'$	$\hat{R}f_q''$	$\hat{R}f_q$	$\hat{R}f_q'$	$\hat{R}f_q''$
1	155	1.0	1.92	1.94	16.93	−	+	+
2		1.0	1.85	1.95	16.95	−	+	+
3		0.6	1.89	1.97	16.96	−	+	+
4		0.7	1.93	2.19	15.31	−	+	+
5		0.5	1.91	1.78	16.77	−	−	+
6		0.5	2.14	1.79	16.78	−	+	+
7		0.4	1.14	1.90	16.89	−	−	−
8		0.4	2.08	1.73	16.71	−	−	−

4 Conclusion

This paper presents a very efficient simplified modification of the algorithm to detect leakages in transient states.

The algorithm' effectiveness has been proved by the obtained results gathered on its implementation on the pilot pipeline, when for the given change of pipeline's transmission conditions, related to a pump's rotation velocity change, by around 10 % of the nominal flow rate, it was possible to detect even 0.5 % volume single simulated leakages in transient states.

Acknowledgments The research work is supported by Bialystok University of Technology, as the research project Nr S/WM/1/2016.

References

1. Begovich, O., Pizano-Moreno, A., Besancon, G.: On-line implementation of a leak isolation algorithm in a plastic pipeline prototype. Lat. Am. Appl. Res. **42**, 131–140 (2012)
2. Colombo, A.F., Lee, P., Karney, B.W.: A selective literature review of transient-based leak detection methods. J. Hydro Environ. Res. **2**(4), 212–227 (2009)
3. Haghighi, A., Covas, D., Ramos, H.: Direct backward transient analysis for leak detection in pressurized pipelines: from theory to real application. J. Water Supply Res. Technol. AQUA **63** (3), 189–200 (2012)
4. Isermann, R.: Leak detection of pipelines. In: Fault-Diagnosis Applications: Model-Based Condition Monitoring: Actuators, Drives, Machinery, Plants, Sensors and Fault-tolerant Systems, pp. 181–204. Springer, Berlin (2011)
5. Kowalczuk, Z., Gunawickrama, K.: Detecting and locating leaks in transmission pipelines. In: Korbicz, K.J., Koscielny, J.M., Kowalczuk, Z., Cholewa, W. (eds.) Fault Diagnosis: Models, Artificial Intelligence, Applications, pp. 822–864. Springer, Berlin (2004)
6. Ostapkowicz, P., Bratek A.: Leak detection in liquid transmission pipelines during transient state related to a change of operating point. In: Kowalczuk, Z. (ed.), Advanced and Intelligent Computations in Diagnosis and Control, Advances in Intelligent Systems and Computing, vol. 386, pp. 253–265. Springer International Publishing, Switzerland (2016)

Modular Multidisciplinary Models for Prototyping Energy Harvesting Products

Jan Smilek, Ludek Janak and Zdenek Hadas

Abstract Energy harvesting systems provide an alternative solution to conventional energy sources by utilizing some form of energy from the ambient environment followed by its conversion to useful electric energy. Designing an efficient energy harvester requires prior knowledge and analysis of the conditions in the location of its intended installation. Therefore, an original energy harvesting solution is needed for every application. The development process of energy harvesters could be sped up by using modular fully parametric models of the system components. These models are developed from basic physical principles in such a way, that they can be freely combined into the multidisciplinary model of the system. The modules are intended to be used for the fast prototyping, comparison of different topologies and transducer types and for prediction of the possible output energy levels. The main motivation for a derivation of such models is their future implementation to the energy harvesting course for mechatronic engineering students. That will allow students to focus on the practical issues of designing an energy harvester for real-life applications using real input data instead of building, debugging and optimizing generic models of the energy harvesting transducers.

Keywords Energy harvesting · Modelling · Kinetic energy · Engineering education · Development methodology

J. Smilek (✉) · L. Janak · Z. Hadas
Faculty of Mechanical Engineering, Brno University of Technology,
Brno, Czech Republic
e-mail: smilek@fme.vutbr.cz

L. Janak
e-mail: janak@fme.vutbr.cz

Z. Hadas
e-mail: hadas@fme.vutbr.cz

© Springer International Publishing AG 2017
R. Jabłoński and R. Szewczyk (eds.), *Recent Global Research and Education:
Technological Challenges*, Advances in Intelligent Systems and Computing 519,
DOI 10.1007/978-3-319-46490-9_53

1   Introduction

Developing energy harvesters for various applications requires the knowledge of
the application environment, constraints placed on the size and performance of the
harvester and knowledge of the possible employable design topologies and energy
transduction methods. During the modelling and simulation phase however, only
few basic types of models are usually used, differing merely in the parameters of
their governing equations.

Most of the kinetic energy harvesters for instance, use a proof mass with one [1]
or two [2] degrees of freedom, dynamic behavior of which is described by the
motion equations. The transduction of energy from kinetic to electric is usually
modelled employed the equations for viscous or electrostatic damping depending
on the type of the harvester [3].

Thermoelectric generators (TEGs) and their modelling can be considered as
another typical example. The most typical model of TEG can be built using a
combination of three types of equations. Thermal domain equations expressing the
effect of heat conduction, convection and radiation are used as a representation of a
thermal part of the thermoelectric energy harvesting system. The Seebeck equation
is used as a model for thermoelectric energy conversion. Finally, the Ohm's and
Kirchhoff's laws are applied to describe the electric equivalent circuit [4].

All the above-mentioned principles of energy harvesting are taught as a part of
Energy Harvesting course for master's engineering students at our university [5].

2   Generic Energy Harvester

Every energy harvesting system can be divided into the following building blocks
(modules) regardless on its operation principle or application constrains:

- Ambient energy—energy to be harvested (energy input). Parameters of this
  input are obtained by measurement.
- Raw energy domain—module of the energy harvesting model ensuring the
  energy treatment before entering the energy conversion. This module can e.g.
  represent mechanical constrains in kinetic energy harvesters or thermal contact
  of thermoelectric generator to its surroundings.
- Conversion to electric energy—the crucial module of the energy harvester
  which performs the conversion of raw energy to the useful electric energy. It can
  be ensured by e.g. electromagnetic, piezoelectric, thermoelectric, pyroelectric, or
  photovoltaic effects.
- Electric energy domain—module which can be described by the electric
  equivalent circuit. This part of model encompasses ideal source of electric
  voltage or current, variable internal impedance and generally variable electric
  load defined by supplied application. Both internal and load impedance can have

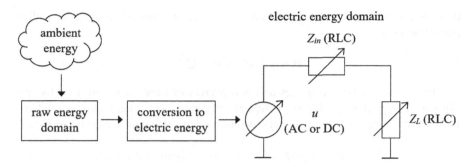

Fig. 1 Diagram of the generic energy harvesting system

the Resistive (R), Inductive (L) and Capacitive (C) behavior or their combination.

Diagram of the generic energy harvesting system is depicted in the following Fig. 1.

3 Modelling Approach for Prototyping Energy Harvesters

A generic set of models and subsystems is developed in such a way, that most of the common designs and topologies can be easily built by combining the ready-made subsystems without having to derive the model governing equations and debug them every single time. This leads to speeding up the modelling phase of the development, allowing the saved time to be used for optimization of the parameters etc. The generic models and subsystems are built as freely combinable with each other and allow for employing the data obtained from analyses in other software.

4 Specific Example: Kinetic Energy Harvesters' Models

Developed models of kinetic energy harvesters include generic models of harvesters with 1, 2 and 3 bodies, where each body has one degree of freedom [6]. Furthermore, several stiffness and damping subsystems, describing different possible characteristics of the elements were created. The harvester models are built in MATLAB/Simulink using differential equations derived from Lagrange equation of the second kind. Number of generalized coordinates and necessary differential equations is equal to the number of degrees of freedom.

Due to using generalized coordinate q and generalized mass, damping and stiffness properties m^*, b^* and k^*, respectively, it is necessary to take care of the correct units of these parameters, depending whether a torsional or linear movement is being investigated. Motion of the harvester with one degree of freedom can be described by the motion equation with generalized parameters and time-dependent excitation $Q(t)$:

$$m^*\ddot{q} + b^*\dot{q} + k^*q = Q(t) \tag{1}$$

The harvester with two bodies and two degrees of freedom is described by two differential equations with two independent generalized coordinates q_2 and q_1. Indexes of the parameters denote the position of the elements according to Fig. 2.

$$m_1^*\ddot{q}_1 + b_1^*\dot{q}_1 - b_{12}^*(\dot{q}_2 - \dot{q}_1) + k_1^*q_1 - k_{12}^*(q_2 - q_1) = Q_1(t) \tag{2}$$

$$m_2^*\ddot{q}_2 + b_2^*\dot{q}_2 + b_{12}^*(\dot{q}_2 - \dot{q}_1) + k_2^*q_2 + k_{12}^*(q_2 - q_1) = Q_2(t) \tag{3}$$

Model of the 3dof harvester exploits following three 2nd order differential equations:

$$\begin{aligned} m_1^*\ddot{q}_1 + b_1^*\dot{q}_1 - b_{12}^*(\dot{q}_2 - \dot{q}_1) - b_{13}^*(\dot{q}_3 - \dot{q}_1) \\ + k_1^*q_1 - k_{12}^*(q_2 - q_1) - k_{13}^*(q_3 - q_1) = Q_1(t) \end{aligned} \tag{4}$$

$$\begin{aligned} m_2^*\ddot{q}_2 + b_2^*\dot{q}_2 + b_{12}^*(\dot{q}_2 - \dot{q}_1) - b_{23}^*(\dot{q}_3 - \dot{q}_2) \\ + k_2^*q_2 + k_{12}^*(q_2 - q_1) - k_{23}^*(q_3 - q_2) = Q_2(t) \end{aligned} \tag{5}$$

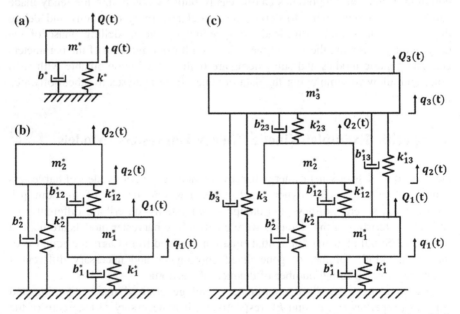

Fig. 2 Spring mass damper systems used to describe the behavior of the harvester with a 1dof, b 2dof, c 3dof

$$m_3^* \ddot{q}_3 + b_3^* \dot{q}_3 + b_{13}^*(\dot{q}_3 - \dot{q}_1) + b_{23}^*(\dot{q}_3 - \dot{q}_2)$$
$$+ k_3^* q_3 + k_{13}^*(q_3 - q_1) + k_{23}^*(q_3 - q_2) = Q_3(t) \tag{6}$$

As all the models contain stiffness and damping parameters between all the bodies and frame, and also between each pair of the bodies, it is possible to obtain all possible different variants of the topology by simply setting the unwanted parameters to 0 or deleting their subsystems from the model.

In order to be able to easily model different types of nonlinear behavior in stiffness, subsystems with different characteristics were developed. The simplest subsystem consists of a constant and represents linear stiffness:

$$F_s^*(q) = k^* q \tag{7}$$

where $F_s^*(q)$ denotes the generalized damping force. Nonlinear behavior of the duffing oscillators can be investigated employing submodel with generalized force dependency:

$$F_s^*(q) = k_1^* q + k_3^* q^3 \tag{8}$$

Depending on the settings of constants k_1^* and k_3^* both hardening and softening or bistable characteristics of the oscillator can be achieved. The effect of mechanical bumpers or step change of the stiffness depending on the position of the oscillator is modelled by the subsystems employing following equation:

$$F_s^*(q) = \begin{cases} k^* q + k_b^*(q - q_{low}) & q < q_{low} \\ k^* q & q_{low} \leq q \leq q_{high} \\ k^* q + k_b^*(q - q_{high}) & q > q_{high} \end{cases} \tag{9}$$

Last stiffness submodel describes the behavior of the pendulum-type harvester with no additional spring (10). L in the Eq. (10) stands for the distance of the center of gravity of the pendulum from the pivotal point. All the developed submodels can be freely combined to achieve unique combinations of characteristics in order to characterize and simulate any investigated topology. A submodel consisting of look-up table was also created to allow for easy exploitation of the stiffness data obtained e.g. from measurements or FEM analyses.

$$F_s^*(q) = m^* g L \cdot \sin q \tag{10}$$

Subsystems for characterizing the damping forces on the system were built in similar manner, resulting in models of viscous (11) and coulomb (12) damping, usable for both mechanical damping in the system due to mechanical losses and electrical damping due to the energy transduction.

$$F_b^* = b^* \cdot \dot{q} \tag{11}$$

$$F_b^* = b^* \cdot sign(\dot{q}) \tag{12}$$

One or more of the damper subsystems will thus consist of two parts—electrical and mechanical damping b_e^* and b_m^*, respectively:

$$b^* = b_e^* + b_m^* \tag{13}$$

Voltage induced on the electrical load is then a function of the electrical damping, respective velocity on the damper and electrical load Z_L:

$$u(i) = f(b_e, \dot{q}, Z_L) \tag{14}$$

5 Implementation to Mechatronic Education

The energy harvesting device is a typical example of mechatronic system. It contains one or even more types of energy conversion and integrates mechanical, electrical, and (if needed) intelligent control algorithm. Development of such device can take a benefit from V-cycle methodology. The following steps have been accommodated to the educational course:

- application scenario—power supply requirements, ambient energy in surroundings;
- identification of target conditions by measurement—determination of ambient energy levels (e.g. acceleration of vibrations);
- model setup—input of application-related data only instead of debugging, etc.;
- simulation—running the model with target conditions from measurement;
- detailed design—developing the components, material characteristics, etc.

Each development step is constrained to the previous one by validation/verification feedback. By meeting all the presented development steps while using the presented model, students are able to understand the physical behavior of the energy harvester, Moreover, the mechatronic approach to engineering a system is presented.

6 Conclusions

The energy harvesting systems are complex engineering products combining a couple of physical domains, design requirements and features. Fortunately, the generic structure of energy harvesting system allows us to advantageously use their modular multidisciplinary models. It has been shown that such models are able to significantly speed up the energy harvester development process.

In this paper, the model development strategy was presented on an example of kinetic energy harvesting system with 1–3 degrees of freedom. The derived model is implementation-independent. However, its implementation for educational purposes was performed in MATLAB/Simulink.

The modules are intended to be used for the fast prototyping, comparison of different topologies and transducer types and for prediction of the possible output energy level. The model will allow students to focus on the practical issues of designing an energy harvester for real-life applications using real input data instead of building, debugging and optimizing generic models of the energy harvesting transducers. The presented method will form an integral part of Energy Harvesting course for master's students at the Faculty of Mechanical Engineering, Brno University of Technology.

Acknowledgments This work has been funded by the FME, Brno University of Technology, project FV 16-13 Multidisciplinary models for prototyping independent power sources.

References

1. Foisal, A.R.M., Hong, C., Chung, G.-S.: Multi-frequency electromagnetic energy harvester using a magnetic spring cantilever. Sens. Actuators A Phys **182**, 106–113 (2012). doi:10.1016/j.sna.2012.05.009
2. Kim, I.-H., Jung, H.-J., Lee, B.M., Jang, S.-J.: Broadband energy-harvesting using a two degree-of-freedom vibrating body. Appl. Phys. Lett. **98**, 214102 (2011). doi:10.1063/1.3595278
3. Beeby, S.P., Tudor, M.J., White, N.M.: Energy harvesting vibration sources for microsystems applications. Meas. Sci. Technol. **17**, R175–R195 (2006). doi:10.1088/0957-0233/17/12/R01
4. Janak, L., Ancik, Z., Hadas, Z.: Simulation modelling of MEMS thermoelectric generators for mechatronic applications. In: Mechatronics 2013 (Recent Technol. Sci. Adv.) pp. 265–271 (2014)
5. Hadas, Z., Singule, V.: New subject energy harvesting for education of mechatronics. In: 13th International Symposium on Mechatronika, 2010, pp. 16–17 (2010)
6. Hadas, Z., Vetiska, V., Vetiska, J., Krejsa, J.: Analysis and efficiency measurement of electromagnetic vibration energy harvesting system. Microsyst. Technol. (2016). doi:10.1007/s00542-016-2832-4

A Multi-attribute Classification Method to Solve the Problem of Dimensionality

A.R. Várkonyi-Kóczy, B. Tusor and J.T. Tóth

Abstract Classification is one of the most important areas of machine learning. However, there are numerous applications where the quantity of attributes is very large, rendering the usage of conventional classifiers very slow or even impossible. The classifier method in this paper is proposed for such problems. Using the assumption that very large problem spaces are typically sparse as well (considering the stored knowledge), it maps the multi-dimensional problem space into a sequential combination of two-dimensional subdomains. The classifier is easy to implement, fast, and capable of recognizing patterns that are similar to known ones.

Keywords Supervised learning · Classification · Fuzzy classification · Problem of dimensionality

A.R. Várkonyi-Kóczy (✉) · J.T. Tóth
Department of Mathematics and Informatics, J. Selye University,
Komarno, Slovakia
e-mail: varkonyi-koczy@uni-obuda.hu

J.T. Tóth
e-mail: tothj@selyeuni.sk

B. Tusor
Integrated Intelligent Systems Japanese-Hungarian Laboratory,
Óbuda University, Budapest, Hungary
e-mail: balazs.tusor@gmail.com

B. Tusor
Institute of Mechatronics and Vehicle Engineering, Óbuda University,
Budapest, Hungary

B. Tusor
Doctoral School of Applied Informatics and Applied Mathematics,
Budapest, Hungary

© Springer International Publishing AG 2017
R. Jabłoński and R. Szewczyk (eds.), *Recent Global Research and Education:
Technological Challenges*, Advances in Intelligent Systems and Computing 519,
DOI 10.1007/978-3-319-46490-9_54

403

1 Introduction

Classification is an invaluable tool to solve problems where the system needs to associate input patterns with different classes. They are widely used in most fields in physics, like in photonics, material science, nanotechnology, biotechnology, environmental engineering, etc.

For example, [1] applies an Artificial Neural Network [2] to estimate the average size of titanium dioxide nanoparticles based on near infrared diffuse reflectance spectra. In [3], Support Vector Machines [4] are used as a nonlinear equalizer in coherent optical orthogonal frequency-division multiplexing.

Another example is [5] that utilizes radial basis networks [6] in order to predict wastewater outputs based data gained and denoised by fuzzy-rough sets methods [7].

However, there are numerous applications where the quantity of attributes is very large, rendering the usage of conventional classifiers very slow or even impossible.

The classifier method in this paper is proposed for such problems. Its base idea comes from fuzzy hypermatrices [8] (which are specialized versions of look-up tables [9]) that realize nearest-neighbor classification in order to recognize patterns similar to known ones. It is done by mapping the problem space into the memory in form of multidimensional matrices, so the class of the input data can be gained instantly in the evaluation phase. The downside of the base method is that the memory requirements scale exponentially with the number of attributes and the size of each attribute domains. The proposed classifier solves this problem by assuming that the bigger a problem space, the sparser it is (considering the stored knowledge). Therefore, only a part of the domain is needed to be stored. Instead of using a single multidimensional matrix, the proposed classifier consists of a layered structure, breaking the multi-dimensional problem to a sequential combination of 2D fuzzy matrices.

The rest of the paper is as follows: Sect. 2 describes the architecture, functionality, and training of the system and presents experiments to show the efficiency of the classifier, while Sect. 3 concludes the paper and presents future work.

2 The Architecture of the Multi-dimensional Classifier

The proposed classifier has a layered architecture that consists of N layers (where N is the amount of attributes, i.e. the dimension of the problem). It is divided to 3 parts: a top layer, $N - 2$ intermediary layers, and a bottom layer.

The purpose of the top layer is to combine the first two attributes. In the first intermediary layer this combination is combined with the attribute associated with the layer, gaining the combination of the first three attributes. This is repeated for all intermediary layers until the bottom layer is reached, where the class belonging to the combination of the whole attribute vector is determined.

The topology of the classifier can be seen in Fig. 1. The top layer and each intermediary layers have an index matrix and a fuzzy matrix (Λ^L and M^L, where L is the given layer). Former ones contain the index number of the given combination, determining which row is examined in the layer below it (so for each unique element (so-called marker) a unique row is assigned in the next layer). The fuzzy matrices assign fuzzy membership function values to the elements of the index matrices (in the same layer). This way a certain level of generalization ability is added to the system: the classifier is capable of recognizing not only the data it has been taught with, but unseen data as well that is similar to the learned ones. The size of the matrices in the top layer is the combined size of the first two attributes, while for each intermediary layer the size of the matrices depends on the domain size of the attribute associated with the given layer and the number of markers in the layer directly above it. Let us mark the amount of markers in layer L with γ^L.

The bottom and each intermediary layer have additional statistical information as well, in the form of a 2D matrix G^L and a 1D array υ^L. In the former, the number of class occurrences (given a C-class problem) is noted for each row (noting the class

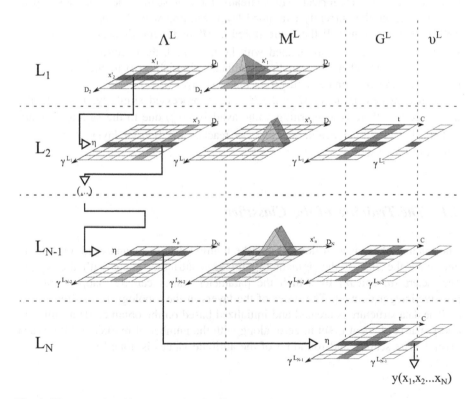

Fig. 1 The general architecture of the classifier

distributions with respect to the known value combination), and in the latter the most commonly occurring class is stored (also for each row).

Determining the order of attributes directly influences the topology of the structure, and through that, the accuracy and size of the classifier. Since one of the primary objectives is reducing the size of the stored problem space, an adequate ordering can be acquired by examining the available training data (attribute by attribute), and learn how much of the individual 1D domains are covered by the known values. Sorting the layers into an ascending order of the coverage can ensure that the structure is of minimal size.

In case of integer or categorical attribute values, the attributes can be used directly as matrix coordinates. In case of floating point values however, they have to be transformed (scaled with an appropriate linear function and rounded).

The evaluation of the classifier is as follows: First, input attribute vector X is transformed:

$$x_i' \cong \max(0, \min(a_i x_i + b_i, S_i)) \tag{1}$$

where S_i is the size of the domain of attribute i, x_i is the ith value of X, a_i and b_i are the coefficients of the linear function for the given attribute. If the problem only has positive integer or categorical values already, then the scaling step can be skipped.

Let L denote the currently examined layer. Starting with the top layer ($L = 1$), each layer is examined. If the value stored in M^L under coordinates (x_i', x_j') (where i and j are the attributes associated with layer L) is 0, then return with *unknown class* as answer. Otherwise, set the combination variable (η) to the value in A^L under the same coordinates and go on the next layer ($L = L + 1$).

Each intermediary layer ($L = 2...N - 1$) is processed similarly, but with the difference that there is a returning value available (v_η^L) due to the stored auxiliary statistical data. If the evaluation process reaches the bottom layer ($L = N$), the output of the system is v_η^N.

2.1   The Training of the Classifier

The training phase is as follows: First, the training data set is analyzed, in order to determine the size of the domain (S_i) for each attribute, along with the scaling parameters (a_i and b_i) by which the parameter values can be mapped into the positive integer domain. The order of the layers in the topology is also made.

The data structure is created and initialized based on the obtained data. Initially, every matrix element is set to zero, along with the number of markers (γ^L) for each layer L. The scaling and rounding of the attribute values is done by

$$x_i' \cong a^L x_i - b^L \tag{2}$$

where a^L and b^L are the appropriate coefficients in layer L (thus indirectly associated with attribute i).

Let L be the top layer, i and j the attributes associated with it. If the value of M^L under coordinates (x_i', x_j') is 100 (character arrays are used, since they need less space than floating point arrays), then set η to the value in Λ^L under the same coordinates and go onto the next layer ($L = L + 1$). If the value of M^L is lower than 100 (there is no marker under those coordinates yet) then calculate the 2D fuzzy membership function values in the area around the given coordinates in M^L and update them where they are larger than the existing values. At the same coordinates (where M^L is updated), set the index values in Λ^L to the current number of markers (γ^L). Set η to γ^L and increment γ^L.

For the intermediate layers ($L = 2 \ldots N - 1$) the first step is incrementing the value of the statistical matrix G^L under coordinates (η, t), where t corresponds to the class of the sample. Let k mark the attribute associated with the layer. The rest of the processing of the intermediary layers is essentially the same as that of the top layer:

- if M^L under coordinates (η, x_k') is 100 then set η to Λ^L under the same coordinates and move on to the next layer,
- if it is not 100, then calculate the 1D fuzzy membership function values in the row marked by η, for the area around x_k' in M^L, and update the appropriate row elements in the same areas in Λ^L to (γ^L). Set η to γ^L and increment γ^L.

When the bottom layer is reached, then its statistical matrix G^L is updated similarly to that of the previous layers. After all training samples have been processed, the most common class is determined for each η, for each layer L:

$$v_\eta^L = \arg \max_t (G_{\eta,t}^L) \tag{3}$$

2.2 Experimental Results

To illustrate the capabilities of the classifier, it has been tested on the Wisconsin Breast Cancer data set (the configuration of the PC is Intel® Core™ i5-2450M CPU, @2.50 GHz, 4 GB RAM). The available data consists of 683 samples, each having 9 attributes, with domain size 10 each. It is a 2-class problem. The system is trained with 500 randomly selected samples, the rest are used for testing.

The results for using different fuzzy set width values can be seen in Table 1. This particular parameter affects the generalization ability of the system, they are measured in the diameter of area covered in the domains (e.g. 25 % covers half of the respective domains of the attributes). The experiments have shown that in this

Table 1 Average training and testing speed and accuracy for the WBC data set

Fuzzy set width	Average training time (s)	Average testing time (s)	Accuracy	%
25 % coverage	0.145	0.014	167/183	91
20 % coverage	0.1612	0.018	167/183	91
15 % coverage	0.1456	0.016	171/183	93
10 % coverage	0.1372	0.017	166/183	90
5 % coverage	0.1484	0.017	160/183	87

particular problem 15 % is the most efficient width value with which 93 % of the testing data is correctly classified. Larger than that performs slightly worse, lower than that performs progressively worse. The average training times range from ~ 0.13 to ~ 0.16 s, while the average testing times range from ~ 0.014 to ~ 0.018 s.

Size of the structure is also interesting to note in this experiment. In comparison, Fuzzy Hypermatrices would require the storage of 10^9 matrix elements in 9D arrays, which are not only hard to handle, but the whole classifier takes ~ 3 GB to store in the memory.

The size of the proposed structure (S_M) for any given problem can be calculated by:

$$S_M = (S_{T_A} + S_{T_M}) \left(D^1 D^2 + \sum_{L=2}^{N-1} \gamma^L D^{L+1} \right) + S_{T_G} \left(\sum_{L=2}^{N} \gamma^L C + \gamma^L \right) \quad (4)$$

where S_{T_A}, S_{T_M}, and S_{T_G} are the size of the variable types used to store matrices A, M, and G; D^L is the size of the domains, N stands for the number of attributes and C corresponds to the number of classes. Thus, the memory required to store the problem ($S_{T_A} = 2$, $S_{T_G} = 2$, $D^L = 10$, $N = 9$, $C = 2$, $\gamma^L = \sim 222$ for all L) is ~ 64 KB, which is a significant reduction.

3 Conclusions

In this paper, a quick classifier method is presented in attempt to solve the problem of dimensionality. The classifier is built on two ideas: the pre-calculation and storage of outputs, and the sparsity of large problem spaces. The proposed system only calculates the outputs for known places of the problem space and stores them in the form of a sequential combination of 2D matrices.

The run-time speed depends on only the number of attributes (linearly). Due to using fuzzy sets, the system is capable of recognizing inputs that are similar to the training samples. Moreover, additional statistical information is used to improve the robustness of the classification.

In future work, the performance of the system will be further investigated and improved.

Acknowledgments This work has partially been sponsored by the Hungarian National Scientific Fund under contract OTKA 105846 and the Research and Development Operational Program for the project "Modernization and Improvement of Technical Infrastructure for Research and Development of J. Selye University in the Fields of Nanotechnology and Intelligent Space", ITMS 26210120042, co-funded by the European Regional Development Fund.

The breast cancer databases has been obtained from the University of Wisconsin Hospitals, Madison from Dr. William H. Wolberg.

References

1. Garmarudi, A.B., Khanmohammadi, M., Khoddami, N., Shabani, K.: Near infrared spectrometric analysis of titanium dioxide nano particles for size classification. In: 10th IEEE Conference on Nanotechnology (IEEE-NANO), 2010, Seoul, pp. 451–453 (2010)
2. Rumelhart, D.E., McClelland, James: Parallel Distributed Processing: Explorations in the Microstructure of Cognition. MIT Press, Cambridge (1986)
3. Nguyen, T., Mhatli, S., Giacoumidis, E., Van Compernolle, L., Wuilpart, M., Mégret, P.: Fiber nonlinearity equalizer based on support vector classification for coherent optical OFDM. IEEE Photon. J. **8**(2), 1–9 (2016)
4. Cortes, C., Vapnik, V.: Support-vector networks. Mach. Learn. **20**(3), 273–297 (1995)
5. Liang, J., Luo, F., Yu, R.H., Xu, Y.G.: Wastewater effluent prediction based on fuzzy-rough sets RBF neural networks. In: International Conference on Networking, Sensing and Control (ICNSC), Chicago, IL, pp. 393–397 (2010)
6. Broomhead, D.S., Lowe, D.: Multivariable functional interpolation and adaptive networks. Complex Syst. **2**, 321–355 (1988)
7. Dubois, D., Prade, H.: Rough fuzzy sets and fuzzy rough sets. Int. J. Gen. Syst. **17**, 91–209 (1990)
8. Várkonyi-Kóczy, A.R., Tusor, B., Tóth, J.T.: A Fuzzy hypermatrix-based skin color filtering method. In: Proceedings of the 19th IEEE International Conference on Intelligent Engineering Systems, INES2015, Bratislava, Slovakia, 3–5Sep 2015, pp. 173–178 (2015)
9. Campbell-Kelly, M., Croarken, M., Robson, E. (eds.): The History of Mathematical Tables From Sumer to Spreadsheets, 1st ed. New York, USA (2003)

Performance Enhancement of Fuzzy Logic Controller Using Robust Fixed Point Transformation

Adrienn Dineva, Annamária Várkonyi-Kóczy, József K. Tar
and Vincenzo Piuri

Abstract Despite its excellent performance as a controller for linear and non-linear systems, the fuzzy logic controller has certain limitations. For instance, large-scale complex fuzzy systems like multi-input, single-output, or multi-output systems are used in various applications with large number of rules. Furthermore, the results also depend on the selected membership functions, etc. This paper presents a novel framework that instead of reducing the number of rules for a fuzzy logic controller, combines it with a fixed point transformation based adaptive control. The adopted approach is based on the Mamdani-type fuzzy controller and enhanced by the Sigmoid Generated Fixed Point Transformation control strategy to cope with modeling inaccuracies and external disturbances that can arise. The general procedure is applied to a nonlinear Kapitza pendulum. Numerical simulations are validating the applicability of the proposed scheme and demonstrating the controller's performance.

Keywords Adaptive control · Iterative learning control · Sigmoid Generated Fixed Point Transformation · Fuzzy logic

A. Dineva (✉)
Doctoral School of Applied Informatics and Applied Mathematics,
Óbuda University, Budapest, Hungary
e-mail: dineva.adrienn@bgk.uni-obuda.hu

A. Dineva
Department of Information Technologies, Doctoral School of Computer Science,
Università degli Studi di Milano, Crema, Italy

A. Várkonyi-Kóczy
Department of Mathematics and Informatics, J. Selye University, Komarno, Slovakia

J.K. Tar
Antal Bejczy Center for Intelligent Robotics (ABC IRob), Óbuda University, Budapest,
Hungary

V. Piuri
Department of Information Technologies, Università Degli Studi Di Milano, Crema, Italy

© Springer International Publishing AG 2017
R. Jabłoński and R. Szewczyk (eds.), *Recent Global Research and Education:*
Technological Challenges, Advances in Intelligent Systems and Computing 519,
DOI 10.1007/978-3-319-46490-9_55

411

1 Introduction

In adaptive nonlinear control Lyapunov's 2nd or Direct method have become very powerful tool in control design because it does not require to analytically solve the equations of motion [7, 8]. In spite of its simple geometric interpretation, the practical application of this technique is difficult due to its high computational need. An other constraint, that often finding the appropriate Lyapunov function candidate is far difficult. Still most of the existing control solutions are take their source from this method see, for e.g. [6, 11, 12]. In certain cases it may be difficult or impossible to derive an accurate mathematical model of the system under control and also the formulas could be more complex than an affordable controller can manage. Hence, in a modern control solution modeling, control, and diagnostic elements are simultaneously present. In the field of control engineering the fuzzy logic controllers has gained much attention in the last decades thanks to their simpler implementation, lower memory requirements and the ability to avoid impractical formula-based methods, etc. However, a difficult aspect is designing the rule bases and extracting the knowledge from the human operators. For e.g., on this reason the fuzzy Lyapunov synthesis method have been introduced by [9]. Furthermore, in some cases the real time operation can be a great difficulty. Although, the increasing number of rules can provide more precise control but with growing complexity. Originally the idea of Robust Fixed Point Transformation (RFPT) have been introduced for replacing Lyapunov's direct method has derived in [14, 15]. This Fixed Point Transformation-based control design based on the transformation of the control signals calculation problem into the task of finding an appropriate fixed point of a contractive map. This fixed point can be obtained by generating an iterative sequence using this contracting map. The method strictly separates the kinematic and dynamic aspects of the control task and predefines the desired system response by the use of purely kinematic terms. It also assumes the existence of the approximate dynamic model of the system under control is available, which is used in the estimation of the necessary dynamic terms that is needful for the realization of the kinematic specifications. The undesired effects of modeling imprecisions and unknown external disturbances are compensated by the adaptive deformation of the *desired response* r^{Des} as the input of the approximate model in order to obtain the desired response as a realized one. Later the *Sigmoid Generated Fixed Point Transformation (SGFPT)* has been proposed for nonlinear adaptive control of *Single Input—Single Output (SISO)* as well as *Multiple Input—Multiple Output (MIMO)* systems [4]. Also the applied sigmoid function have been reconsidered, which has crucial significance in the adaptive deformation [5]. In this paper we investigate possible enhancement of the classical fuzzy controllers performance with the SGFPT approach in the control of the Kapitzas pendulum system. The applicability and effectiveness are validated with numerical simulations.

2 Fuzzy Controller Performance Enhancement with SGFPT

2.1 The System Under Consideration

Controlling the inverted pendulum with horizontal displacement is a widely documented problem. Furthermore, an ample number of papers have been published involving the design of a fuzzy controller for stabilizing the classical inverted pendulum. However, only few papers discussing the control of the inverted pendulum with vertical vibration of the basement, i.e. the so-called Kapitza's pendulum system [1, 3, 10, 13, 17, 18]. For the investigations we consider the simplified model depicted in Fig. 1 in which the *generalized coordinates* denoted by $q_1 [m]$, and $q_2 [rad]$, while the *generalized forces* are $Q_1 [N]$ and $Q_2 [Nm]$. In the case of *underactuated model* $Q_2 \equiv 0$ so we can prescribe the *"nominal trajectories"* for q_2 and adjusting the driving force value Q_1 by using the equations of motion given by (1).

$$\begin{bmatrix} (M+m) - mL\sin q_2 \\ -mL\sin q_2 \; mL^2 \end{bmatrix} \begin{bmatrix} \ddot{q}_1 \\ \ddot{q}_2 \end{bmatrix} + \begin{bmatrix} -mL\cos q_2 \dot{q}_2^2 + (m+M)g \\ -mgL\sin q_2 \end{bmatrix} = \begin{bmatrix} Q_1 \\ Q_2 \end{bmatrix} \quad (1)$$

2.2 Principles of Fixed Point Transformation

As it has been mentioned before the undesired effects of modeling imprecisions, etc. are compensated by the adaptive deformation that is carried out by some kinematically expressed trajectory error reduction, i.e. by the comparison of the *desired response* r^{Des} and the *actually observed response* r^{Act}. Than, the *actually observed response* is deformed due to the exact system under control, formally $r^{Act} = f(r^{Des}, \ldots)$ where the symbol "..." stand for the unknown state variables.

Fig. 1 The scheme of the vertically excited pendulum system

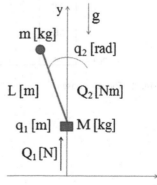

Because of the presence of additive noises $r^{Act} \neq r^{Des}$. Thus, the input of the response function from r^{Des} to $r_{\star}$ is deformed using *Banach's Fixed Point Theorem* [2] in order to obtain $r^{Act} \equiv f(r_{\star}) = r^{Des}$. Details of the mathematical background is well studied an can be read for e.g. in [15, 16].

2.3 Fuzzy Controller Design

For controlling the system a Mamdany-type fuzzy controller was applied. The inputs of the system are the sensed values of the angle of pole relative to the vertical axis $q_2[rad]$ and the derived values $\ddot{q}_2[rad/s^2]$ while the output is the $Q_1[N] \in [250; -250N]$. For the fuzzification of the variables 7–7 sigmoid membership function were used. The evaluation of the membership functions of the output signal of the fuzzy system carried out with the max-dot algorithm. For the defuzzification the center of sums method was used. This design can ensure acceptable performance but with weak disturbance rejection ability. In order to improve the performance instead of increasing the complexity with a growing number of membership functions and rules, at the borders of the range defined with the rules it switches to the SGFPT control strategy. The rule bases can be seen below. In the next sequel we will demonstrate the operation of the purely fuzzy - based method with comparison of the fuzzy controller supported with SGFPT, both in disturbed and disturbance-free cases (Fig. 2).

		input1						
		BN	**MN**	**SN**	**Z**	**SP**	**MP**	**BP**
	BN	switch to SGFPT	switch to SGFPT	BN	BN	MN	MN	0
	MN	switch to SGFPT	BN	BN	MN	SN	0	SP
	SN	BN	BN	MN	SN	0	SP	MP
	Z	BN	MN	SN	0	SP	MP	BP
	SP	MN	SN	0	SP	MP	BP	BP
	MP	SN	0	SP	MP	BP	BP	switch to SGFPT
input2	**BP**	0	SP	MP	BP	BP	switch to SGFPT	switch to SGFPT

Fig. 2 The rule bases

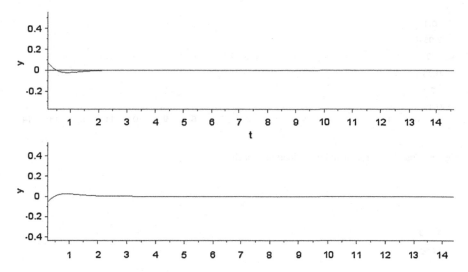

Fig. 3 Tracking error of the fuzzy controller in the disturbance free case

3 Simulation Results

The following *exact model parameters* were used in the simulation: $m = 2$ [kg], $M = 1$ [kg], where m represents the mass attached the end of the pendulum and M is the mass at the bottom of the pivot. The length of the beam was stated in $L = 1$ [m] with neglecting its mass. The *approximate model parameters* were set as $\tilde{m} = 2.2$ [kg], $\tilde{M} = 1.5$ [kg], $\tilde{L} = 0.8$ [m], and $g = 9.81$ [m/s²]. Figures 3 and 4 the operation of the purely fuzzy based control can be seen. With comparing Figs. 4 and 5 the performance enhancement can be seen (Fig. 6).

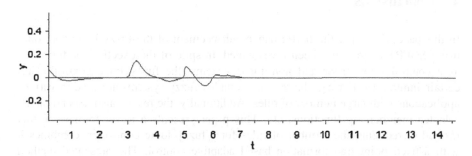

Fig. 4 Tracking error of the fuzzy controller with disturbing noise

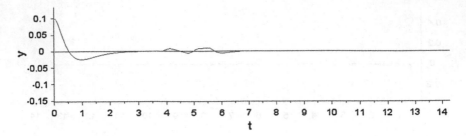

Fig. 5 Tracking error of the combined approach

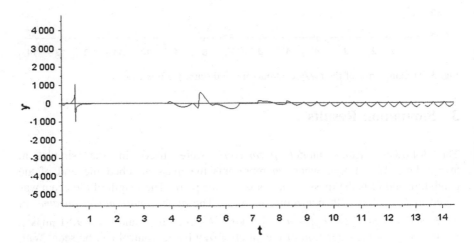

Fig. 6 Q[N] values for the combined approach

4 Conclusions

In this paper the paper the performance enhancement of the fuzzy logic controller using SGFPT method has been investigated. In spite of the excellent performance as a controller for linear and non-linear systems, the fuzzy logic controller has certain limitations. For e.g., large-scale complex fuzzy systems are used in various applications with large number of rules. Additionally, the results also depend on the selected membership functions, etc. This paper presents a novel framework that instead of reducing the number of rules for a fuzzy logic controller, combines it with a fixed point transformation based adaptive control. The presented method strategy is based on the Mamdani-type fuzzy controller and enhanced by the Sigmoid Generated Fixed Point Transformation control strategy. The Vertically excited inverted pendulum, i.e. the Kapitza pendulum system served as a benchmark problem. Simulation results have shown, that the purely fuzzy logic based solution can ensure acceptable performance, but with weak disturbance rejection ability. The combined design is able to cope with modeling inaccuracies and external disturbances. Numerical simulations are validating that the proposed scheme enhances the performance and far promising.

Acknowledgments This work has been sponsored by the Hungarian National Scientific Fund (OTKA 105846). This publication is also the partial result of the Research and Development Operational Programme for the project "Modernisation and Improvement of Technical Infrastructure for Research and Development of J. Selye University in the Fields of Nanotechnology and Intelligent Space", ITMS 26210120042, co-funded by the European Regional Development Fund.

References

1. Aracil, J., Gordillo, F., Acosta, J.: Stabilization of oscillations in the inverted pendulum. In: Proceedings of the 15 IFAC World Congress, pp. 261 (2002)
2. Banach, S.: Sur les opérations dans les ensembles abstraits et leur application aux équations intégrales (About the operations in the abstract sets and their application to integral equations). Fund. Math. **3**, 133–181 (1922)
3. Belman, R., Bentsman, J., Meerkov, S.M.: Vibrational control of a class of nonlinear systems: vibrational stabilization. IEEE Trans. Aut. Control **32**(8), 710–716 (1986)
4. Dineva, A., Tar, J., Varkonyi-Koczy, A., Piuri, V.: Generalization of a sigmoid generated fixed point transformation from siso to mimo systems. In: IEEE 19th International Conference on Intelligent Engineering Systems (INES2015), 3–5 Sept 2015, Bratislava, Slovakia. pp. 135–140 (2015)
5. Dineva, A., Tar, J., Varkonyi-Koczy, A., Piuri, V.: Adaptive control of underactuated mechanical systems using improved sigmoid generated fixed point transformation and scheduling strategy. In: IEEE 14th International Symposium on Applied Machine Intelligence and Informatics, 21–23 Jan 2016, Herlany, Slovakia. pp. 193–197 (2016)
6. Hosseini-Suny, K., Momeni, H., Janabi-Sharifi, F.: Model reference adaptive control design for a teleoperation system with output prediction. J. Intell. Robot Syst. pp. 1–21 (2010). doi:10.1007/s10846-010-9400-4
7. Khalil, H.: Nonlinear Systems, 2nd ed. Upper Saddle River, Hall (1996)
8. Lyapunov, A.: A general task about the stability of motion (in Russian). Ph.D. Thesis, University of Kazan, Tatarstan (Russia) (1892)
9. Margaliot, M., Langholz, G.: New Approaches to Fuzz Modeling and Control: Design and Analysis. World Scientific Publishing, Singapore (2000)
10. Miroshnik, I., Odinets, N.: Stabilization of pendulum oscillations around upper position. In: 6th IFAC Syposium on Nonlinear Control Systems (NOLCOS 2004), vol. 3 (2004)
11. Nguyen, C., Antrazi, S., Zhou Jr., Z.L., Campbell Jr., C.: Adaptive control of a stewart platform-based manipulator. J. Robot. Syst. **10**(5), 657–687 (1993)
12. Slotine, J.J.E., Li, W.: Applied Nonlinear Control. Prentice Hall International Inc., Englewood Cliffs (1991)
13. Stephenson, A.: XX. On induced stability. Philos. Mag. Ser. 6 **15**(86), 233–236 (1908)
14. Tar, J.: Replacement of Lyapunov function by locally convergent robust fixed point transformations in model based control, a brief summary. J. Adv. Comput. Intell. Intell. Inf. **14**(2), 224–236 (2010)
15. Tar, J.: Towards replacing Lyapunov's 'direct' method in adaptive control of nonlinear systems (invited plenary lecture). In: Proceedings of the Mathematical Methods in Engineering International Symposium (MME), Coimbra, Portugal (2010)

16. Tar, J.: Adaptive control of smooth nonlinear systems based on lucid geometric interpretation (DSc dissertation). Hungarian Academy of Sciences, Budapest (2012)
17. Yabuno, H., Goto, K., Aoshima, N.: Swing-up and stabilization of an underactuated manipulator without state feedback of free joint. IEEE Trans. Robot. Autom. **20**, 259–365 (2004)
18. Yabuno, H., Tsumoto, K.: Experimental investigation of a buckled beam under high-frequency excitation. Arch. Appl. Mech. **77**(5) (2007)

Hip Articulation in Orthotic Robot

Marcin Zaczyk, Dymitr Osiński and Danuta Jasińska-Choromańska

Abstract The paper describes a concept of a hip articulation of the 'Veni-Prometheus' System for Verticalisation and Aiding Motion, developed at the Division of Design of Precision Devices, Faculty of Mechatronics, Warsaw University of Technology. The device is to be attached to the hip belt of the orthotic robot and allows movements of lower limb in sagittal and transverse planes to be realized. Such motion can enable the orthotic robot to perform turns by allowing a hip rotation of lower limb.

Keywords Orthotic robot · Exoskeleton · Lower extremity · Hip articulation · Turning system · Turning module

1 Introduction

1.1 The 'Veni-Prometheus' Orthotic Robot

A system for verticalisation and aiding motion intended mainly for the disabled and handicapped people has been designed at the Division of Design of Precision Devices, Faculty of Mechatronics, Warsaw University of Technology (Fig. 1) [1, 5]. The orthotic robot clasps the lower limbs of the user, enabling an articulated movement of flexion and extension of hip and knee joints. The ankle joints can also

M. Zaczyk · D. Osiński (✉) · D. Jasińska-Choromańska
Faculty of Mechatronics, Warsaw University of Technology, Warsaw, Poland
e-mail: d.osinski@mchtr.pw.edu.pl

M. Zaczyk
e-mail: m.zaczyk@mchtr.pw.edu.pl

D. Jasińska-Choromańska
e-mail: danuta@mchtr.pw.edu.pl

© Springer International Publishing AG 2017   419
R. Jabłoński and R. Szewczyk (eds.), *Recent Global Research and Education: Technological Challenges*, Advances in Intelligent Systems and Computing 519,
DOI 10.1007/978-3-319-46490-9_56

Fig. 1 The
'Veni-Prometheus' system for
verticalization and aiding
motion

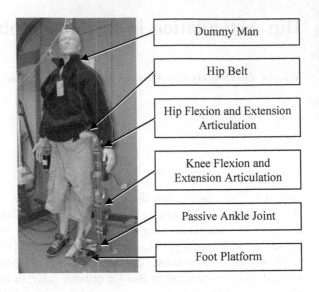

Dummy Man

Hip Belt

Hip Flexion and Extension
Articulation

Knee Flexion and
Extension Articulation

Passive Ankle Joint

Foot Platform

move in the sagittal plane, but this movement is not powered, i.e. it is passive. The
exoskeleton has been constantly developed to further increase its capabilities. The
presented efforts are focused on creating a capability of turning, as an addition to
current set of functions that already incorporates the capabilities of walking, sitting
down as well as standing up and traversing stairs.

1.2 State of the Art and the Intended Development

There already exists a wide range of various exoskeletons and orthotic robots that
aid the gait cycle, which are designed for civilian, medical and military users [3].
A survey of those machines has been performed, focusing on their ability to turn
[7]. Only a few of the surveyed exoskeletons, which could be used by an impaired
person, are capable of turning. Among those are REX Exoskeleton [4, 9], IHMC
Mobility Assist Exoskeleton [6] and Mindwalker Exoskeleton [2, 10]. As results
from the survey, the issue of turning in orthotic robots and exoskeletons is significant and worth of further development. Thus, a mechanism enabling turning by
powered rotation of the hip joint have been proposed [8]. One of the solutions, a
2-degree of freedom hip articulation allowing a movement in sagittal and transverse
planes to be performed, is described in this paper.

2 Conception of the Hip Articulation for the Orthotic Robot

The drive module of the hip joint (Fig. 2) is located at the side parts of the hip belt, above the articulations of hip flexion and extension of the lower limbs. The device contains two articulations enabling flexion and extension movements (within the sagittal plane) as well as a hip rotation (within the transverse plane). Both articulations are driven by DC motors coupled with reducers. Drive of the flexion and extension movements includes a worm gear, just as the drive of the rotation, yet the latter is additionally equipped with cams. Owing to such solution, it is possible to reduce the rotational speed and increase at the same time the driving torque. Besides, orientation of the axes, about which the rotations are performed, is changed, keeping a small volume of the drive and its high efficiency. Output movements of both drives are realized only within a section of the full angle, which approximately corresponds to the anatomic range of motion. None of the mechanisms is self-locking, what makes it possible to harvest energy in the case when the user performs passive movements (generator operation). Thus, the energetic balance is positively influenced; moreover, the user can perform free movements when the aiding system is inactive.

3 Discussion

The device featuring the presented kinematics enables realization of a turn with a whole lower limb by means of its hip rotation; at the same time the rotation axis is close to the anatomic axis of hip rotation. Owing to that, it is possible to realize independent and driven movements of extension and flexion as well as hip rotation

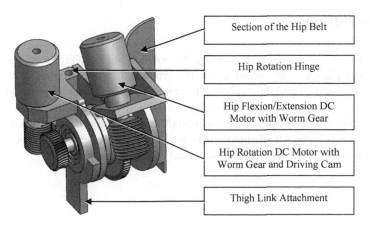

	Section of the Hip Belt
	Hip Rotation Hinge
	Hip Flexion/Extension DC Motor with Worm Gear
	Hip Rotation DC Motor with Worm Gear and Driving Cam
	Thigh Link Attachment

Fig. 2 Conception of the hip articulation (perspective view of a single module)

of the lower limb. The movements are useful while performing a gait cycle as well as motion involved while a human wearing an orthotic robot makes a turn. Besides, even in the case of a flexion of the lower limb in the hip joint (within the sagittal plane) and its rotation, the anatomic motion is till performed while spreading the legs. Performing a motion that is not anatomic while spreading the legs, may result in an injury of the user. The motion of spreading the legs at a sitting position becomes helpful while putting on and off the orthotic robot. It should be emphasized that the drives for hip rotation are located at the side part of the hip belt. Such location should not disturb walking (including performance of turns) or sitting having the system for aiding motion put on. Motion of the user's hands could remain relatively free as well. However, an impediment may be armrests or a backrest of chairs, on which the mechanism could be caught. Location of the drives at the hip belt seems to be more advantageous than location at the lower limbs, since the drives do not influence the mass and the mass moments of inertia of the limbs, which perform a motion of a considerable amplitude. It is obvious that the hip belt also moves while walking, however this motion is less significant and takes place in vicinity of the human mass center in the orthotic robot. An advantage of the proposed solution is its modular structure, making it possible to configure the drives according to the motor capabilities of an individual user of the exoskeleton. A man completely devoid of a capability of controlling the lower limbs could use the pair of drives to realize a gait cycle and turning, whereas a user who has use of his legs could employ the drives of extension and flexion of the lower limbs in the gait cycle, making turns by himself. A shortcoming of such solution is a displacement of the rotation axis within a transverse plane, out of the natural motion of the head of the hip bone at the pelvis, outside the outline of the human body. A parallel connection will be created then: the orthotic robot and the human body. Thus, each rotary movement will feature also a translational component, what will result in displacements of the human body with respect to the frame of the orthotic robot.

4   Conclusion

The proposed solution is characteristic due to merging capabilities of performing within two planes independent aided movements. It is realized by a small and modular unit attached to the hip belt. The concept has been developed as one of few solutions of the problem of performing turns in the 'Veni-Prometheus' orthotic robot; further related works are foreseen, starting with modeling and simulations, to tests of a prototype. All of these works are aimed at implementation of a chosen version in the system for verticalisation and aiding motion.

References

1. Bagiński, K., Wierciak, J.: Forming of operational characteristics of an orthotic robot by influencing parameters of its drive systems. In: Szewczyk, R., Zieliński, C., Kaliczyńska, M. (eds.) Advances in Intelligent Systems and Computing, vol. 351. Progress in Automation, Robotics and Measuring Techniques, vol. 2, pp. 1–9. Springer, Berlin (2015) (Robotics)
2. Gancet, J., et al.: MINDWALKER: going one step further with assistive lower limbs exoskeleton for SCI condition subjects modeling, design, and optimization of mindwalker series elastic joint. In: 4th IEEE RAS and EMBS International Conference on Biomedical Robotics and Biomechatronics, pp. 1794–1800 (2012)
3. Herr, H.: Exoskeletons and orthoses: classification, design challenges and future directions. J. NeuroEng. Rehabil. **6**(1) (2009)
4. Irving, R., Little, R.: Mobility aid Europe. Patent, EP 2231096 (2013)
5. Jasińska-Choromańska, D., et al.: Mechatronic system for verticalization and aiding the motion of the disabled. Bullet. Pol. Acad. Sci. Techn. Sci. **61**(2), 419–431 (2013)
6. Kwa, H. et al.: Development of the IHMC mobility assist exoskeleton. In: IEEE International Conference on Robotics and Automation, IEEE ICRA, pp. 2556–2562 (2009)
7. Osiński, D., Jasińska-Choromańska, D.: Survey of turning systems used in lower extremity exoskeletons. Chall. Autom. Robot. Measurement Tech. Adv. Intell. Syst. Comput. **440**, 447–457 (2016)
8. Osiński, D., Zaczyk, M., Jasińska-Choromańska, D.: Conception of turning module for orthotic robot. Adv. Mechatronics Solutions Adv. Intell. Syst. Comput. **393**, 147–152 (2016)
9. REX Bionics, http://www.rexbionics.com/
10. Wang, S., et al.: Design and control of the MINDWALKER exoskeleton. IEEE Trans. Neural Syst. Rehabil. Eng. **23**(2), 277–286 (2014)

Application of Artificial Neural Networks for Early Detection of Breast Cancer

Krzysztof Urbaniak and Krzysztof Lewenstein

Abstract Objects placed in mammograms and their features have high diagnostic value. The method of extraction and calculation parameters of the images is a key element in the whole process of assessment of individual medical cases. This applies both to detect the location of microcalcifications, as well as the accurate assessment of the degree of malignancy [1–4, 8, 9]. The analysis of the literature shows that in many cases, to assess them as neural classifiers used neural networks unidirectional scholars algorithm back propagation of error. The large popularity of this type of network stems from the rich knowledge about them supported a wide range of applications [11–13, 17, 18]. This does not mean that this type of network is the best tool for the identification of cancer. In order to verify the correctness of the research, it was decided to compare the two types of neural networks: one-way multi-layer network MLBP [17, 18] and a network of independently building the architecture i.e. Fahlmana network [5, 17]. Both networks were used for the same task, the task of verification and classification of medical data based on the feature vectors. The study attempted to assess the mammograms for detecting the locus and the evaluation grade of microcalcifications [19, 23]. According to the stated objective, the structure of the neural networks were designed so that it was possible to indicate location and performance evaluation grade localized potential microcalcifications. Way of describing mammogram—a set of characteristics and size of the database are the factors that were taken as those that have a key impact on the quality of the diagnostic computer tools [21–23].

Keywords Microcalcifications · Extraction method · Fahlman neural network · MLBP neural network

K. Urbaniak (✉)
Wydział Administracji i Nauk Społecznych, Politechnika Warszawska, Warsaw, Poland
e-mail: k.urbaniak@ans.pw.edu.pl

K. Lewenstein
Wydział Mechatroniki, Politechnika Warszawska, Warsaw, Poland
e-mail: k.lewenstein@mchtr.pw.edu.pl

© Springer International Publishing AG 2017
R. Jabłoński and R. Szewczyk (eds.), *Recent Global Research and Education: Technological Challenges*, Advances in Intelligent Systems and Computing 519,
DOI 10.1007/978-3-319-46490-9_57

425

1 Introduction

Mammographic images and specific information encoded within, may be a source of information—features (parameters) which can be successfully used for their evaluation. Objects located in mammographic images and their features have significant diagnostic value [1–10]. The method of extracting and calculating image parameters is a crucial element and at the same time a task to be realized in the process of the evaluation of particular medical cases. It relates to both, detecting the location of microcalcifications and accurate evaluation of the degree of their malignancy. Information included in digital mammographic images, as well as their nature directed the development of computer methods supporting the extraction of characteristic features of the objects used for identifying typical elements of the images and evaluation of their "condition" [14–16, 21–23].

Obstacles encountered in the evaluation of this issue, as well as the analysis of the available literature made the Author formulate the following conclusions [19–21, 23]:

- computer tools supporting data analysis and using the algorithms of artificial neural networks to data evaluation may be effectively applied for the analysis of information included in mammogram and can successfully support the process of detecting microcalcifications,
- detection of microcalcifications is divided into two stages: the analysis of mammographic image allowing for defining the location of microcalcifications' concentration and an attempt to assess the degree of malignancy of microcalcifications,
- improvement of the results of microcalcifications' detection is possible due to searching and change of the parameters describing a mammographic image,
- in the majority of works presenting the application of artificial neural networks for the analysis of mammographic images, feed-forward networks of MLBP type are presented.

From the literature analysis appears, that in the majority of cases feed-forward neural networks, taught by the algorithm of back propagation error were applied for the evaluation as neuron classifiers. Huge popularity of this type of networks results from the extensive knowledge on them supported by a wide spectrum of applications. It does not however mean that this type of network is the best tool for diagnosing cancer. In order to verify the correctness of the conducted studies, the Author decided to compare two types of neural networks: multilayer feed-forward MLBP network and networks independently designing architecture, i.e. Fahlman network [12, 13, 17, 18]. Both types of networks were used for the same task—verification and classification of medical data on the basis of the features' vectors. In the course of the studies an attempt was made to evaluate mammograms in terms of detecting the place of occurrence and the evaluation of the degree of malignancy of microcalcifications.

According to the assumed goal, the structure of neural networks was designed in such a way as to enable indicating the place of occurrence of microcalcifications (division into segments) and performing the evaluation of the degree of malignancy of the identified, potential microcalcifications [15, 21]. The manner of describing mammogram—a set of features and the extensity of database are factors which were treated as those, which had crucial influence on the diagnostic quality of computer tools [16, 23].

2 Feed-Forward Network

The basic, best known and most frequently used type of neural network is a feed-forward, multilayer neural network, often described as MLBP—multilayer perceptron. In this work, particularly during the analysis of the results, full name of the network will be replaced by MLBP. Now, continuous activation functions are usually used which are the approximation of step function, which was initially applied in McCulloch-Pitts model [17, 18]. As far as multilayer neural networks are concerned, neuron output signal is propagated to the next neurons of the hidden layers. Activation functions are most frequently defined as sigmoid function or hyperbolic tangent. Trained neural network has the abilities to generalize knowledge, which was possessed during the training. The degree of training is verified by way of a test consisting in recognizing a test set. Training and test sets should be separate.

For optimization of the values of network weights (learning) gradient methods of learning are frequently used. A classic algorithm used for their modification is error back propagation algorithm [17, 18]. This method was applied in the realization of the studies when training MLBP feed-forward networks.

3 Fahlman Cascade Correlation Network

An advantage and characteristic feature of Fahlman neural network [5] is its ability to automatically generate architecture depending on the problem being solved. Another feature of the algorithm is the nature of architecture resembling a cascade [17]. It is caused by the specificity of the method of including subsequent hidden neurons in the network, together with a cascade of weight connections. Each subsequent added neuron is a connection with input junctions and all already existing, included hidden neurons [5]. All neuron outputs, together with input signals of a network directly feed decision-output neurons (Fig. 1).

Taking the properties of Fahlman neural network into consideration, the Author decided to verify the correctness of inference through conducting each mammogram examination with the use of neural network which independently designs its architecture.

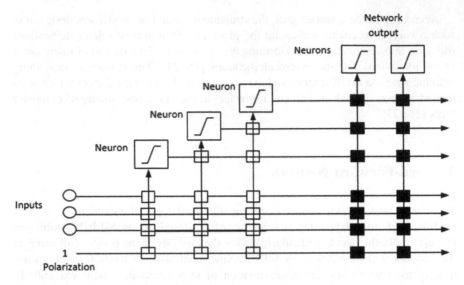

Fig. 1 General form of Fahlman cascade correlation [5, 17]

4 Simulation of the Use of Neural Networks to Identification of Microcalcifications' Location and for the Evaluation of Microcalcifications' Malignancy

The results of medical examinations, together with feature vectors, could have been used for training and testing neural networks and comparing the results achieved this way. The result of processing 600 mammographic images was obtaining a high number of segments allowing for representative assessment of particular cases. When an image was divided by a network of 32 × 32 pixels, 2000 randomly chosen cases were selected with microcalcifications and 2000 segments containing normal tissue [15, 21]. Further, another division was made into 4 subsets containing the following numbers of segments. As it was mentioned earlier, each segment separated from the mammogram had its features determined [16, 23]. For each image and determined segment the required parameters' values were calculated. As a result the Author had to his disposal feature vectors for all images from database. These data participated in training the neural networks. Those mammographic images were used as training data, which did not take part in the process of training. Verification of the network consisted in assigning test vectors to two affiliation classes: 0—no microcalcifications, 1—the appearance of microcalcifications. Having network results to the Author's disposal, as well as knowing the results of medical diagnoses, he considered the usefulness of neural networks in terms of indicating the places where microcalcifications occur. When such place was detected, it was possible to enlarge the "threatened" area, filtering it and further analysis in terms of the degree of malignancy was possible. In order to determine the accuracy of a neural network, all segments included in test images were used. In the case of segmentation by field of size a = x pixels, test set was T = (512/a) 2 * Ltest of segment where: T—the number of all vector pairs (segments), a—the size of a segment, Ltest—the number of images in a test base [16, 21–23]. Neural networks were taught and tested repeatedly. For training and testing, previously created sets were used containing 4000 records on the whole for segments of 32 × 32 pixels and 1200 segments of 64 × 64 pixels. After finishing teaching at the first subset, verification of the degree of learning of the networks at the test set was conducted. After obtaining the results, the process of teaching and training was repeated until the set was empty. Detecting microcalcifications, as well as the description of their shape is the beginning of the process of diagnosing the degree of the tumor's malignancy. In order to get a full evaluation it is necessary to prepare a classification of microcalcifications and their concentrations [8, 10]. Such studies are very difficult, and unfortunately in many cases also ambiguous. It is probably caused by a very initial phase of existence of the objects, namely—microcalcifications. This means that a huge amount of them, noticeable in the form of lighter points, may, after some time, be absorbed by other tissues. Even those classified as not bringing the risk of tumor, may also cause its formation after some time. Summing up, it needs to be emphasized that microcalcifications may be a very early symptom of breast cancer formation and studies on their unambiguous evaluation are unquestionably necessary. In order to evaluate the degree of their malignancy [6, 14], the Author used the same method of examination due to which microcalcifications' location was identified.

5 The Results of the Studies

Neural system classified images (areas with marked microcalcifications) into two groups: malign tumor (binary value 1) and benign tumor (binary value 0). A series of tests was carried out with a different number and type of features [16, 22, 23]. The results are presented in Figs. 2, 3, 4, 5, 6, 7, 8 and 9.

ACCURACY [%]	Benign microcalcifications	Malignant microcalcifications
	64%	67%

Fig. 2 The accuracy of the system (MLBP network). Features: V, CV, MAX, MIN

ACCURACY [%]	Benign microcalcifications	Malignant microcalcifications
	70%	69%

Fig. 3 The accuracy of the system (Fahlman network). Features: V, CV, MAX, MIN

ACCURACY [%]	Benign microcalcifications	Malignant microcalcifications
	76%	73%

Fig. 4 The accuracy of the system (MLBP network). Features: V, CV, AFP, LFP, MAX, MIN

ACCURACY [%]	Benign microcalcifications	Malignant microcalcifications
	75%	76%

Fig. 5 The accuracy of the system (Fahlman network). Features: V, CV, AFP, LFP, MAX, MIN

ACCURACY [%]	Benign microcalcifications	Malignant microcalcifications
	74%	72%

Fig. 6 The accuracy of the system (MLBP network). Features: V, CV, AFP, LFP, LK, ZM, LMK

ACCURACY [%]	Benign microcalcifications	Malignant microcalcifications
	72%	74%

Fig. 7 The accuracy of the system (Fahlman network). Features: V, CV, AFP, LFP, LK, ZM, LMK

ACCURACY [%]	Benign microcalcifications	Malignant microcalcifications
	83%	79%

Fig. 8 The accuracy of the system (MLBP network). Features: V, CV, AFP, LFP, LK, ZM, LMK, MAX, MIN

ACCURACY [%]	Benign microcalcifications	Malignant microcalcifications
	84%	82%

Fig. 9 The accuracy of the system (Fahlman network). Features: V, CV, AFP, LFP, LK, ZM, LMK, MAX, MIN

6 Summary

The obtained results indicated that the number and type of features influence the accuracy of evaluation of the degree of malignancy. When choosing such features as variance, variance coefficient, average variance, maximal and minimal greyscale [11, 16, 23], the accuracy of diagnosis was the lowest. Expanding feature vector by angular and Fourier power spectrum [16], MLBP network accuracy increased to more than 70 %. If a full set of features was used for training neural networks, the accuracy increased to about 80 %. It needs to be emphasized that such parameters as the number of clusters, microcalcification outline, the number of microcalcifications in a cluster, minimal and maximal greyscale had a significant influence on the evaluation of the degree of microcalcifications malignancy with the use of artificial neural networks.

Summing up, it has to be highlighted that with the use of Fahlman network max. 84 % of accuracy for malignant microcalcifications and 82 % of accuracy for benign microcalcifications was achieved as a result of the tests. These studies confirmed already existing conclusions indicating Fahlman neural network as a more efficient tool than MLBP network.

The most important feature and advantage of Fahlman network is the ability to design the architecture of artificial neural network [11, 17]. The level of complexity of the created structure is contingent upon and dependent on the level of difficulty of the analyzed medical problem. Network structure created in the stage of training is adjusted to a given level of complexity of the problem being solved.

References

1. Arbach, L., Bennett, D.L., Reinhardt, J.M., Fallouh, G.: Classification of mammographic masses: comparison between backpropagation neural network (BNN) and human readers. Department of Biomedical Engineering, The University of Iowa, Iowa City, IA 52242 USA. Department of Radiology, The University of Iowa, Iowa City. Department of Biomedical Engineering, Damascus University, Syria (2003)
2. Arodz, T., Kurdziel, M., Popiela, T.J., Sevres, E.O.D., Yuen, D.A.: A 3D visualization system for computer-aided mammogram analysis. University of Minnesota, Research report UMSI 2004/181. Submitted to Computer methods and programs in biomedicine, Elsevier (2004)

3. Burke, H.B., Goodman, P.H., Rosen, D.B., Henson, D.E., Weinstein, J.N., Harrell, F.E., Marks, J.R., Winchester, D.P., Bostwick, D.G.: Artificial neural networks improve the accuracy of cancer survival prediction. Am. Cancer Soc. (1997)
4. Collins C.: Breast cancer detection aided by new technology installed at Magee-Womens Hospital of University of Pittsburgh Medical Center. Pittsburgh, 16 Jan 2002. Magee Womens Hospital (2002)
5. Fahlman, S.E., Labiere, C.: The cascade-correlation learning architecture advances in NIPS2. Touretzky, D. (ed.) pp. 524–532 (1990)
6. Kim, J., Park, H.: Statistical textural features for detection of mikrocalcifications in digitized mammograms. IEEE Trans. Med. Imag. **18**, 231–238 (1999)
7. Kopans, D.: (1998) Breast Imaging. Lippincott-Raven
8. Kuo, W., Chang, R.F., Moon, W.K., Lee, C.C., Chen, D.R.: Computer-aided diagnosis of breast tumors with different US systems. Acad. Radiol. **9**(7) (2002)
9. Lado, M.J., Tahoces, P.G., Mendez, A.J., Souto, M., Vidal, J.J.: A wavelet based algorithm for detecting clustered microcalcifications in digital mammograms. Med. Phys. **26**(7), 1294–1305 (1999)
10. Lado, M.J., Tahoces, P.G., Mendez, A.J., Souto, M., Vidal J.J.: Evaluation of an automated wavelet-based system dedicated to the detection of clustered microcalcifications in digital mammograms. Departamento de Radiologia Faculted de Medicina, Universidad de de Santiago de Compostela C/San Francisco, 1. 15782 Santiago, Spain (2003)
11. Lewenstein, K., Urbaniak, K.: Detection and evaluation of breast tumors on the basis of microcalcification analysis. In: Advances in Intelligent Systems and Computing, vol. 393, pp. 159–167. Springer International Publishing, Switzerland (2016)
12. Lewenstein, K., Urbaniak, K.: Interpretation of mammograms in system with Fahlman's neural network and picture segmentation. In: Embec VI International Conference on Medical Physics, pp. 181–187. Monduzzi Editore (1999)
13. Lewenstein, K., Urbaniak, K., Łubkowski, P., Pałko, T.: The comparison of computer aided discrimination of breast cancer based on the analysis of mammo-grams transformed digitally in three different ways. In: Proceedings of IX Mediterranean Conference on Medical and Biological Engineering and Computing—Medicon, Pula, Croatia, 12–15 June 2001, part I pp. 534–538 (2001)
14. Li, L., Clark, R.A., Thomas, J.: Improving algorithm robustness for mass detection in digital mammography. Department of Radiology, Collage of Medicine, H. Lee Moffit Cancer Center and Research Institute at University of South Florida, Uniformed Services University of the Health Sciences. The U.S. Army Medical Research and Materiel Command under DAMD 17-00-1-0245 (2000)
15. Nieniewski M. (2005) Segmentacja obrazów cyfrowych. Metody segmentacji wododziałowej. Akademicka Oficyna Wydawnicza EXIT. Warszawa (2005)
16. Ogawa, K., Fukushima, M., Kubota, K.: Computer-aided diagnostic for diffuse liver diseases with ultrasonography ny neural networks. IEEE Trans. Nucl. Sci. **45**(6) (1998)
17. Osowski, S.: Sieci neuronowe w ujęciu algorytmicznym. WNT (1996)
18. Osowski, S.: Sieci neuronowe do przetwarzania informacji. OWPW (2006)
19. Pardela, M.: Współczesne rozpoznawanie i leczenie guzów sutka u kobiet. Śląska Akademia Medyczna, Katowice (1997)
20. Ripley, B.D., Ripley, R.M.: Neural networks as statistical methods in survival analysis. In: Dybowski, R., Gant, V. (eds.) Artificial Neural Networks: Prospects for Medicine. Landes Biosciences Publishers (1998)

21. Rudnicki, Z.: Wybrane metody przetwarzania i analizy cech obrazów teksturowych. Informatyka w Technologii Materiałów. Nr 1, Tom 2, pp. 1–18 (2002)
22. Urbaniak, K., Lewenstein, K., Łubkowski, P., Chojnacki, M.: Influence of the digital transformation of the mammograms on computer discrimination of breast cancer. In: 3rd International Conference Mechatronics, Robotics and Biomechanics 2001 Trest, pp. 181–189 (2001)
23. Wróblewska, A., Przelaskowski, A.: System automatycznej detekcji i klasyfikacji mikrozwapnień w cyfrowej mammografii. Elektronizacja, 3:8–11, Mat. IV Sympozjum'Techniki Przetwarzania Obrazu', pp. 299–305 (2002)

New Ways of Selection of Vibroacoustic Isolation Selection for Utilization in Checkweighting Systems

Paweł Nowak, Marcin Kamiński and Roman Szewczyk

Abstract Paper presents a methodology of selection of vibroisolation parameters on the example of checkweighter system. Due to the high degree of complexity of analysed system some simplifications were made—main modules are represented by the concentrated masses, connected by the universal models of bonds. Paper presents two methods of determination of bonds parameters. First way requires physical presence of analysed system and can be utilized for applying vibroisolation on developed system in order to minimize vibration transmission from the environment to the system. Second way can be applied during system design which significantly reduces the cost of the further tests. Utilization of Finite Element Method analyses allows to design proper shape of bonds between modules. Presented method utilizes numerical simulations for different parameters of vibroisolation. Based on the results of multi-parameter optimization significant reduction of vibration transmission to the crucial elements of checkweighter system can be achieved. Limiting the transmission of the vibrations to the main weighting module may result with the significant improvement of the mass measurement accuracy.

Keywords Vibroisolation efficiency · Mechanic modeling · Spectral analyses

1 Introduction

Vibroisolations are utilized for minimization of vibrations transfer from environment to device ("displacement vibroisolation") or inversely—to decrease the amplitude (or completely attenuate) of the vibes from working device to its sur-

P. Nowak (✉) · R. Szewczyk
Faculty of Mechatronics, Warsaw University of Technology,
sw. A. Boboli 8, 02-525 Warsaw, Poland
e-mail: p.nowak@mchtr.pw.edu.pl

M. Kamiński
Industrial Research Institute for Automation and Measurements,
Al. Jerozolimskie 202, 02-486 Warsaw, Poland

© Springer International Publishing AG 2017
R. Jabłoński and R. Szewczyk (eds.), *Recent Global Research and Education:
Technological Challenges*, Advances in Intelligent Systems and Computing 519,
DOI 10.1007/978-3-319-46490-9_58

roundings ("force vibroisolation") [1, 2]. Considering checkweighter, which is a sensitive mass measurement system, which does not generate substantial vibes itself, vibroisolation has to be chosen in order to reduce and influence of external vibrations. Properly selected parameters of isolators—dumping coefficient (c_{izol}) and stiffness coefficient (k_{izol}) should result with increment of measurement accuracy by substantial attenuation of dynamic impact transmitted to the mass measurement system.

2 Simplified Checkweighter Model

Analyzed checkweighter [3] is, from the mechanics point of view, a system with many degrees of freedom. In order to simplify the analyses during vibroisolation selection, analyzed system was reduced to a model with significantly decreased number of degrees of freedom. Analyzing checkweighting process, most important interactions occur between the weighting module and conveyor belt utilized to transport weighted objects. This is due a direct transmission of all dynamic interactions between those elements to a measurement signal. Transporter and weighting module are connected with wishbone, which can be represented as parallel elastic-dissipation system with and dumping c_{p2} and stiffness k_{p2}. Due to the fact, that transmitter stiffness is significantly greater than wishbones (which principle of operation is based on elastic deflection), conveyor can be simplified as a concentrated mass m_{p2}. Weighting module is placed on a thick aluminum plate connected with supporting structure of checkweighting system. During analysis those elements were simplified by one concentrated mass m_w. Two transporters (which are delivering and receiving weighted elements) are connected to the aluminum plate with square shaped brackets. Those objects were replaced in model, as well as main transporter connected with weighting module, with concentrated mass m_{p1} placed on the ideal spring with stiffness k_{p1} parallel with suppressor with dumping c_{p1}.

Whole checkweighter system is placed on vibroisolators, which parameters—stiffness k_{izol} and dumping c_{izol} [4, 5] are to determine. Figure 1 presents simplified model of checkweighter.

2.1 Analytical Description

Based on created checkweighter model, kinematic Eqs. (1), (2) and (3) were derived.

$$m_{p1}\ddot{x}_{p1} + c_{p1}\left(\dot{x}_{p1} - \dot{x}_w\right) + k_{p1}\left(x_{p1} - x_w\right) = 0 \tag{1}$$

$$m_{p2}\ddot{x}_{p2} + c_{p2}\left(\dot{x}_{p2} - \dot{x}_w\right) + k_{p2}\left(x_{p2} - x_w\right) = 0 \tag{2}$$

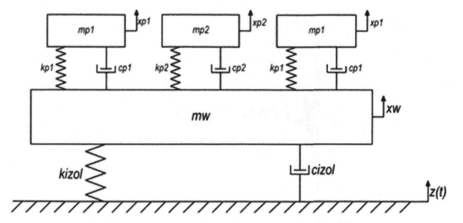

Fig. 1 Simplified mechanic scheme of checkweighter system

$$m_w \ddot{x}_w + 2c_{p1}\left(\dot{x}_w - \dot{x}_{p1}\right) + 2k_{p1}\left(x_w - x_{p1}\right) + 2c_{p2}\left(\dot{x}_w - \dot{x}_{p2}\right)$$
$$+ 2k_{p2}\left(x_w - x_{p2}\right) + c_{izol}(\dot{x}_w - \dot{z}) + k_{izol}(x_w - z) = 0 \qquad (3)$$

Without vibroisolation between checkweighter and ground, (3) would be formed as (4). Equations (1) and (2) would remain analytically the same. On the other hand both consists displacement x_w, so transmitter movement would be different.

$$m_w \ddot{x}_w + 2c_{p1}\left(\dot{x}_w - \dot{x}_{p1}\right) + 2k_{p1}\left(x_w - x_{p1}\right) + c_{p2}\left(\dot{x}_w - \dot{x}_{p2}\right) + k_{p2}\left(x_w - x_{p2}\right) = 0 \qquad (4)$$

2.2 Initial Parameters Determination

Values of coefficients of stiffens k_{p1} and dumping c_{p1} occurring in (1), (3) and (4) were determined. To determine those parameters (5), (6) and (7) were utilized.

$$k_{p1} = \omega_{p1}^2 m_{p1} \qquad (5)$$

$$c_{p1} = 2\xi_{p1}\sqrt{k_{p1}m_{p1}} \qquad (6)$$

$$\xi_{p1} = \ln\frac{A_i}{A_{i+1}} \cdot 2\pi^{-1} \qquad (7)$$

In order to obtain parameters ω_{p1}, m_{p1} and A_i/A_{i+1} for those calculations, transient characteristic of system had to be determined. Thus impulse response of vibration on block under weighting module was measured. Results are presented in

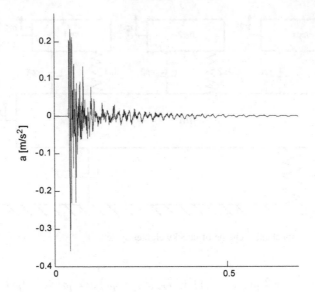

Fig. 2 Impulse response of vibrations on block under weighting module

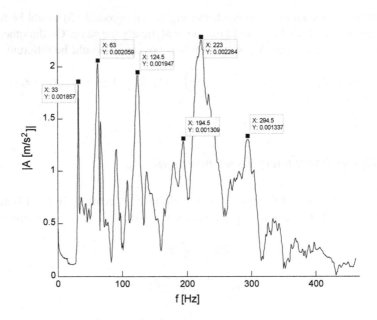

Fig. 3 Frequency analysis of impulse response

Fig. 2. During analyses, FFT was applied on measured response. Results are presented in Fig. 3 and clearly indicate that initial resonance frequency of checkweighter setup equals 33 Hz.

Also impulse extortion generates large amount of signal harmonics. Those frequencies are not caused by resonance of block itself. Their source are resonances of elements connected to the block.

Basing on systems characteristics presented on Figs. 2 and 3 required parameters results were obtained and equal:

$$\omega_{p1} = 2\pi f_{p1}\big|_{f_{p1}=33\ \text{Hz}} \approx 207\ \frac{\text{rad}}{\text{s}}$$

$$m_{p1} = 8\ \text{kg}$$

$$\frac{A_i}{A_{i+1}} \approx 1.1$$

Then, basing on (5), (6) and (7) values of k_{p1} and c_{p1} were calculated. Their values equal $k_{p1} = 343{,}936$ N/m and $c_{p1} = 5.2 \cdot 10^{-6}$.

Similar determination of analogous coefficients k_{p2} and c_{p2} for transmitter placed over the weighing module is impossible due to the fact that this module works with electromagnet, which prevents its free movement. Theoretically those coefficients can be extract from the measurement data but due to the complexity of the signal it was proved impossible as well.

Thus stiffness coefficient k_{p2} was determined with utilization of finite element method on transmitter model. Obtained value is $k_{p2} = 26{,}455$ N/m, where mass of conveyor over weighting module is $m_{p2} = 5.5$ kg. Values of vibroisolation parameters (c_{izol}, k_{izol}) are object of analysis.

3　Numeric Modeling

Equations (1), (2), (3) and (4) as well as values of coefficients were implemented in Matlab-Simulink software. Group of simulations was conducted [6], in order to obtain spectral characteristics of vibration transmission from the base (z) to the block under the weighting module (x_w), auxiliary transmitters (x_{p1}) and weighting transmitter (x_{p2}). Those simulations were conducted for different values of vibroisolation parameters. Exemplary obtained characteristics is presented in Fig. 4.

After analysis of results of simulation, optimal parameters of the vibroisolation coefficients were selected. Optimal attenuation of vibrations was obtained for coefficients $c_{izol} = 0.2$ and $k_{izol} = 2000$ N/m. Figure 5 present spectrum of vibrations on the models of analyzed checkweighter elements.

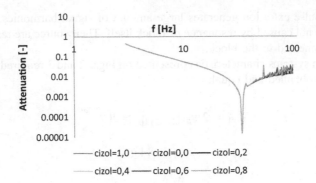

Fig. 4 Efficiency of vibroisolation on weighting transmitter x_{p2} for different values of vibroisolation dumping cizol, for stiffness coefficient kizol = 2000 N/m

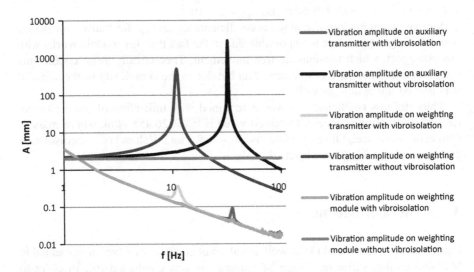

Fig. 5 Spectrum of vibrations on the models of analyzed checkweighter elements

It is clearly visible that utilization of optimized vibroisolations significantly decreases amplitudes of transmitted vibrations. The most important is suppression of checkweighter resonance frequency (33 Hz). Due to the crucial role of weighting module in checkweighter operation, efficient isolation of vibration propagation is crucial during improvement of measurement accuracy. Significant suppression of main harmonics frequency (10 Hz) is visible.

Due to significant mass of the block under conveyor belts, which in simplified model represents supporting structure of checkweighter, its resonance frequency is not visible in analyzed spectrum. On the other hand significant attenuation of vibrations amplitude in whole analyzed spectrum is visible.

4 Conclusion

In paper exemplary method of vibroisolation selection was described. Due to considerable complexity of analyzed system, development of simplified model was required. Reduced model contained four concentrated masses, connected by universal models of bonds.

Most universal bond model contains parallel connection of ideal spring and ideal suppressor. Coefficients of those models determine their behavior. Thus determination of those parameters is crucial during system modelling. Two ways of parameters determination was presented in paper. First was based on spectral analyses of impulse response of initial model. This method is more reliable but requires physical presence of analyzed object, thus cannot be utilized during system design. Second method of determination of bonds coefficients utilized Finite Element Method analysis conducted on model of the element. Results obtained with this method may be utilized during system design, in order to optimize its resonance frequency.

Based on created model numerical simulations of vibration transition were conducted. Influence of vibroisolation parameters was tested, and based on that optimal stiffness and dumping coefficients were selected.

Conducted simulations for optimized vibroisolations confirmed significant attenuation of vibrations transmission. Resonance frequency of all modules of checkweighter system were suppressed. The most important, for the system purpose, is suppression of vibrations on weighting module. Due to efficient vibroisolation improvement of mass measurement accuracy can be achieved.

Basing on numerical procedure vibroisolators can be selected from commercial elements, without conducting expensive and time-consuming tests or real object.

Acknowledgment This research is result of the project co-financed from the European Regional Development Fund under the Operational Program Innovative Economy 2007–2013 (contract number POIG.01.03.01-14-086/12) "Grants for innovation".

References

1. Lewandowski, R., Chorążyczewski, B.: Identyfikacja parametrów lepkosprężystego tłumika drgań. Zeszyty Naukowe Politechniki Rzeszowskiej **258**, 203–214 (2008)
2. Dobry, M.W.: Efficiency of the constant interaction force vibroisolation (WOSSO). J. Theor. Appl. Mech. **52**, 1083–1091 (2014)

3. Ugodziński, R., Gosiewski, Ł., Szewczyk, R.: FPGA based processing unit for a checkweighter. Adv. Intell. Syst. Comput. **267**, 713–719 (2014)
4. Jaworowski, H., Kasprzyk, S., Wapiennik, J.: Vibroisolation parameter selection method and stability determination of a discrete-continuous system of the (∞, 1)-type. ZAMM-J. Appl. Math. Mech. **60**, 212–215 (1980)
5. Goliński, J.: Vibroisolation of Machines and Devices. WNT (1979)
6. Hath, M.R.: Vibration Simulation using MATLAB and ANSYS. CRC Press (2000)

Thermoanemometric Flowmeter of Biofuels for Motor Transport

Igor Korobiichuk, Olena Bezvesilna, Andrii Ilchenko
and Yuri Trostenyuk

Abstract For vehicles, especially those running on the biofuel, the thermo-anemometric flowmeters are considered promising. The thermo-anemometric flowmeters, the design of which is based on dependence between the amount of heat expressed by the heating element located in the fluid flow and its mass consumption, are of particular interest. The thermo-anemometric flowmeter operating as a part of the hardware and software to measure the biofuel consumption by motor vehicles was developed, manufactured and customized. All its 13 channels were calibrated. The dependences of the "voltage-temperature" for sensitive elements of each channel were obtained.

Keywords Thermoanemometric flowmeter (TAF) · Fuel consumption · Sensitive element · Motor transport

I. Korobiichuk (✉)
Institute of Automatic Control and Robotics,
Warsaw University of Technology, Warsaw, Poland
e-mail: kiv_igor@list.ru

O. Bezvesilna
National Technical University of Ukraine "Kyiv Polytechnic
Institute", Kyiv, Ukraine
e-mail: bezvesilna@mail.ru

A. Ilchenko · Y. Trostenyuk
Zhytomyr State Technological University, Zhytomyr, Ukraine
e-mail: avi_7@rambler.ru

Y. Trostenyuk
e-mail: mix_ua@meta.ua

© Springer International Publishing AG 2017
R. Jabłoński and R. Szewczyk (eds.), *Recent Global Research and Education:
Technological Challenges*, Advances in Intelligent Systems and Computing 519,
DOI 10.1007/978-3-319-46490-9_59

1 Introduction

To measure the fuel consumption on transport, especially when using biofuels, thermal flowmeters are considered promising. They have a range of advantages compared, for example, to rotary, piston, diaphragm flowmeters and so on. When using ultrasonic flowmeters [1, 2], there is the problem of the loss of desired precision of low consumption measurement that occurs in the operating of vehicles. It is believed that in this case the accuracy of determining the low fuel consumption can be increased by using a larger number of sensitive elements (SE) (Fig. 1).

Thermoanemometric flowmeters (TAF) [3, 4] are of the greatest interest for the use in vehicles. In thermal flowmeters, in most cases, a nichrome spiral is used as the heating element (HE). At low fuel consumption this can lead to overheating and spontaneous combustion of fuel, the consumption of which is being measured.

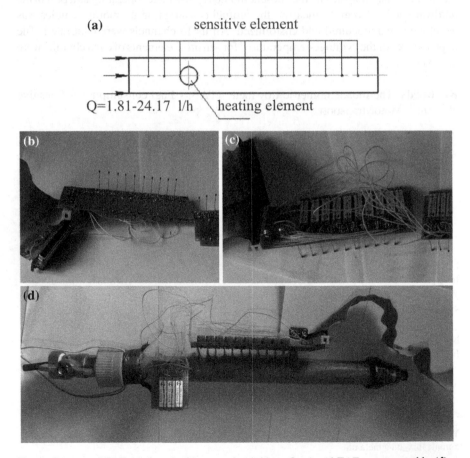

Fig. 1 Scheme of TAF (**a**), board with measuring bridges (**b**, **c**) and TAF as an assembly (**d**)

As shown by previous studies [3–7], flowmeters based on the principle of heat transferring is quite promising for the use in transport.

The aim of this study is to develop and configure hardware of the software-hardware complex, which includes TAF to measure the biofuel consumption; to produce a working model of the hardware part of the complex; to obtain the dependences of the output voltage of the measuring bridges on the temperature of sensing elements (it is proposed to use thermistors as sensing elements), and to calibrate the measuring bridges.

2 The Main Results of the Study

Experimental setup for calibration and configuration of TAF includes power supplies HE (PS(HE)) and measuring bridges (PS (TAF)), a tank with liquid, a pump, a measuring cylinder, an analog-digital converter (ADC) and a personal computer (PC) (Fig. 2). Thermistors NTC 2K2 1 % 50 mW DSTU 2815-94 were used as SE. This design is connected to the ADC using an external analog port DRB-37M ADC L-Card E14-440. It gives the opportunity to receive signals from 16 individual channels simultaneously. Voltage sampling, measured by ADC, is 0.1×10^{-3}V. In its turn, the ADC is attached to a PC via a USB port. Information processing is performed using software systems LGraph2 [8] and PowerGraph, which makes it possible to record data array "voltage-time", and then to represent it in the form of dependence "temperature-time".

For evaluation and comparison of TAF operating parameters, its computer model was studied by means of CFD-complex of COSMOSFloWorks [9].

Pre-calibration, obtaining the dependencies "temperature-voltage", as well as for research safety, the temperature distribution depending on the liquid flow rate and

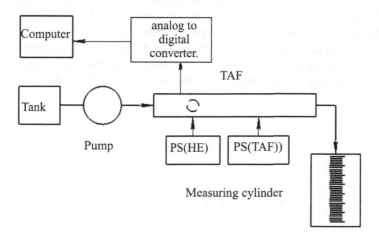

Fig. 2 Diagram of the setup for calibration and configuration of TAF

Fig. 3 The dependence of the temperature on the distance to HE for given consumption, obtained experimentally in the TAF calibration process

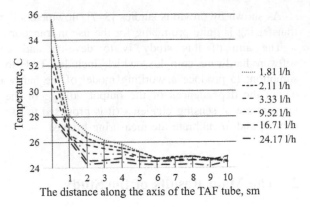

The distance along the axis of the TAF tube, sm

Fig. 4 The dependence of the temperature on the distance to HE for given consumption, obtained by computer simulation of the TAF operating

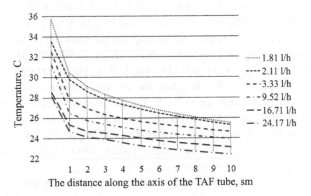

The distance along the axis of the TAF tube, sm

distance from HE were obtained for water. These dependencies require further correction taking into account the differences of the properties of the fuel, for which TAF is being developed. As the sensitive elements in this design of TAF is proposed to place on the axis at the intervals of 1 cm, the distance from HE to each sensitive element corresponds to its serial number (Figs. 3 and 4).

The obtained data show that the discrepancy between the experimental results and computer simulation is possible due to different coefficients of thermal conductivity of biofuel and its properties in the computer model (Tables 1 and 2).

Table 1 The data of dependence of the temperature on the distance to HE for given consumption, obtained by computer simulation of the TAF operating

Distance from HE (sm)	Consumption					
	1.81 l/h		2.21 l/h		3.33 l/h	
	Error		Error		Error	
	Absolute (°C)	Relative (%)	Absolute (°C)	Relative (%)	Absolute (°C)	Relative (%)
0	0.07	0.19	0.28	0.85	0.35	1.06
1	2.39	7.86	2.59	8.67	1.35	4.83
2	2.37	8.13	2.29	8.03	1.14	4.24
3	2.19	7.72	2.06	7.38	0.89	3.37
4	1.89	6.85	1.78	6.5	0.68	2.59
5	1.99	7.33	1.65	6.13	0.81	3.15
6	2.12	7.93	1.78	6.71	0.67	2.65
7	1.60	6.07	1.3	4.99	0.39	1.55
8	1.19	4.58	0.92	3.58	0.08	0.33
9	1.11	4.31	0.82	3.22	0.11	0.44
10	1.02	3.98	0.76	3.02	0.13	0.51

Table 2 The data of dependence of the temperature on the distance to HE for given consumption, obtained by computer simulation of the TAF operating

Distance from HE (sm)	Consumption					
	9.52 l/h		16.71 l/h		24.17 l/h	
	Error		Error		Error	
	Absolute (°C)	Relative (%)	Absolute (°C)	Relative (%)	Absolute (°C)	Relative (%)
0	0.67	2.16	0.35	1.21	0.1	0.36
1	0.07	0.25	1.08	4.26	1.41	5.72
2	0.58	2.27	0.1	0.42	0.18	0.73
3	0.32	1.25	0.06	0.22	0.48	2.02
4	0.01	0.03	0.47	1.95	0.88	3.72
5	0.01	0.03	0.53	2.21	1.03	4.42
6	0.17	0.69	0.76	3.22	1.21	5.26
7	0.27	1.11	0.99	4.22	1.38	6.05
8	0.46	1.88	1.2	5.13	1.59	7.02
9	0.52	2.16	1.34	5.78	1.63	7.22
10	0.94	3.93	1.82	7.9	2.06	9.21

3 Conclusions

(1) TAF operating as a part of hardware and software complex for measuring the biofuels consumption for motor transport was designed, manufactured and configured.

(2) Calibration of all 13 channels of TAF was made. The dependences "voltage-temperature" for their sensitive elements were obtained.

References

1. Soldatov, A.I., Tsekhanovsky, S.A.: Ultrasonic flowmeter with acoustic waveguide tract, Izvestiya YuFU. Eng. Sci. 163 (2008)
2. Mashchenko, T.G., Akoyev, R.I.: Methods of control of moving fluids flow. Bull. NTU "KhPI". 15, 97 (2014)
3. Rumyantsev, A.V., Shevchenko, P.R., Guskov, K.V.: High temperature gas flow meter, bulletin of RSU named after I. Kant. Phys. Math. Sci. 4, 70 (2006)
4. Korobiichuk, I., Bezvesilna, O., Ilchenko, A., Shadura, V., Nowicki, M., Szewczyk, R.: A mathematical model of the thermo-anemometric flowmeter. Sensors 15, 22899–22913 (2015). doi:10.3390/s150922899
5. Ilchenko, A.V., Romanova, A.O.: Taking into account properties of two-component fuels in the measurement of their consumption by thermoanemometric flowmeter. Proc. Automobile Road Inst. 1(4), 104 (2007)
6. Korobiichuk, I., Shavursky,Yu., Nowicki, M., Szewczyk, R.: Research of the thermal parameters and the accuracy of flow measurement of the biological fuel. J. Mech. Eng. Autom. 5, 415–419 (2015). doi:10.17265/2159-5275/2015.07.006
7. Bezvesilna, E.N., Ilchenko, A.V., Trostenyuk, Yu.V.: Calorimetric flowmeter of motor fuels with high precision of consumption measurement. In: Perspective Directions of the Motor Complex P 27 Development: Collection of Articles of VI International Scientific—Production Conference/IRC PGSKHA, 14 p. RIO PGSKHA, Penza (2013)
8. Software systems LGraph2. http://www.lcard.ru/support/lgraph
9. Alyamosky, A.A., Sobachkin, A.A., Odintsov, Ye.V., Kharitonovich, A.I., Ponomarev, N.B.: Solidworks 2007/2008. Computer modeling in engineering practice/SPb, 1040 p. BHV, Petersburg (2008)

Research on Automatic Controllers for Plants with Significant Delay

Igor Korobiichuk, Dmytro Siumachenko, Yaroslav Smityuh and Dmytro Shumyhai

Abstract The article represents the possibility of using the modified algorithm of automatic controllers with higher order derivatives to control complex plants with a significant delay. The mathematical models of control systems are developed and comparative analysis of quality indexes of derived transient responses is carried out.

Keywords Significant time delay · Multiparameter controllers · PID-controller · Optimization of controller's parameters

1 Introduction

Time delay (or delay) is always observed in industrial heat and mass transfer processes and lead to late transfer of current state information of the plant to the controller, which can lead to the loss of stability and efficiency of closed loop control system. The control complexity of the plant with delays is characterized by the ratio of the delay to the constant time of the plant: if ratio is larger than it is more difficult to achieve the required quality control.

The time during which the change of the input signal does not effect on output value is called delay. The linear system with delay is a system that includes at least one link, which has the delay time τ in control and measurement channels, that has

I. Korobiichuk (✉)
Institute of Automatic Control and Robotics, Warsaw University of Technology, Warsaw, Poland
e-mail: kiv_igor@list.ru

D. Siumachenko · Y. Smityuh · D. Shumyhai
National University of Food Technology, Kiev, Ukraine
e-mail: dmitrij-syumachenko@yandex.ru

Y. Smityuh
e-mail: smityuh1@gmail.com

D. Shumyhai
e-mail: shumygai@gmail.com

influence on final signals speeding rate of information in the plant (transport delay) [1]. This leads to increased order of differential equation of the plant, in other words, to a large number of sufficiently small time constants of the plant.

Control quality of such plants can be improved in two ways:

- reducing delay in the plant by making structural changes;
- using more complex structure of control system that reduces the negative impact of delay [2, 3].

Synthesis capabilities of control systems of complex plants were extended due to development of software, numerical methods and computation.

The presence of the delay in the plant increases the overcontrol and oscillation in the typical single-loop system [4]. The carried out analysis shows that the common disadvantage of such systems is the inability to provide high performance with a small overcontrol in systems with a significant delay [5].

That problem is solved in double-loop control systems proposed by Veronesi [6] and in systems with delay compensation. The idea of building such systems is to insert the model of the plant in a regulation loop and to obtain signal without delay from the plant. That signal is required to build a satisfactory control system [7]. The disadvantage of this approach is the requirement that parameters and model structure must conform to real plant, that requires precise priori information about it. This creates difficulties in the construction of control system.

Problem statement. The problem of comparative analysis of transient responses of control systems for plants with delay is solved using both classical PID controller and its modifications with second and third order derived (PIDD2 and PIDD2D3 regulators). Juice heater with transfer function of first order aperiodic link with delay is set as the first plant. Second plant is deaerator that is described by the second order aperiodic link with delay. The structure of control system should be created using Matlab software and controller settings should be optimized according to the selected quality criteria.

2 Task Solution

2.1 Computer Simulation

Plant is presented as a mathematical model for analyzing the plant, designing the optimal control system and studying the impact of control actions. The input signals transmission is analyzed during the study, in other words, the system response to step signal is determined. System properties were estimated in the time and frequency ranges.

Juice heater is described as a mathematical model of the transfer function (aperiodic links of the first order with delay):

$$W(p) = \frac{0.25}{156p + 1} \cdot e^{-27p} \tag{1}$$

Mathematical model of the second plant (deaerator) is second order aperiodic link with delay:

$$W(p) = \frac{3.5}{(42p + 1) \cdot (10p + 1)} \cdot e^{-20p} \tag{2}$$

The structures of control systems based on PID (Fig. 1), PIDD2 (Fig. 2) and PIDD2D3 (Fig. 3) controllers were created in Simulink toolbox of Matlab software. The proportional component of the controller provides the necessary accuracy of the system in the midrange frequencies, integrated amplifies signals at low frequencies and provides high accuracy of the system in this area. Included integral part of the system allows to provide the unlimited growth of amplitude-frequency response with decreasing frequency. This means that static error is zero and all system errors are generated by other causes, such as static sensor error [8].

Differential component should rise amplitude-frequency response at high frequencies, resulting in the achieved stability of the system and very high quality control, in other words resulting in slight overshoot or its complete absence [9]. Harmful effects from differentiation can occur in excessive strengthening the noise of feedback signal. Another reason of the negative impact of differentiation may include excessive increase of the area of amplitude-frequency characteristics in

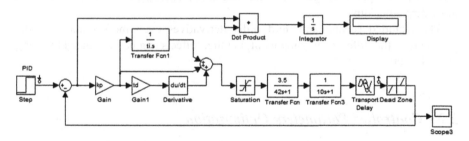

Fig. 1 Ideal PID-controller (standard, noninteractive, ISA algorithm)

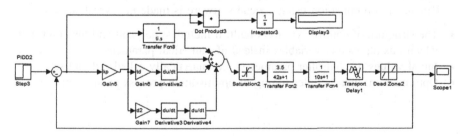

Fig. 2 PIDD2-controller (with second order derivative)

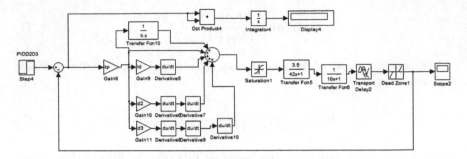

Fig. 3 PIDD2D3-controller (with second and third order derivative)

which the phase shift is so great that differentiation does not improve the control quality. This is especially true for high-order plants and plants with delay.

The component that is proportional to the second derivative or to the deviation acceleration of controlled parameter with time constant is added in PIDD2 algorithm. By analogy, we can talk about the third derivative, which characterizes the rate of acceleration.

The transfer function of multiparameter PIDD2D3 controller is:

$$W(p) = K_p \cdot \left(1 + \frac{1}{T_i \cdot p} + T_{d1} \cdot p + T_{d2} \cdot p^2 + T_{d3} \cdot p^3\right), \qquad (3)$$

where K_p—proportional gain, T_i—integral time, T_{d1}, T_{d2}, T_{d3}—first second and third derivative time.

The included second and third order derivatives even more complicate the problems of the differential component, but their effects can be minimized by using special filters.

2.2 Controller Parameters Optimization

Multiparameter controllers have more configuration options, so they need for optimization to provide the necessary quality of transient response.

Parameters optimization using special software is made in several stages:

- The structure of control system is built. It includes the plant and the controller, which has alphabetic variables instead of numerical parameters.
- Initial values of controller parameters are analytically calculated and set.
- Optimization, that includes tuning of optimization process parameters.

Moreover, control systems with optimal controllers have the following requirements:

1. Control systems and controllers must meet the technological and operational requirements and production modes of the plant.
2. The system should maintain stable when the plant is influenced by the most characteristic disturbances.
3. The set quality of the control process, operational reliability and economic efficiency should be ensured in dynamic and static modes of the system.

Method CHR (Chien-Hrones-Reswick) [10] is used to determine the initial values of controllers settings, which uses the criterion of maximum increasing rate with no overshoot. The criterion is based on plant response to a step signal. Criterion provides a higher stability factor than the method of Ziegler-Nichols MP [11].

Optimization is done in software Matlab using toolbox NCD (The Nonlinear Control Design Blockset), which is a graphical user interface (GUI) to implement a Simulink design control systems in the time domain. The optimal settings for nonlinear system model in Simulink can be found with NCD. Those settings will provide the necessary response process that is limited graphically in the time domain [12]. We can draw an analogy with the interface "box" of V. V. Solodovnikova.

The toolbox NCD automatically turns limits in the time domain to the limits of optimizing problem of the control system parameters and then solves the problem by using algorithms and procedures laid down in the optimization package (sequential quadratic programming algorithm SQP and quasinewton gradient search method [13]). The integral quadratic criterion is used to estimate the quality of control in numerical optimization procedures:

$$I = \int\limits_{t_n}^{t_\kappa} (e(t))^2 \, dt, \tag{4}$$

where $e(t)$–tracking error, t_n–start time, t_k–end time.

3 Results of Research

3.1 Analysis of the PID, PIDD2, PIDD2D3 Controllers Comparison During Juice Heater Control

PIDD2D3 controller was the best for controlling the plant, as the control variable got to the set point for the minimal time (92 s) (Fig. 4). PIDD2 controller had minimal dynamic error. PIDD2 controller was better for plant with changeable

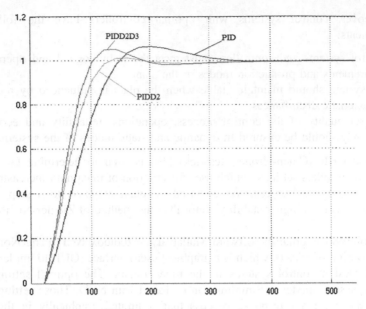

Fig. 4 Transient responses in heating control systems

Table 1 Controllers comparison

Controller	K_p	T_i	T_d	T_{d2}	T_{d3}	t, s	$A1$	ψ (%)	I^2
PID	6.1511	109.6879	−21.1596	–	–	200	0.05	100	59.28
PIDD2	9.7494	159.27	−3.156	−4.275	–	109	0.005	100	49.09
PIDD2D3	11.0361	159.2525	−3.3277	4.2810	0.9975	92	0.048	100	46.29

parameters. It provided an extra margin of stability with either changeable time delay or changeable transfer coefficient. The table presents a values comparison of controller coefficients (K_p—proportional gain, T_i—integration time, T_d—differentiation time of the first order, T_{d2}—differentiation time of second order T_{d3}—differentiation time of third order) and quality of response (t—settling time $A1$—dynamic error, ψ—decay ratio, I^2—integral quadratic criterion of quality) (Table 1; Fig. 5).

3.2 Analysis of the PID, PIDD2, PIDD2D3 Controllers Comparison During Deaerator Control

The best performance for the plant with transfer function (1) was obtained using PIDD2D3 controller, but PID controller was better for the plant with changeable

Fig. 5 Influence of the time delay on decay ratio for control systems with different controllers

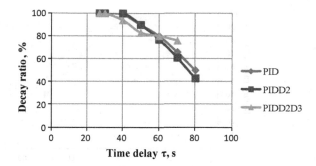

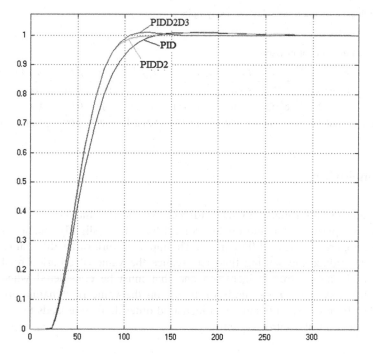

Fig. 6 Transient responses of deaerator control systems

parameters. Dynamic error and decay ratio hardly changed. Systems with PIDD2 and PIDD2D3 controllers reduce the settling time and at the same time decrease margin of stability (Figs. 6 and 7; Table 2).

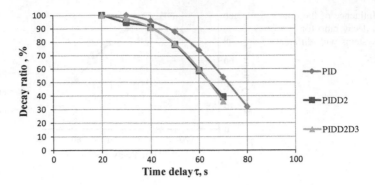

Fig. 7 Influence of the time delay on decay ratio

Table 2 Controllers comparison

Controller	K_p	T_i	T_d	T_{d2}	T_{d3}	t, s	AI	ψ (%)	I^2
PID	0.2650	53.1323	8.1730	–	–	106	0.01	100	45.99
PIDD2	0.2922	53.6707	7.4808	40.1451	–	89	0.01	100	43.52
PIDD2D3	0.3053	56.3750	7.2048	−1.6772	0.0254	89	0.007	100	43.18

4 Conclusions

Multiparameter controllers (PIDD2, PIDD2D3) showed the best results. They provide high performance and quality control but have a slightly smaller stability margin compared to usual PID control. The input of third-order differential component can reduce the settling time and reduce the value of criterion quality I2. However, dynamic error is increased and that must be considered when other plants, that may have more rigid constraints on the value of dynamic error, are controlled. Figures 4 and 6 show that increased order derivative leads to a stability margin reduction of control system.

References

1. Dorf, R.: Sovremennyye sistemy upravleniya / R. Dorf, R. Bishop.; Per. s angl. B. I. Kopylova. M.: Laboratoriya bazovikh znaniy YUNIMEDIASTAYL, p. 832 (2002)
2. Yermolovich, D. A.: Upravleniye ob"yektami s bol'shim zapazdyvaniyem/D. A. Yermolovich, A. P. Movchan // Materialy konf. Yevropeyskaya nauka KHKHÍ veka, 2010 / Natsional'nyy tekhnicheskiy universitet Ukrainy «KPI». Kiyev (2010)
3. Korobiichuk, I.: Mathematical model of precision sensor for an automatic weapons stabilizer system. Measurement **89**, 151–158 (2016)
4. Rotach, V. Y. A.: Teoriya avtomaticheskogo upravleniya / V. Y. A. Rotach; Izdatel'stvo MEI. — M. p. 396 (2008)

5. Ladanyuk, A. P.: Arkhanhel's'ka K. S., Vlasenko L. O. Teoriya avtomatychnoho keruvannya tekhnolohichnymy ob"yektamy: Navch. posib. K.: NUKHT, p. 274 (2014)
6. Veronesi M.: Performance improvement of Smith predictor through automatic computation of dead time. Yokogawa tech. rep. **35**, 25–30 (2003)
7. Hongdong, Z., Ruixia, L., Huihe, S.: Control for integrating processes based on new modified smith predictor / Control 2004. University of Bath: UK (2004)
8. Vagia, M. (ed.): PID Controller design approaches—theory, tuning and application to frontier areas. Croatia: InTech (2012)
9. Denisenko, V. V.: PID regulyatory: voprosy realizatsii / V. V. Denisenko // Sovremennyye tekhnologii avtomatizatsii. **4**, 86–97 (2007)
10. Zhmud' V. A. O.: metodakh rascheta PID-regulyatorov/V. A. Zhmud' // Avtomatika i pro-grammnaya inzheneriya. **3**(4), 118–124 (2013)
11. Äström, K. J.: Revisiting the Ziegler-Nichols step response method for PID control / K. J. Äström, T. Hägglund // J. Proc. Cont. **14**, 635–650 (2004)
12. D'yakonov, V. P. M.: Polnyy samouchitel' / V. P. D'yakonov. M.: DMK Press, p. 768 (2012)
13. Chau, P. C.: Process control—a first course with MATLAB // Cambridge University Press: Cambridge (2002)

Early Support of Technical Education

Jaromir Hrad, Tomas Zeman, Boris Simak, Daniela Spiesova
and Dusan Maga

Abstract An international project realized by 9 partners from 4 EU countries (CZ, ES, SK, DE) will be introduced in the paper. The main objective of the project is to increase the quality of education at schools using a form that will be highly attractive. Topics are based on 20 different subjects (telecommunications, electronics, automation and related areas). The following types of new-generation learning objects have been developed for students and teachers: explanatory modules, worksheets, tests, explanatory dictionary, and multimedia translation dictionary. These learning objects are described in the paper. Also the quality management of a large international team, as well as the social impact and expectations will be discussed.

Keywords Technical education · Learning objects · VET

1 Introduction

The introduced project reacts to the pan-European decrease of interest in studying technical disciplines, which can be observed already at secondary schools. From the strategic point of view, the fields related to electrical and information engineering

J. Hrad (✉) · T. Zeman · B. Simak · D. Maga
Czech Technical University in Prague, Technicka 2, 166 27 Prague, Czech Republic
e-mail: hrad@fel.cvut.cz

T. Zeman
e-mail: zeman@fel.cvut.cz

B. Simak
e-mail: simak@fel.cvut.cz

D. Maga
e-mail: dusan.maga@fel.cvut.cz

D. Spiesova
Czech University of Life Sciences Prague, Kamycka 129, 165 21 Prague, Czech Republic
e-mail: spiesova@pef.czu.cz

© Springer International Publishing AG 2017   459
R. Jabłoński and R. Szewczyk (eds.), *Recent Global Research and Education:*
Technological Challenges, Advances in Intelligent Systems and Computing 519,
DOI 10.1007/978-3-319-46490-9_61

are vital for sustainable European competitiveness. However, very often we can witness the situation that many students do not understand even the basic principles of modern electronic devices that all of them use daily. Our motivation is to contribute to higher attractiveness of study, to foster the students' interest in the given field and involve them in the international competition. On the other hand, we also want to increase the motivation of secondary school teachers, so that the best ones would not leave the sector of education.

According to the preliminary analysis, vocational secondary schools are missing:

(a) Really modern learning aids that would be not only efficient in pedagogical sense, but also well received by the students,
(b) Study materials for many areas, especially those covering new topics that have not been processed in the form of textbooks or other learning aids yet.

Therefore the main objective of the project is to increase the quality of education at vocational secondary schools using a form that will be highly attractive for their students. Project web portal can be found at www.techpedia.eu.

2 Types of Learning Objects

The designed learning system consists of five types of mutually related objects. The core of the system are the learning modules. They are supported by four other functionalities—multilingual and explanatory dictionary, tests and worksheets. The used standards are verified by international team (the managing problems and quality issues are discussed in [1]). The structure of the team (universities complemented by secondary schools) guaranties the topicality and the professional content of the designed materials, including the acceptance of the target group (secondary school students).

2.1 Multilingual Dictionary

The primary goal of the project is to design a new technology for electronic support of education and deliver its content; nevertheless, thanks to the international partners of the project, another interesting issue has been identified and solved. The project partners are from four different European countries speaking 4 different languages (German, Spanish, Czech and Slovak). These languages have been complemented with additional 4 ones (English, French, Slovenian and Swedish—these are based on the interest and help of other partners outside the consortium). It is worth mentioning that the web services are also offered in 5 languages (based on the partners' countries plus English). Therefore the idea of a professional multilingual translation dictionary (including multimedia features) originated. Today the dictionary has about 3200 entries and their number still grows.

2.2 Explanatory Dictionary

The abovementioned service is supplemented with unique professionally elaborated encyclopedic entries. The content of these objects is inspired by Wikipedia-like services; however, the aim, purpose and value of the information is entirely different. The major difference is that the content cannot be edited by community—it has been developed and maintained by experts, so that the professional value is guaranteed. Nowadays, more than 500 entries have been processed here. This part of the material is multilingual as well—it covers all 5 languages (example of the explanatory service can be seen in Fig. 1). Therefore, this functionality has been labeled as "a jewel among the learning objects" [2].

2.3 Learning Modules

The core of the developed system is based on learning modules. They are realized as independent chapters dealing with specific professional topics defined by project partners' needs, as one of the major aims is to build a substantial support for technically oriented education at secondary schools specialized in electrical engineering. The topics (as described in [2]) are focused mainly on telecommunications and ICT (wireless networks, IoT, optical systems and networks, LTE, high-speed

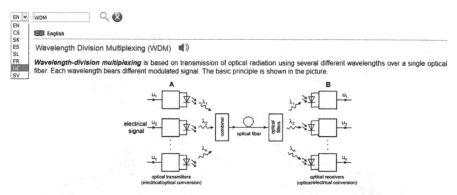

The signal from n optical transmitters (laser diodes) working on wavelengths λ_1 to λ_n is coupled to the fiber in device A and then transmitted to device B, where optical filters are used to decompose it back to n individual optical signals, which are converted into electrical ones. Thus, n independent signals can be transmitted over a single fiber. The picture illustrates one direction of transmission (A–B). For the opposite direction, another fiber would be required, with the same circuitry (the function of a two-fiber communication circuit corresponds to that of a 4-wire metallic line, i.e. 2 pairs).

Fig. 1 An example from explanatory dictionary

Internet access, etc.), automation (home automation, systems for intelligent buildings and homes, etc.) and security issues (cryptography, cybercrime, security systems, etc.). Project portal offers the content in several most important formats (e.g. ePub, Amazon Kindle, PDF or SCORM; selected parts are also available as XML files).

2.4 Tests

Each module is supported by tests. The regular self-evaluation tests are realized as a set of single-choice questions. The number of questions may differ from module to module, but the minimum number is 20. They should help the students to validate their knowledge. There is also an "Olympic" test prepared for each module—these tests are multiple-choice ones and they are prepared for the international Technical Olympiad.

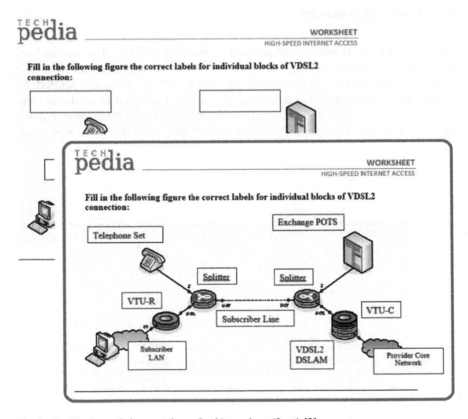

Fig. 2 TechPedia worksheet: students (back), teachers (front) [3]

2.5 Worksheets

Worksheets can be used to verify the students' understanding of the given topic in a different way. Typical sets of professional tasks are presented in this form, based not only on texts, but also on schemes, puzzles, crosswords, figures and graphs, etc., which makes the students' work interesting and more natural. The graphical interpretation also helps the students to memorize important data and find a good way in solution-searching procedures. The worksheets accompany the complete range of modules. The students' and teachers' versions differ, naturally, in hiding or displaying the correct solution and/or procedures. Typical view of the students'/ teachers' worksheet can be seen in Fig. 2.

3 Pilot Testing

We consider the pilot testing of all developed learning objects as a component of outputs quality control. The core of the pilot testing is an extensive pilot run, both for teachers and for students of secondary schools. In order to increase the attractiveness of the pilot run for students (and the willingness to participate in it), we have included a motivational element—an international competition called "Technical Olympiad" [4]. An integral part of the pilot testing is a system of feedbacks and evaluation, which leads to the elimination of (potential) weak points in the developed learning objects.

The first part of the learning objects testing consists in the pilot run for teachers. This pilot run has two purposes. The first one is the promotion—the teachers can learn about the advantages and the quality of the developed learning objects, which will motivate them to use the project outputs it in their lessons. The second purposes is to master the learning environment (see above) and to get familiar with the learning objects.

The pilot run for teachers begins with an all-day face-to-face tutorial. Within the tutorial the teachers learn how to get the offered learning objects and how to use the learning environment efficiently. They also get familiar with the support that they will receive during the pilot run. The topics of the individual learning objects and their scope will be introduced. Practical demonstrations of using the learning objects in classes with students are provided as well.

The all-day tutorial is followed by a 14-day self-study period. This part of the pilot run for teachers is aimed at making the teachers familiar with new topics, i.e. the factual content.

The objective of the pilot run for students is the maximum utilization of the developed learning objects. The pilot run for students will therefore be organized within regular lessons—the teachers, at their own discretion, will include the appropriate chapters of the newly developed materials. It is not the purpose of the project to force the teachers to use the full range of learning objects; on the contrary, we want the teachers to focus on using that part of the learning objects, which is of real value for their subjects from their own point of view. All materials are arranged in this way.

The pilot run will be organized on a large scale in the four countries involved in the project and at many schools. We expect the involvement of more than a thousand students. The duration of the pilot run for students will be 7 months.

4 Expected Social Impact

The long-term excess of demand over supply of labor force is typical for the ICT sector, which significantly increases the salary levels there. The situation does also bring negative effects—too fast growth of salaries raises the price of labor faster than the labor productivity, and thus the companies may have a problem with global competitiveness. The professional staff fluctuation is also growing within the sector, reducing the motivation to work on the individuals' personal development.

Based on National Training Fund studies, ICT professions are characterized by high percentage of university-educated staff. Tertiary education has more than one third of employees, but only one sixth of them graduated from a college or university with a relevant professional focus.

At present, companies in the Czech Republic are missing up to 20,000 ICT professionals. The qualification discrepancies are discussed as well. Similar situation is observed at the European level (as shown by a study carried out under the Digital Single Market project, European Commission priorities). According to this study, more than 825,000 ICT professionals will be missing in 2020.

New investors place the demand for thousands of jobs annually. ICT professions are also often searched by both industrial companies and services. Currently, while the tertiary education in the Czech Republic supplies about two thousand graduates per year (and although this number will grow), that still will not be enough for the labor market.

From this point of view, the project Techpedia is a very good way to increase the interest in technical fields. It encourages secondary school students to choose technical universities, and thus it contributes to the growing number of ICT experts. Increasing this will help to solve of the negative aspects observed within the ICT sector. So, the introduced activity might support the increase of the competitiveness of companies and the national economy in the global market.

From the social point of view, the questionable factor of the Techpedia project is that students receive the information from heterogeneous sources perfectly processed. This might result in their decreased abilities to work with various information sources and handle them individually and correctly for their own needs.

5   Conclusions

The project introduced in this paper follows the idea of improving the level of education at secondary schools focused on engineering area and their collaboration with technical universities. Taking into account the professional quality guaranteed by the applied quality management system, the developed learning objects together with the learning portal could become a basis for a continuously developing and sustainable system that will be used for support of routine lessons [5], not only at the schools involved in the project implementation, but throughout the participating countries or even other ones, thanks to the internationalized content (offered in several major languages [6, 7]).

Acknowledgments This paper has originated thanks to the grant support obtained from the Erasmus + program within the project No. 2014-1-CZ01-KA202-002074 "European Virtual Learning Platform for Electrical and Information Engineering—TechPedia".

References

1. Zeman, T., Hrad, J.: Quality management for development of large-scale electronic learning objects. In: Proceedings of eLearning 2015, pp. 184–187. Gaudeamus, Hradec Kralove (2015). ISBN 978-80-7435-615-5
2. Zeman, T., Hrad, J.: Improved chances for entering technical universities. In: Proceedings of the 24th EAEEIE Annual Conference, pp. 16–21. Technical University of Crete, Chania (2013). ISBN 978-1-4799-0043-5
3. TechPedia. [Online], http://www.techpedia.eu/
4. Maga, D., Dudak, J., Pavlikova, S., Hajek, J., Simak, B.: Support of technical education at primary and secondary level. In: Proceedings of 15th Mechatronika 2012, pp. 195–198. CVUT, Praha (2012)
5. Bauer, P., Dudak, J., Maga, D., Hajek, V.: Distance practical education for power electronics. Int. J. Eng. Educ. **23**(6), 1210–1218 (2007). ISSN 0949-149X
6. Maga, D., Sitar, J., Bauer, P.: Automatic control, design and results of distance power electric laboratories. In: Recent Advances in Mechatronics 2008–2009, pp. 281–286. Springer, Berlin (2009). ISBN 978-3-642-05021-3
7. Bauer, P., Staudt, V.: Remote controlled practical education for power electronics. In: European Conference on Power Electronics and Applications, EPE 2007, p. 10. EPE Association, Brussels (2007). ISBN 978-92-75815-10-8

Part VII
Metrology, Sensors and Devices

Uncertainty Analysis as the Tool to Assess the Quality of Leak Detection and Localization Systems

Mateusz Turkowski, Andrzej Bratek and Paweł Ostapkowicz

Abstract The new method of uncertainty analysis application for the assessment of leak detection systems is presented. It enables to assess the sensibility of the system, giving it the possibility of detecting leaks without excessive number of false alarms. It is also possible to assess the accuracy of leak localization. Experimental research confirmed the effectivity of the proposed method.

Keywords Leak detection · Leak localization · Leak localization uncertainty

1 Introduction

Even if the pipeline has been designed and built very carefully, there is always a potential for leaks. If a leak occurs, its effects can be minimized only by fast detection and localization, enabling reaction of the dispatcher (stopping pumps, closing valves), so pipeline leak detection systems are crucial for the minimization of the leaks impact. A wide range of technologies is available today [1–3].

The leak detection and localization systems are sensitive to the irregularities of input data, e.g. the noise, uncertainty and systematic errors of the instruments. Low quality of the data influences the leak detection systems and generate false alarms.

M. Turkowski (✉)
Faculty of Mechatronics, Warsaw University of Technology,
Warszawa, Poland
e-mail: m.turkowski@mchtr.pw.edu.pl

A. Bratek
Industrial Research Institute for Automation and Measurements,
Warszawa, Poland
e-mail: abratek@piap.pl

P. Ostapkowicz
Faculty of Mechanical Engineering, Bialystok University of
Technology, Białystok, Poland
e-mail: p.ostapkowicz@pb.edu.pl

© Springer International Publishing AG 2017
R. Jabłoński and R. Szewczyk (eds.), *Recent Global Research and Education:
Technological Challenges*, Advances in Intelligent Systems and Computing 519,
DOI 10.1007/978-3-319-46490-9_62

An attempt to analyze the problem in a rational and systematic way and obtain objective, quantitative criteria of the quality of the leak detection system is the task of the paper. The authors claim, the issue can be resolved using methods known in metrology, described in the Guide, so called GUM [4] and in standard ISO 5168 [5].

2 Analytical Methods of Leak Localization

For the purposes of leak detection and localization systems, the following process variables should be permanently measured:

- pressure at the inlet p_0, at the end p_n, and pressures p_i at several or more locations x_i (at the valve stations, metering stations, terminals),
- flow rate, at least at the inlet or/and at the outlet of the pipeline,
- temperature preferably at the inlet and outlet and ground temperature,

Other variables influencing the system (viscosity, density) are measured directly or are calculated from measured temperature and pressure.

The measurements are compared with the data calculated from the analytical model of the system, and the differences between the modeling results and the measurements results are used to identify and localize the leak. Transients usually influence the model, decreasing its accuracy and increasing the uncertainty of leak localization.

In contemporary systems sophisticated, but accurate dynamic models, such as real time transient modeling, are used [6], which can produce results at almost the same accuracy level of modeling as the steady state modeling. Dynamic models use the same data as static models, encumbered with the same uncertainties, so the use of a static model for the purposes of uncertainty analysis is justified.

For uncertainty analysis was therefore used the formula representing pressure drop Δp across the pipeline segment for the modeling of liquid flow through the pipe

$$\Delta p = \lambda \frac{\rho V^2}{2} \frac{L}{D} \tag{1}$$

The pressure distribution along the pipeline before and after leak is shown in Fig. 1.

In no-leak conditions the pressure along the pipeline is represented by straight line, pressure gradient $G_0 = dp/dx$ is constant over the whole pipeline length L. If the leak occurs at the localization x_l the following phenomena can be observed:

- change of the pressure at outlet Δp_{II},
- increase of the pressure gradient ΔG_I between inlet and leak coordinate x_l,
- decrease of the pressure gradient ΔG_{II} between leak coordinate x_l and pipeline outlet.

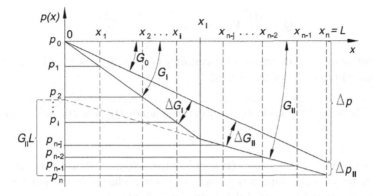

Fig. 1 Pressure distribution along the pipeline during no-leak conditions and after leak occurrence

3 Leak Identification

The first task of the system is to identify if there is, in fact, a leak. Usually, some indicating parameters are calculated based on the differences between the model and measurements results. The threshold values are determined during the system tuning in order to identify real leaks without the generation of false alarms.

The question is how to determine the quality of the system before installation.

The sensitivity for leaks depends on the uncertainty of both measured pressure p_m and pressure p_c calculated from the model. If the extended uncertainty fields of calculated pressures $U_{p,c}$ and measured pressures $U_{p,m}$ overlap, the existence of a leak is doubtful.

If, however, at some points the uncertainty fields do not overlap, there is high probability that a leak has really occurred. In other words, a leak is probable if ($p_c - p_m$) is greater than the sum of uncertainties of calculated and measured pressure, so

$$|p_c - p_m| \geq \sqrt{U_{p,c}^2 + U_{p,m}^2} \qquad (2)$$

The expanded uncertainty of directly measured pressure $U_{p,m}$ can be evaluated from manufacturer of pressure transmitters data.

The calculated pressure is instead calculated from the formula

$$p_c = p_0 - \lambda \frac{\rho V^2}{2} \frac{L}{D} \qquad (3)$$

A more convenient parameter is, however, volumetric flow rate. It is usually measured directly at the inlet and sometimes also at the outlet of the pipeline. It equals

$$Q = \frac{\pi D^2}{4} V, \quad \text{so} \quad p_c = p_0 - \frac{8\lambda\rho Q^2 L}{\pi^2 D^5} \tag{4}$$

According to [3, 4] the standard uncertainty $u(p_c)$ of the leak coordinate x_l can be calculated with the use of the formula

$$u(p_c) = \sqrt{[u(p_0)c(p_0)]^2 + [u(\lambda)c(\lambda)]^2 + [u(\rho)c(\rho)]^2 + [u(Q)c(Q)]^2 + [u(D)c(D)]^2 + [u(L)c(L)]^2} \tag{5}$$

The sensitivity coefficients $c(p_0)$, $c(\lambda)$, $c(\rho)$, $c(Q)$, $c(D)$, $c(L)$ are defined as change in the output estimate x_l produced by a unit change in the input estimates and can be obtained by partial differentiations of (5), e.g.

$$c(p_0) = \frac{\partial p_c}{\partial p_0} = 1, \ c(\lambda) = \frac{\partial p_c}{\partial \lambda} = \frac{8\rho Q^2 L}{\pi^2 D^5}, \ c(\rho) = \frac{\partial p_c}{\partial \rho} = \frac{8\lambda Q^2 L}{\pi^2 D^5} \text{ etc.} \tag{6}$$

The expanded uncertainty of the calculated pressure $U_{p,c}$ equals standard uncertainty multiplied by a coverage factor usually $k = 2$, providing a level of confidence of approximately 95 %, so $U_{p,c} = 2u(p_c)$.

The uncertainties of the pressure and flow measurements can be evaluated from instruments manufacturers data. Density is measured directly, taken from tabularized data or, for liquid hydrocarbons, calculated according to API standard. The length of pipeline and distances between pressure transmitters can be measured with the use of contemporary geodesic measurements with low uncertainty and are stored in GIS systems. The parameters above can be measured with very low uncertainty.

The pipeline internal diameter can be hard to measure directly, in particular if the leak detection system has to be installed on an existing pipeline. We can only use the nominal data and the uncertainty can be evaluated from ironworks working tolerances. They are very high—1 % on diameter and 15 % on wall thickness. A similar problem arises in the case of the friction factor λ. It can be calculated either from formula

$$\frac{1}{\sqrt{\lambda}} = -2\log\left(\frac{\varepsilon}{3.7D} + \frac{2.523}{Re\sqrt{\lambda}}\right) \tag{7}$$

or, based on measurements of pressure drop Δp across the pipe section of length L

$$\lambda = \frac{\Delta p \pi^2 D^5 L}{8\rho Q^2} \tag{8}$$

In both cases D value, which is very inaccurate, is used as input data. In Colebrok-White formula (8) the roughness height ε, which also cannot be measured accurately affects the calculation accuracy. Moreover, during the pipe life ε can change in an uncontrolled way. This is why the formulae (8) or (9) can be used only for the initial calculations of λ.

These are the main reasons that the uncertainty calculations directly from (6) give overestimated uncertainties, so that the condition (3) would hardly be fulfilled.

Fortunately, in fact, in most leak detection systems the procedure of tuning is implemented. In short, it consists of modifying the λ value in cases when there are differences between pressures calculated from the model and measured pressures (residuals). Tuning is conducted to get the residuals down to zero.

We cannot directly improve the measurement accuracy of D, so it is handled as a constant, nominal value (although the corrections for pressure and temperature are usually applied). But the effects due to the difference between real and nominal values of D are incorporated in λ, so the uncertainty of D can be neglected.

Also an imprecise roughness height ε, even if it changes during pipeline operation, is not important, also these changes are included in λ while system tuning.

As the friction coefficient λ is tuned against pressure transmitters, its uncertainty after tuning is of the same order of magnitude as the uncertainty of pressure measurement.

4 Leak Localization

The classical method of leak localization [7, 8] consists of the calculation of the abscissa of the intersection point of two lines representing pressure distribution along the pipe upstream and downstream of the leak (Fig. 1).

The coordinate of the leak x_l lies at the intersection of these two lines, so in point

$$x_l = \frac{p_n - p_0 - G_{II}L}{G_I - G_{II}} \tag{9}$$

The uncertainty of x_l can be calculated according the same procedure as in case of uncertainty of calculated pressure $u(p_c)$.

5 Experimental Part

The experimental research in real fuel or crude oil pipelines by simulation of the leak releasing product to the tank truck is extremely difficult and dangerous. It was therefore decided to use the laboratory model of the pipeline built in the Bialystok University of Technology in Poland and described widely in [9] (Fig. 2).

Fig. 2 Laboratory model of the pipeline used during experiments

The model consists of pump, valves, tanks at the inlet and outlet, 380 m of polyethylene pipeline of 34 mm diameter. The stand enables the measurement of pressure distribution along the pipe. Solenoid valves with interchangeable orifices to set the leak intensity, enable the stimulation of leaks of desired flow rate in 9 places distributed along the pipe. The flow rate can be measured at both ends of the pipe with the use of electromagnetic and Coriolis meters.

The simulated leaks were localized 115, 155, 195 and 275 m from the pipe inlet and the leak flow rates were 1, 2, 4, 8 and 9 % of the flow rate in the pipe.

The calculated value of the term $\sqrt{U_{p,c}^2 + U_{p,m}^2}$ was 1.206 kPa.

All the leaks of 2 % or more were detectable. The condition (3) in all these cases was fulfilled.

The leaks of 1 % were undetectable so criterion (3) should not to be fulfilled for these cases. However, in 1 case (for 8 experiments) the criterion (3) was fulfilled.

In the next step the assessment of the accuracy of the leak localization was determined for the leak rates 2 % and more. The localization algorithms used were gradient method and detection of the pressure shock wave. For 80 experiments the calculated leak localization uncertainty was greater than localization error in 76 cases, so only in the 4 cases the localization error is greater than calculated uncertainty, which is 5 % of total number of leak detection experiments. On the other hand, the extended uncertainty was calculated with a confidence level of 95 %, so theoretically also 5 % of calculated uncertainties can be greater than the real error. The agreement between experiment and calculation is therefore good.

6 Conclusions

The uncertainty analysis can be the precise tool for the assessment of the reliability and accuracy of leak detection systems. It enables the assessment of both sensitivity of the system to leaks and leak localization accuracy. After some modification, uncertainty analysis can be helpful to determine the necessary accuracy of field instruments to achieve the determined level of the system sensitivity and accuracy.

Acknowledgments This research is funded from the budget of The Polish Ministry of Science in the years 2008–2011 as the research project Nr O R00 0013 06 and partly by statutory funds.

References

1. Glen, N., Ryan, J., Findlay, T.: A review of pipeline integrity systems. In: Proceedings of 24th international north sea flow measurement workshop paper 3.1 24–27 October (2006)
2. Glen, N.: Leak detection based pipeline integrity systems. Flow Measurement Guidance Note no. 49. NEL, Glasgow (2005)
3. Turkowski, M., Bratek, A.: Methods of decreasing the influence of the factors disturbing the reliability of leak detection systems. J Autom. Mobile Rob. Intell. Syst. **3**(2), 33–37 (2012)
4. Joint Committee for Guides in Metrology 100:2008, Evaluation of the measurement data—Guide to the expression of uncertainty in measurement
5. ISO 5168:2005 Measurement of fluid flow—evaluation of uncertainties
6. den Hollander, H., et al.: Field data of an E-RTTM based leak detection system. In: Proceedings of 24th international north sea flow measurement workshop paper 3.3, 24–27 October (2006)
7. Siebert, H., Isermann, R.: Leckerkennung und—Lokalisierung bei Pipelines durch On-Line-Korrelation mit einem Prozeßrechner, *Regelungstechnik*, 25. Jahrgang 1977, heft 3
8. Siebert, H.: Ein einfaches Verfahren zur Erkennung und Ortung kleiner Lecks in Gaspipelines", *Regelungstechnik*, 29. Jahrgang 1981, heft 6
9. Bratek, A., Turkowski, M.: Analytical system of leak detection and localization for long range liquid pipelines. In: Proceedings of diagnostics of processes and systems (2011)

Development of a Microfluidic Device System Using Adhesive Vinyl Template to Produce Calcium Alginate Microbeads for Microencapsulation of Cells

Chin Fhong Soon, Hiung Yin Yap, Mohd Khairul Ahmad, Kian Sek Tee and Siew Hwa Gan

Abstract Microfabrication technique based on microelectronic technology is commonly used to produce microfluidic devices but this technique involves with costly and toxic chemicals. In this paper, we proposed the use of patterned adhesive vinyl template to produce a poly-dimethylsiloxane (PDMS) microfluidic device that was applied to generate microbeads of calcium alginate for microencapsulation of cells. In the microfluidic system, an infusion pump of high flow rate (2000 µl/min) and a commercial syringe pump were used to emulsify the continuous and disperse phases of liquids in forming the microbeads. Microbeads of calcium alginate in the range of 438 ± 38–799 ± 20 µm were successfully produced using this environmental friendly technique.

Keywords Microfluidic · Vinyl adhesive template · Infusion system · Calcium alginate · Microbeads

1 Introduction

Microfluidic device is a widely use lab-on-chip that integrates a network of micro sized channels to execute special functions such as diluting [1], mixing and splitting of fluids [1, 2]. The main applications of the microfluidic devices are in the field of biomedical engineering, molecular biology and drug delivery [3]. Among the advantages of microfluidic device include processing accuracy, efficiency and

C.F. Soon (✉) · H.Y. Yap · M.K. Ahmad · K.S. Tee
Faculty of Electrical and Electronic Engineering, Universiti Tun Hussein Onn Malaysia, 83000 Batu Pahat, Johor, Malaysia
e-mail: soon@uthm.edu.my

C.F. Soon
Biosensor and Bioengineering Laboratory, MiNT-SRC, Universiti Tun Hussein Onn Malaysia, 83000 Batu Pahat, Johor, Malaysia

S.H. Gan
School of Medical Sciences, Universiti Sains Malaysia, Kubang Kerian, Kelantan, Malaysia

© Springer International Publishing AG 2017
R. Jabłoński and R. Szewczyk (eds.), *Recent Global Research and Education: Technological Challenges*, Advances in Intelligent Systems and Computing 519,
DOI 10.1007/978-3-319-46490-9_63

minimum consumption of reagents [1]. The flow control of microfluidic device is predominantly involved with physics and engineering principles [4].

In soft lithography, the fabrication of the microfluidic device through the micropatterning of a silicon wafer requires clean room facility, costly masking methods and toxic chemicals. In our work, a technique was proposed to produce microfluidic device on a work bench based on adhesive vinyl template. The application of the microfluidic device in cell engineering was reported.

2 Materials and Methods

2.1 Design and Fabrication of the PDMS Microfluidic Device

The microfluidic pattern was designed in the COMSOL Multi physics software version 4.2 based on the requirements to perform emulsification at the crossed junction in which, the continuous (paraffin oil) and disperse (sodium alginate) phases were intersected. During emulsification, the minute droplets of the sodium alginate will be finely dispersed in the miscible oil phase. To perform emulsification in the microfluidic device, an inlet (inlet 1) with two split channels were assigned to induce two continuous flows of paraffin oil. At a crossed junction, the extensions of the two split channels then intercepted with the second inlet (inlet 2) assigned for disperse flow (sodium alginate). The three input channels were connected to an output channel leading to the outlet of the microfluidic device.

The design of the microfluidic template was plotted using Autocad® software (Autodesk®, service pack 1, USA). Then, the patterns on the vinyl sheet was plotted and cut using a vinyl plotter machine (Roland STIKA SV-12, Japan). Subsequently, a single layer of the patterned adhesive vinyl template was attached to a petri dish. Ten layers of similar patterned vinyl template were stacked to a thickness of 300 μm. The elastomer (Sylgard 184, Dow Corning, Michigan, USA) and curing agent (Dow Corning, Michigan, USA) were mixed at a ratio of 10:1 to produce poly-dimethylsiloxane (PDMS) gel. Then, the petri dish containing the vinyl template was filled up with the PDMS gel before curing in an oven at 70 °C for 1 h. After curing, the PDMS microfluidic device was peeled off and adhered to a glass slide. Holes were punched for the inlets and outlet of the microfluidic device.

2.2 Cell Culture and Preparation of Cell Suspensions

The human keratinocyte cell lines (HaCaT) were cultured in a 25 cm^2 culture flask with Dulbecco's Modified Eagle's Medium (DMEM, Sigma Aldrich, Dorset, UK) supplemented with L-Glutamine (2 mM, Sigma-Aldrich, Dorset, UK), Penicillin

(100 units/ml, Sigma Aldrich, Dorset, UK), Streptomycin (100 mg/ml, Sigma-Aldrich, Dorset, UK), Fungizone (2.5 mg/l, Sigma-Aldrich, Dorset, UK) and 10 % Fetal Calf Serum (Promocell, Heidelberg, Germany). The cells were incubated in an incubator at 37 °C and harvested using standard cell splitting procedure.

2.3 Cells Encapsulation in Calcium Alginate Microbeads

The sodium alginate and calcium chloride were prepared at a concentration of 2 and 5 %w/v, respectively. After the experimental setup as shown in Fig. 1, the HaCaT cells at a density of 82.8×10^4 cells/ml were mixed with 0.2 ml of sodium alginate and DMEM. The HaCaT cells containing in the sodium alginate solution were filled in a 5 ml syringe and infused into the microfluidic device as the disperse phase. Similarly, paraffin oil filled in a 5 ml syringe was infused to the microfluidic device as the continuous phase (Fig. 1). The calcium alginate microbeads encapsulated with HaCaT cells were polymerised in a sterilised petri dish containing the calcium chloride solution. The microbeads containing the cells were then incubated at 37 °C in a 5 % CO_2 humidified incubator and monitored in an inverted phase contrast microscope (TS-100, Nikon, Tokyo, Japan). Images of the microcapsules were captured using a Go-3 camera (QImaging, Surrey, Canada). The cells in encapsulation were stained using 2 µg/ml of 4',6-diamidino-2-phenylindole (DAPI, Sigma Aldrich, Dorset, UK) and then, observed in a fluorescence microscope (BX53, Olympus, Tokyo, Japan).

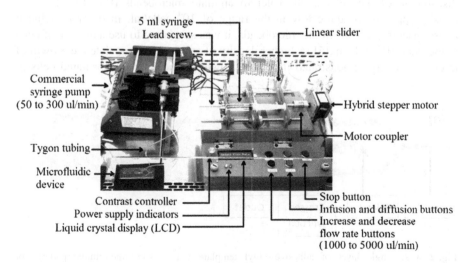

Fig. 1 The microfluidic infusion system setup for fabrication of microbeads in a Class II biological safety cabinet

3 Results and Discussions

3.1 The Microfluidic Device

This is the first demonstration of producing microfluidic template using adhesive vinyl template as shown in Fig. 2a. However, the resolution limit was subjected to the resolution of the plotter at approximately 500 μm. Figure 2b shows the PDMS microfluidic device fabricated based on the patterned adhesive vinyl template for microencapsulation. The patterned adhesive vinyl template is reusable up to 20 times to fabricate PDMS microfluidic devices.

3.2 Size Distribution of the Microbeads

At a continuous flow rate of 2000 μl/min and disperse flow rates of 50 to 300 μl/min, the diameter of the microbeads produced were ranging from 438 ± 38 μm to 799 ± 20 μm (Fig. 3). The inset in Fig. 3 shows the microbeads of calcium alginate produced using the microfluidic device.

3.3 Encapsulation and Analysis of HaCaT Cells

Figure 4 shows the cells encapsulated and randomly scattered in a microcapsule with a diameter of 437 μm. DAPI staining provided a clearer indication on the distribution of the cells in the calcium alginate microbeads (Fig. 4b). The cell density applied was quite low in the range of 10^4 cells/ml. In order to achieve denser packing of cells in the microbeads, it was suggested to use a density of cells higher than 10^7 cells/ml [5]. A higher density of cells would stimulate the growth of micro tissues as demonstrated in [5, 6]. However, the microencapsulated cells in

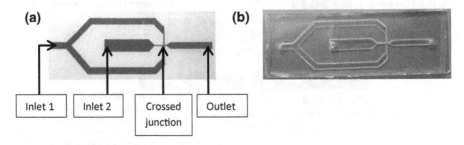

Fig. 2 a A single layer of adhesive vinyl template with design microfluidic pattern for microencapsulation. **b** The PDMS microfluidic device adhered to a glass slide for microencapsulation with 2 inlets and 1 outlet

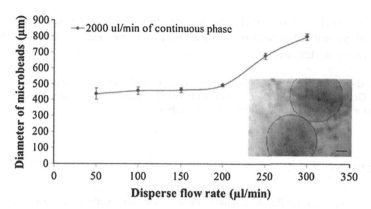

Fig. 3 The diameter of the microbeads fabricated at different flow rates of the disperse phase and a fixed continuous flow rate of 2000 µl/min. The *inset* shows the photomicrographs of microbeads at an average size of 460 µm. (Scale bar 50 µm)

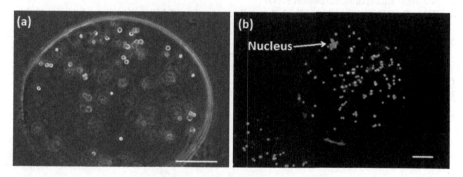

Fig. 4 The phase contrast micrograph of **a** microencapsulation of cells in microbeads of calcium alginate and fluorescence micrograph of **b** DAPI staining of the HaCaT cells. (Scale bar 100 µm)

calcium alginate were successfully demonstrated using the developed microfluidic system.

4   Conclusion

A PDMS microfluidic device was designed based on adhesive vinyl template to perform emulsification of the continuous and disperse phases of two immiscible liquids. This technique enabled the fabrication of PDMS microfluidic without the use of toxic chemicals or expensive clean room facility. At a continuous phase fixed at 2000 µl/min and a disperse flow rates ranging from 50 to 200 µl/min, the microcapsules produced were determined at a minimum size of 438 ± 38 µm which was due to the size of the crossed junction of the microfluidic device. The developed microfluidic system was successful in microencapsulation of HaCaT cells in calcium alginate.

Acknowledgments The authors would also like to thank financial support from Research Incentive Grant Scheme (IGSP Vot. U251) and Research Acculturation Grant Scheme (RACE Vot No. 1515) awarded by Malaysia Ministry of Education.

References

1. Hung, L.-H., Lee, A.P.: Microfluidic devices for the synthesis of nanoparticles and biomaterials. J. Med. Biol. Eng. **27**(1), 1–6 (2007)
2. Duncanson, W.J., Lin, T., Abate, A.R., Seiffert, S., Shah, R., David, A.: Microfluidic synthesis of advanced microparticles for encapsulation and controlled release. Lab Chip **12**, 2135–2145 (2012)
3. Nguyen, N.-T., Wereley, S.T.: Fundamentals and Applications of Microfluidics. Artech House, Norwood Incorporation, Boston (2009)
4. Björn, S., GPatrick, r., Göran, S.: A thermally responsive PDMS composite and its microfluidic applications. J. microelectromech. syst. 16 (1), 1057–7157 (2007)
5. Soon, C.F., Thong, K.T., Tee, K.S., Ismail, A.B., Denyer, M., Ahmad, M.K., Kong, Y.H., Vyomesh, P., Cheong, S.C.: A scaffoldless technique for self-generation of three-dimensional keratinospheroids on liquid crystal surfaces. Biotech. Histochem. **91**(4), 283–295 (2016)
6. Wong, S.C., Soon, C.F., Leong, W.Y., Tee, K.S.: Flicking technique for microencapsulation of cells in calcium alginate leading to the microtissue formation. J. Microencapsul. 33(2), 162–172 (2016)

Rutile Phased Titanium Dioxide (TiO$_2$) Nanorod/Nanoflower Based Waste Water Treatment Device

M.K. Ahmad, Adila Fitrah Abdul Aziz, C.F. Soon, N. Nafarizal, Abd Hamed Noor Kamalia, Shimomura Masaru and K. Murakami

Abstract Titanium dioxide (TiO$_2$) nanorod/nanoflower was prepared using hydrothermal method and annealed at different temperatures. The samples were characterized using Field Emission Scanning Electron Microscopy (FE-SEM), X-Ray Diffraction (XRD) and customized waste water treatment device in order to observe the surface morphology, structural properties and photo degradation activity, respectively. The efficiency of the TiO$_2$ nanorod/nanoflower for waste water treatment was investigated using the waste water treatment device. A device with rotating mechanism was designed and developed in order to stir the textile waste water using the TiO$_2$ nanorod/nanoflower. The effectiveness of the treatment process was observed using the Chemical Oxygen Demand (COD) and Biochemical Oxygen Demand (BOD) tests. The objectives of this project were successfully achieved in which annealing temperature of 500 °C presented the most optimum results.

Keywords Titanium dioxide · Hydrothermal · Annealing · Waste water treatment

1 Introduction

Effluents from industrial sector often face a major environmental problem. Semiconductor photocatalysis is one of the promising methods that have high potential to restrain the contaminants. The potential of Titanium Dioxide (TiO$_2$) in degrading organic or inorganic pollutants that is generated from photo activation is

M.K. Ahmad (✉) · A.F.A. Aziz · C.F. Soon · N. Nafarizal · A.H.N. Kamalia
Microelectronic and Nanotechnology – Shamsuddin Research Centre (MiNT-SRC), Faculty
of Electrical and Electronic Engineering, Universiti Tun Hussein Onn Malaysia, Parit Raja,
Batu Pahat 86400, Johor, Malaysia
e-mail: akhairul@uthm.edu.my

S. Masaru · K. Murakami
Department of Engineering, Graduate School of Integrated Science and Technology,
Shizuoka University, 432-8011 Hamamatsu, Shizuoka, Japan

© Springer International Publishing AG 2017 483
R. Jabłoński and R. Szewczyk (eds.), *Recent Global Research and Education:
Technological Challenges*, Advances in Intelligent Systems and Computing 519,
DOI 10.1007/978-3-319-46490-9_64

comprehensively utilized for a photocatalyst in wastewater treatment. TiO_2 is one of the various oxide semiconductor which is have strong oxidizing power, nontoxicity and long term photostability [1]. There are various techniques can be used to synthesize TiO_2 thin film such as sol-gel [2], DC magnetron sputtering [3], spin-coating [4], spray pyrolysis deposition (SPD) method [5] and hydrothermal method [6]. Over these past years, many researchers have been focused in producing TiO_2 fine powder because fine powder has higher photocatalytic effect than TiO_2 thin films. However, fine powder requires stirring during reaction and separation process after the reaction [7]. These problems can be solved by preparing the TiO_2 catalyst on thin film. Titanium dioxide occurs in nature as minerals rutile (tetragonal), anatase (tetragonal) and brookite (orthorhombic). Rutile phased is more stable than anatase and brookite. Rutile absorbed more than anatase in UV light because rutile has a lower band gap which is 3.0 eV than anatase (3.2 eV). Some author claimed anatase is better than rutile and brookite in photocatalytic activity. This is because rutile has higher rates of electron and hole recombination [8]. Some other researcher claimed mixed phased of anatase/rutile TiO_2 gives higher effect on photocatalytic activity. The reason for the synergistic effect of the mixed phased still in debate [9]. It is believed that, rutile in mixed phased TiO_2 help to improve the charge carrier separation through electron trapping and reduced the electron reconmination. Phased structure, crystallite size, pore structure and specific surface area play an important role in photo catalytic activity. In this work, a thin film of nanostructures rutile phased TiO_2 were prepared and used in customized water treatment device to study the effect of annealing temperatures to the photo degradation of prepared TiO_2 films.

2 Experimental

2.1 Preparation of the Nanorod/Nanoflower of Rutile TiO_2

Fluorine-doped SnO_2 coated (FTO) glass was used as substrate with thickness of 0.5 mm. Ethanol, acetone, hydrochloric acid and titanium butoxide (TBOT) were purchased from Sigma Aldrich and all chemicals were used as received. Deionized water was used to prepare all the solutions. The FTO glass was cut into dimension of 10×25 mm and cleaned by sonicating method in acetone, ethanol and deionized water with volume ratio of 1:1:1 for 30 min, and finally dried in air. Rutile-phased titanium dioxide nanorods (r-TNRs) and rutile-phased titanium dioxide nanoflowers (r-TNFs) were synthesized from a chemical solution using the hydrothermal process. The chemical solution was prepared by dissolving 80 mL of concentrated hydrochloric acid (HCl) (36.5–38 %) in 80 mL of deionized (DI) water. The mixture was vigorously stirred at 35 °C and 300 RPM for 5 min before 3 mL of titanium butoxide (TBOT) was added drop wise using a capillary tube. After stirring for nearly 20 min, the solution was placed into steel made

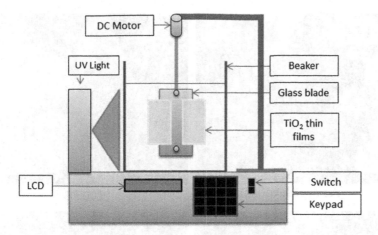

Fig. 1 Schematic diagram of customized waste water treatment device

autoclave with Teflon made liner (300 mL) for hydrothermal process in which the FTO glass substrates were placed with the conducting FTO surface facing upwards. The process was performed at 150 °C for reaction times of 10 h. After that, the autoclave was taken out from the oven and was let to cool to room temperature. The prepared samples were rinsed with DI water and left to dry at room temperature. The morphology and thickness of TiO₂ nanorods were characterized by using a field emission scanning electron microscope (FESEM) JEOL JSM-7600F model operated at 15 kV. The crystal structure and crystallite size are defined by an X-ray diffractometer (XRD) PANalytical X-Part3 Powder model. The scan axis used were 2θ with range of 20–60° and the type of slit used were fixed divergence slit. Schematic diagram of customized waste water treatment device was shown Fig. 1.

3 Results and Discussion

3.1 Surface Morphological Properties

Figure 2 shows the FE-SEM images at times 100,000 magnification scales. When annealing temperatures were increased the nanostructure's surface show decreased in roughness especially at annealed temperature of 500 °C. The nanostructures clearly show that the particles stick closely to one another and caused the surface contact between the particles were increased. When the surface contact was increased, the surface roughness of nanostructures was decreased. Therefore, the electron will easily to jump one another and caused increased in conductivity. When the conductivity increased, the photocatalytic reaction easily occurred.

Fig. 2 FESEM image of TiO$_2$ nanorods annealed at **a** as deposited, **b** 100 °C, **c** 200 °C, **d** 300 °C, **e** 400 °C, and **f** 500 °C

3.2 Structural Properties

Figure 3 shows the XRD patterns of TiO$_2$ nanorod/nanoflower thin films were annealed at different temperatures from 100 to 500 °C. Annealing is a best way to

Fig. 3 XRD pattern of TiO_2 nanorods annealed at **a** as deposited, **b** 100 °C, **c** 200 °C, **d** 300 °C, **e** 400 °C, and **f** 500 °C

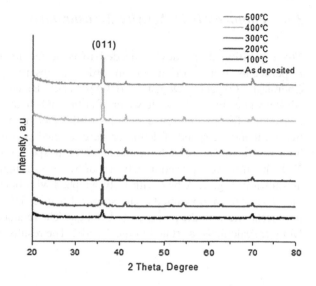

Table 1 The variation of crystallite sizes at different annealing temperatures

Annealing temperatures, (°C)	FWHM, β (radians)	Bragg angle, 2θ (radians)	Crystallite size, (nm)
As deposited	0.2952	36.1676	6.46
100	0.1968	36.1471	9.69
200	0.2460	36.1646	7.76
300	0.2460	36.1582	7.76
400	0.2952	36.1669	6.46
500	0.2460	36.1579	7.76

improve the crystallinity of TiO_2 nanorod/nanoflower thin film. The position of the maximum intensity (011) peak of TiO_2 nanorod/nanoflower thin films whose diffraction features are qualitatively identical, reveals the peak shift $2\theta = 36.16$, 36.14, 36.16, 36.15, 36.16 and 36.15° for annealed samples of as deposited, 100, 200, 300, 400 and 500 °C respectively. As the annealing temperature increases, there is no additional peak observed in the XRD profiles. The peak intensities do not significantly increase either. The XRD pattern of the TiO_2 nanorod/nanoflower thin films confirmed that prepared films were pure rutile phase TiO_2.

The estimated crystallite size of the TiO_2 nanorod/nanoflower films annealed at as deposited, 100, 200, 300, 400 and 500 °C were shown in Table 1. Based on the Table 1, the average of crystallite size was about 7.64 nm.

3.3 Photocatalytic Activity Measurement

The capability of the device needed to be tested in the real practical scenario. Experiments were carried out in order to observe the percentage change of Chemical Oxygen Demand (COD) and Biochemical Oxygen Demand (BOD) values in textile waste water after the TiO_2 treatment. The experiments were carried out for six difference samples of the TiO_2 thin films. The duration for the treatment and the speed of dc motor were selected as constant variable in which one hour per treatment for each sample and 80 rpm speed respectively. Two samples of TiO_2 thin film were placed in between of two glasses that connected to the dc motor as shown in Fig. 1. A box with fully wrapped with black paper was used to cover the device from unwanted light source because TiO_2 thin films only react as photocatalytic with the presence of UV light. The reduction percentage of COD and DO were calculated as state in Eqs. 1 and 2. The results were shown on Table 2 and Fig. 4.

Table 2 The result of textile waste water before and after treatment using COD and BOD tests

Annealing Temperature, °C	COD_i (mg/L)	COD_f (mg/L)	COD reduction (%)	BOD_i (mg/L)	BOD_f (mg/L)	BOD reduction (%)
As deposited	4119	3965	3.74	7.59	7.39	2.64
100	4119	3380	17.94	7.61	7.31	3.94
200	4119	2850	30.81	7.60	7.28	4.21
300	4119	2355	42.83	7.59	7.14	5.93
400	4119	1164	71.74	7.56	6.68	11.64
500	4119	596	85.53	7.57	6.32	16.51

Fig. 4 Graph of reduction in percentage of COD and BOD versus the annealing temperature

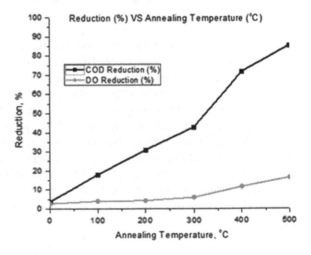

$$\text{Reduction COD}\% = \frac{COD_i - COD_f}{COD_i} \times 100\% \tag{1}$$

$$\text{Reduction DO}\% = \frac{DO_i - DO_f}{DO_i} \times 100\% \tag{2}$$

Figure 4 shows the relation between percentage of reduction for COD and BOD values with different annealing temperatures. It can be observed that, as the annealing temperatures were increased the percentages of reduction were increased as well for the both reading of COD and BOD. Based on the COD reading, the percentage of reduction has the highest reading at temperature 500 °C which was 85.53 % and same goes to BOD reading, at temperature 500 °C has the highest reading which is 16.51 %.

4 Conclusion

In summary, rutile phased TiO$_2$ nanorods films were successfully synthesized on FTO substrate using hydrothermal method at 150 °C and 10 h reaction time. The prepared sample were used in customized waste water treatment device and films that annealed at 500 °C gives the highest percentage of 16.51 %. It is due to better electron mobility in the films.

Acknowledgments The authors would like to thank to Ministry of Education (MOE) Malaysia and Universiti Tun Hussein Onn Malaysia (vot U414) for financial support.

References

1. Tayade, R.J., Surolia, P.K., Kulkarni, R.G., Jasra, R.V.: Photocatalytic degradation of dyes and organic contaminants in water using nanocrystalline anatase and rutile TiO$_2$. Sci. Technol. Adv. Mater. **8**(6), 455–462 (2007)
2. Biju, K.P., Jain, M.K.: Effect of crystallization on humidity sensing properties of sol–gel derived nanocrystalline TiO$_2$ thin films. Thin Solid Films **516**(8), 2175–2180 (2008)
3. Domaradzki, J.: Structural, optical and electrical properties of transparent V and Pd-doped TiO$_2$ thin films prepared by sputtering. Thin Solid Films **497**(1–2), 243–248 (2006)
4. Diebold, U.: Structure and properties of TiO$_2$ surfaces: a brief review. Appl. Phys. A Mater. Sci. Process. **76**(5), 681–687 (2003)
5. Shinde, P.S., Sadale, S.B., Patil, P.S., Bhosale, P.N., Brüger, a, Neumann-Spallart, M., Bhosale, C.H.: Properties of spray deposited titanium dioxide thin films and their application in photoelectrocatalysis. Sol. Energy Mater. Sol. Cells **92**, 283–290 (2008)
6. Bin Ahmad, M.K., Murakami, K.: Influences of Surface Morphology of Nanostructured Rutile TiO$_2$ Nanorods/Nanoflowers as Photoelectrode on the Performance of Dye-sensitized Solar Cell. MAKARA J. Technol. Ser. **17**(2), 73–76 (2013)
7. Yu, J., Jimmy, C.Y., Cheng, B., Zhao, X.: Photocatalytic activity and characterization of the sol-gel derived Pb-doped TiO$_2$ thin films. J. sol-gel Sci. Technol. **24**(1), 39–48 (2002)

8. Wu, J., Lo, S., Song, K., Vijayan, B.K., Li, W., Dravid, V.P., Gray, A.K.: Growth of rutile TiO$_2$ nanorods on anatase TiO$_2$ thin films on Si-based substrates. J. Mater. Res. **26**, 1646–1652 (2011)
9. Mohamed, M.A., Salleh, W.N.W., Jaafar, J., Yusof, N.: Preparation and photocatalytic activity of mixed phase anatase/rutile TiO$_2$ nanoparticles for phenol degradation. J. Teknologi **70**(2), 1 (2014)

Design—Simulation—Optimization Environment of Specialized MEMS

Magdalena A. Ekwińska, Grzegorz Janczyk, Tomasz Bieniek,
Piotr Grabiec, Jerzy Zając and Jerzy Wąsowski

Abstract Independent simulations of mechanical and electrical parts are no longer sufficient. It is necessary to incorporate parameters of both type of elements in a multidomain process in which the joint action is checked and co-optimization is performed. Such an approach allows to reduce cost and shorten time to market. This article presents how complex a multidirectional simulation process is nowadays. Moreover it present methodology and tools applicable for MEMS product development together with results of numerical simulations and optimization.

Keywords MEMS · Modelling · Simulation · Optimization · Co-simulation · FEM

M.A. Ekwińska (✉) · T. Bieniek · P. Grabiec · J. Zając
Division of Silicon Microsystem and Nanostructure,
Institute of Electron Technology, Warsaw, Poland
e-mail: ekwinska@ite.waw.pl

T. Bieniek
e-mail: tbieniek@ite.waw.pl

P. Grabiec
e-mail: grabiec@ite.waw.pl

J. Zając
e-mail: jzajac@ite.waw.pl

G. Janczyk · J. Wąsowski
Department of Integrated Circuits and Systems,
Institute of Electron Technology, Warsaw, Poland
e-mail: janczyk@ite.waw.pl; janczyk@imio.pw.edu.pl

J. Wąsowski
e-mail: jwasowski@ite.waw.pl

G. Janczyk
Institute of Microelectronics and Optoelectronics,
Warsaw University of Technology, Warsaw, Poland

© Springer International Publishing AG 2017
R. Jabłoński and R. Szewczyk (eds.), *Recent Global Research and Education:
Technological Challenges*, Advances in Intelligent Systems and Computing 519,
DOI 10.1007/978-3-319-46490-9_65

491

1 Introduction

The ongoing race in microelectronics results in increasing complexity of micro-electro-mechanical systems (MEMS) being developed. Several subsequent design and fabrication challenges have been solved. One of them is application of the most suitable design-simulation-optimization environment and product engineering methodology. The second one is technology optimization of the applied fabrication processes. Last but not least is the comparison of the achieved parameters of manufactured structure with parameters obtained during simulation process. Development of smart innovative heterogeneous systems incorporating micro-electro-mechanical part (MEMS) supported by dedicated electronic modules (ASICs) does not fit to any standard design methodology, nor design verification flow yet. Therefore modeling and simulation efforts assisted by virtual fabrication run and optimization are one of the crucial development processes [1–4].

2 Product Specification and the Design Concept

The final product is the element that has been designed and manufactured following customer specification. One of the crucial elements of the manufacturing process is modeling simulation and optimization loop applied for an element under development. Typical design flow of the product is presented in the Fig. 1.

One of the most important of the design flow elements is the initial idea and precise product specification necessary to build preliminary model of the device structure and virtually set up its manufacturing process. Such a model is evaluated during the simulation process and undergoes subsequent iterative optimization. Physical manufacturing is the next step if virtual process run leads designers from design to valuable product. Resulting parameters achieved by measurement are compared with initial specification and test-bench simulation results achieved using the optimized design. Two operation schemes are available for consideration: if something goes wrong the product comes back to the design simulation and

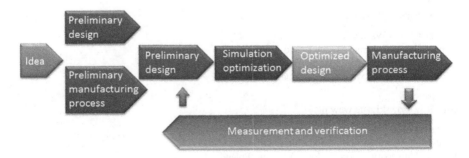

Fig. 1 Design flow

iterative optimization step, until achieved results match the requested specification of the product.

It is well known that physical structure manufacturing is a time consuming and expensive process. Therefore, companies with microfabrication facilities (FAB's) need reduction of development costs and the time to market shortening. The accurate simulation model interactively applicable in product development flow significantly shortens the time necessary to develop, fabricate, measure and introduce final product to the market [5].

3   MEMS Design, Modeling and Simulation

A MEMS device simulation process is only more or less accurate approximation of the real physical phenomenon and behavior of the object. The device simulation model is a simplified mathematical description neglecting some features assumed as less significant, with minor influence on the MEMS behavior MEMS structure model is a combination of mechanical and electrical models, ready to realize electro-mechanical simulations. For a long time separate simulations of mechanical and electrical parts were feasible. Contemporary EDA design software (like MEMS + by Coventor [7]) opens an unprecedented possibility to perform co-simulations—combining several simulation domains (thermal, electrical, mechanical) into some sort of a virtual reality.

There are two types of MEMS simulation: finite element method (FEM) and behavioral. The FEM analysis is the most popular method enabling various types of simulations such as mechanical, thermal, piezo, electrostatic, electro-quasi-static, magnetic, fluidic, and combinations of the aforementioned types (e.g. thermo-electro-mechanical). In all these cases a large problem is subdivided into smaller pieces, whose behavior is modeled by simple equations, forming matrices to describe the behavior of whole big system of equations reflecting number of finite elements. The FEM method uses algorithms from the calculus of variations to approximate a solution by minimizing an associated error function. The FEM disadvantage is the time necessary to perform simulation and get the result. Long simulation time is caused by the high amount of finite elements generated. The time is even longer when optimization is taken into account, due to the fact that it requires on each step building of a new model from the beginning, sharing it on the components and performing set of simulations.

Another approach is the behavioral MEMS simulation method combining number of pieces of known geometrical and mechanical parameters and behavior, aggregated in a library. Number of such pieces (modules, elements) by definition is limited. Behavioral modeling simplifies model parameterization, which simplifies process of MEMS shape optimization. It applies to all multidomain aspects of MEMS modeling. It will be presented on the MEMS microphone design sample where CoventorWare® EDA software environment was used [6].

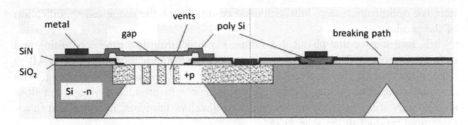

Fig. 2 Model of the mechanical part of the MEMS microphone

Based on the initial MEMS specification model of a capacitive microphone was developed (Fig. 2). The main design issue was to optimize membrane parameters: shape, size and thickness. For that purpose a set of simulations has been performed using all combinations of membrane sizes (500×500 μm, 1000×1000 μm, 1500×1500 μm) and thicknesses (2 and 3 μm) an verified for particular loads of 2 Pa–2 kPa evenly distributed over the whole surface of the membrane.

Achieved simulation results lead designers to a conclusion that optimal solution to be used for further analysis is a 1000×1000 μm membrane with thickness of 2 μm.

Once the size and thickness of the membrane were set, a special suspension, providing parallel movement of the membrane to the back-plate, was modeled and optimized. Final, optimized shape of the membrane suspension is presented in Fig. 2.

Finally a mechanical model of the MEMS microphone has been successfully developed using MEMS + design environment [7], capable to export MEMS model Cadence EDA environment for IC design as well as to perform local co-simulation of the MEMS module with dedicated electronics described as a SPICE or behavioral simulation model. The MEMS + environment used the membrane design imported from the CoventorWare design software [6]. For model validation purposes the comparison of the resonant frequencies modes of the membrane in CoventorWare and MEMS + have been performed.

4   Electronic Modules Design, Modeling and Simulation

Integrated electronic circuits split into at least three main groups: analog, digital and mixed (analog + digital). Analog circuits are modeled by electrical schematics using basic components like transistors, resistors, capacitors, etc. All these components have simulation models associated (for example SPICE models), which are technology-specific. All analog and digital element are characterized and respective simulation models are provided by manufacturing technology vendor. After physical characterization of electronic components, dedicated simulation models are being formulated using particular modeling standards like the SIPCE.

Calibrated simulation models are circulated in form of specialized EDA databases called Design Kits available for designers. Such an approach supports analog circuit design, modeling and simulation by dedicated EDA simulation software like SPICE-like circuit simulator (Cadence SPECTRE simulator). The first analog simulation can be run once the schematic design is completed to find out if all electrical parameters comply with initial specification. In case of analog circuits development, if all simulations are correct and meet requirements, layout designers start to prepare Integrated Circuit Layout which is also directly related to the particular fabrication technology selected for device under development. When layout design is completed, it needs to be verified. Do to so, EDA software contains extraction tools building electrical diagram back from the layout specification down to the basic analog elements enclosed in PDK simulation models database. The electrical post-layout simulation of the extracted circuitry is performed using parasitic RLC elements extracted from the layout. The quality of the simulation depends on quality of component models used.

Digital circuits are modeled functionally using high level modeling languages like Verilog or VHDL. Such a model can be functionally verified using dedicated logic simulator. Final circuit implementation (netlist) for selected technology is automatically generated making use of logic synthesis tool (e.g. Synopsys Design Complier). Electrical and timing parameters of all logic components used (flip-flops, gates, etc.) are provided within PDK. After layout design and interconnections parasitics extraction, post-layout logic simulation is performed for final functional and timing verification.

The block diagram of implemented MEMS microphone system has been presented in Fig. 3. Analog front-end block (AMP) performs microphone signal conditioning including signal amplification and filtering for proper frequency response. Then the resulting signal is transformed to digital domain by means of high resolution Sigma Delta Analog-to-Digital Converter (ADC). The Sigma Delta ADC is composed of two main parts: Sigma Delta Modulator, transforming input analog signal to high data rate bit stream in PDM (Pulse Density Modulation) format and digital decimation filter, transforming PDM bit stream to high resolution parallel output data with near Nyquist data rate. Digital microphone signal is available at serial output. Analog front-end block (AMP) and Sigma Delta ADC with SPI serial interface have been integrated on a single chip. The whole

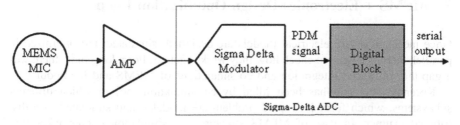

Fig. 3 MEMS microphone system

Fig. 4 MEMS MIC simulation (frequency response)

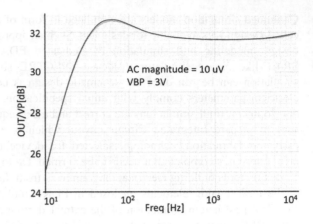

AC magnitude = 10 uV
VBP = 3V

Fig. 5 Sigma delta modulator output power spectral density (PSD)

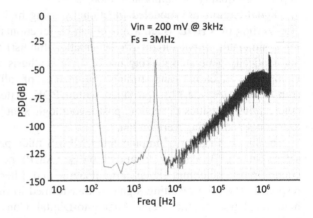

Vin = 200 mV @ 3kHz
Fs = 3MHz

microphone system has been successfully simulated in Cadence environment using SpectreVerilog mixed-signal simulator (Figs. 4 and 5). Required MEMS microphone Symbol view and Spectre simulation model have been imported from MEMS + environment.

5 MEMS + Electronics Design Optimization Loop

Once mechanical and electrical models are designed, simulated and optimized in parallel, there is a need to check how they work together. For a long time, there was a gap that embraced integration and co-simulations of MEMS and IC modules.

Recently, this gap has been filled by co-simulation tool in which different subsystems, which form a coupled problem are modeled and simulated in a distributed manner. In case of MEMS structures, co-simulation engine takes into consideration both: electro-mechanical features of the MEMS module and electronic parameters of the associated readout electronic. For the sake of simplicity, the coupled simulation is carried out by running the subsystems in a black-box manner. The model itself is done on the subsystem level without having the coupled problem in mind. During the simulation the subsystems exchange their data. This enables to perform co-simulations for MEMS and IC modules and validate achievements of the whole design process from the initial idea to co-design, co-simulation and optimization stage. This helps to keep the design in the requested range of parameters following the initial system specification and lower cost of the prototype [8, 9]. The example of such a tool is Cadence, which implements the application of the MEMS + (Coventor®) model and creates digital MEMS microphone. The last one is the object of extensive analysis in order to identify and effectively handle accuracy problems, specific compatibility and integration at the early design stage.

6 Summary

In this article we presented typical design flow of a sensor using new multidomain optimization and co-simulation. It was showed that it is possible to check all parameters of the design sensor as well as solve most of the technological problems and perform test of the electro-mechanical structure using just simulation engines. It was proved that nowadays simulation process is complex, multidirectional and multidomain and incorporate different types of parameters (e.g. electrical and mechanical).

The innovative design flow is a design alternative approach of primary importance which leads smart system to lower design costs and shorten development time e.g. using simulation and optimization process assisted by virtual fabrication run.

References

1. Janczyk, G., Bieniek, T., Szynka, J., Grabiec, P.: Reliability aspects of 3D-Oriented heterogeneous device related to stress sensitivity of MOS transistors. In: Proceedings of the IEEE international 3D systems conference CD 2009, pp. 1–6 (2009)
2. Janczyk, G., Bieniek, T., Grabiec, P., Szynka, J.: Thermo-mechanical aspects of reliability for vertically integrated heterogeneous systems. In: Proceedings of 16th international conference mixed design of integrated circuits and systems (2009)
3. Bieniek, T., Janczyk, G., Janus, P., Ekwińska, M., Szmigiel, D., Domański, K., Grabiec, P., Dumania P.: Efficient scenarios, methodology and tools for MEMS/NEMS product development. In: Proceedings of the 10th international conference on multi-material micro manufacture (4 M 2013), pp. 284–287 (2013)
4. Ortloff, D., Schmidt, T., Hahn, K., Bieniek, T., Janczyk, G., & Brück, R.: MEMS product engineering. In MEMS Product Engineering, pp. 53–83. Springer, Vienna (2014)

5. Bieniek, T., Janczyk, G., Grabiec, P., Szynka, J., Kalicinski, S., Janus, P., Domanski, K., Sierakowski, A., Ekwinska, M., Szmigiel, D., Tomaszewski, D., Holzer, G., Schropfer G.: Customer-oriented product engineering of micro and nano devices. In: Proceedings of the 17th international conference on mixed design of integrated circuits and systems, pp. 81–84 (2010)
6. Coventor® CoventorWare User Guide, more info available at www.coventor.com (2015)
7. Coventor® MEMS + User Guide, more info available at www.coventor.com (2015)
8. Bieniek, T., Janczyk, G., Ekwińska, M., Budzyński, T., Głuszko, G., Grabiec, P., KociubińskiA.: Novel methodology for 3D MEMS-IC design and co-simulation on MEMS microphone smart system example. IEEE, 978-1-4799-8472-5/14 (2014)
9. Scheeper, P.R., Nordstrand, B., Gullov, J.O., Bin, L., Clausen, T., Midjord, L., Storgaard-Larsen, T.: A new measurement microphone based on MEMS technology. J. Microelectromech. Syst. 12(6), 880–891 (2003)

Instability in CdTe Detector Characterized by Real-Time Measurement of Pulse Height and Carrier Transit Time

Hisaya Nakagawa, Tsuyoshi Terao, Tomoaki Masuzawa, Tetsu Ito, Hisashi Morii, Akifumi Koike, Volodymyr Gnatyuk and Toru Aoki

Abstract CdTe radiation detectors have been used for γ-ray and X-ray detection because CdTe have good physical properties. A current issue of CdTe detectors is instability in long-term operation. This instability is called polarization, however, the details of this phenomenon are still being discussed with suggested some mechanism by many researchers. In this study, we measured pulse height and carrier transit time under the polarization condition with aim of estimating mechanism of polarization from carrier transport properties. For evaluation of carrier transport properties, we have developed a new measurement system, which enables real time monitoring of both pulse height and carrier transit time of signal pulse. First, the carrier transit time is later after biasing. However, the carrier transit time is earlier due to reducing depletion layer. As a result, we estimated a mechanism of

H. Nakagawa (✉) · T. Terao · T. Aoki
Graduate School of Science and Technology, Shizuoka University, Shizuoka, Japan
e-mail: nakagawa.hisaya.14@shizuoka.ac.jp

T. Terao
e-mail: terao.tsuyoshi.15@shizuoka.ac.jp

T. Aoki
e-mail: aoki.toru@shizuoka.ac.jp

T. Masuzawa · T. Ito · T. Aoki
Reserch Institute of Electronics, Shizuoka University, Shziuoka, Japan
e-mail: masuzawa.tomoaki@shizuoka.ac.jp

T. Ito
e-mail: ito.tetsu@shizuoka.ac.jp

H. Morii · A. Koike · T. Aoki
ANSeeN Inc, Shizuoka, Japan
e-mail: morii@anseen.com

A. Koike
e-mail: koike@anseen.com

V. Gnatyuk
Institute of Semiconductor Physics, National Academy of Science of Ukraine, Kiev, Ukraine
e-mail: gnatyuk@ua.fm

© Springer International Publishing AG 2017
R. Jabłoński and R. Szewczyk (eds.), *Recent Global Research and Education: Technological Challenges*, Advances in Intelligent Systems and Computing 519,
DOI 10.1007/978-3-319-46490-9_66

polarization and the distribution of electric field including carrier transport properties.

Keywords CdTe semiconductor · Radiation detector · Long-term instability

1 Introduction

In recent years, cadmium telluride (CdTe) compound semiconductor detector have been used for various applications of γ-ray and X-ray detection in many situations. CdTe detector have high detection efficiency for γ-ray and X-ray because CdTe is composed with high density and high atomic number elements. Moreover, CdTe detector can be operated at room temperature because of low thermal noise by its wide band gap. Despite of CdTe detector has good properties, instability in long term operation is reported. This instability is called polarization in this field. The main effect of polarization is shifting energy peaks to lower energy and degradation of count rate. Polarization is often explained by charge accumulation model [1]. But the details of this phenomenon are still being discussed with suggested some mechanism by many researchers [1–4]. These suggested mechanisms were not considered carrier transport properties which is an important factor in radiation detection.

In this study, we measured the pulse height and the carrier transit time of radiation pulse under the polarization with the aim of estimating mechanism of polarization from carrier transport properties. For the evaluating of carrier transport properties, we have developed a new measurement system. This system enables real time measurement of both pulse height and carrier transit time of the signal pulse. As a result, we estimated mechanism of polarization and the distribution of electric field including carrier transport properties.

2 Experiment

In this study, we measured carrier transit time for estimating model including carrier transport properties. The using CdTe Schottky diode detector is based on a p-type crystal and has in as a Schottky electrode and Pt as a Ohmic electrode. The size of CdTe detector is $4 \times 4 \times 0.5$ mm. Figure 1 shows schematic diagram of the measurement system. The CdTe detector was connected to a charge sensitive amplifier (CSA). The output signals from CSA were observed and recorded by our measurement developed system, which enables real time measurement of both pulse height and carrier transit time. To control the interaction position, ^{241}Am (60 keV) and ^{57}Co (122 keV) γ-ray radioisotopes were measured at the room temperature (around 298 K). A ^{241}Am interacts near the incident surface and a ^{57}Co interacts throughout the CdTe detector. To the drift distance of electrons and holes, these

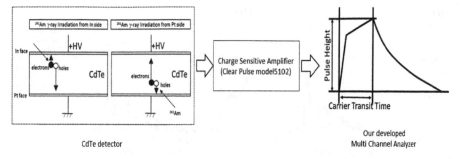

Fig. 1 The schematic diagram of measurement system

sources were irradiated from In side and Pt side of CdTe detector respectively. The reverse bias voltage was applied 100 V during measurement. The measurement time is 5 min and the interval time is each an hour until 6 h.

3 Results

Figures 2 and 3 show energy spectra by pulse height of each ^{241}Am and ^{57}Co γ-ray sources in passage time after biasing voltage. It was confirmed that these spectra are shifting peaks to lower energy and degradation of count rate with passage time. It is seemed that polarization was occurred with time in both energy cases.

Figure 4 shows the degradation of counts the changing ratio from pulse height spectrum of ^{241}Am source irradiating from both sides. The counts changing rate become indicator of thickness reducing of depletion layer in CdTe detector. The depletion layer of CdTe detector did not change with time when the ^{241}Am sources were irradiated from In side. In the case of irradiating from Pt side, the depletion layer changed after 120 min from biasing voltage. Assuming that the depletion layer forms fully in the CdTe detector, it was decreasing about 0.1 mm after 360 min.

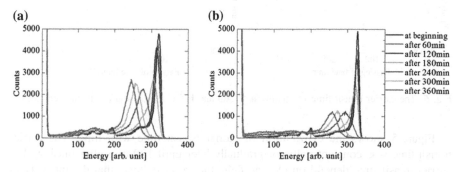

Fig. 2 Energy spectra of ^{241}Am γ-ray source irradiating from **a** In side and **b** Pt side

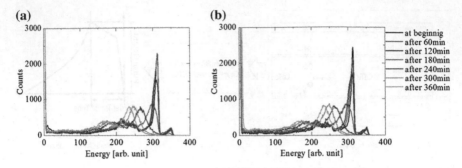

Fig. 3 Energy spectra of ^{57}Co γ-ray source irradiating from **a** In side and **b** Pt side

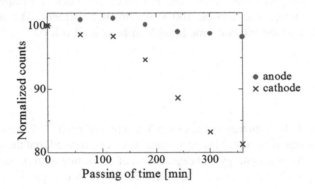

Fig. 4 The number of count changing when ^{241}Am γ-ray sources is irradiated from each sides

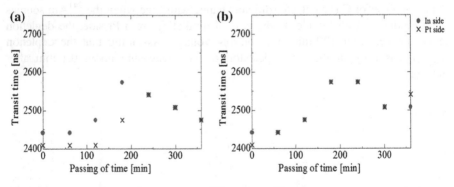

Fig. 5 The carrier transit time irradiating **a** ^{241}Am and **b** ^{57}Co γ-ray sources from each sides

Figure 5 shows the distribution of carrier transit time with time. The carrier transit time was constant value or gradually later until 120 min after biasing. The carrier transit time depend on electric field, the result indicate that the intensity of electric filed was reduced. The transit time took the latest time from 180 to 240 min. Moreover, the transit time is gradually earlier after 240 min.

4 Discussions

We considered that the relationship exists between the carrier transit time and the thickness of the depletion layer. The carrier transit time is shown as Eq. (1)

$$T = \frac{W}{\mu E} \tag{1}$$

where W is thickness of the depletion layer, μ is mobility and E is electric field. The cause of becoming later transit time is the reducing of electric field between keeping depletion layer fully in CdTe detector. After the decreasing the depletion layer, the carrier transit time is earlier. Therefore, the effect of the decreasing of depletion layer is bigger than the effect of reducing electric field.

We estimated the model of polarization from carrier transport properties. Figure 6 shows the estimated model of polarization. The depletion layer forms and the distribution of electric field are applied uniformly throughout CdTe detector (Fig. 6a). From Fig. 4, the depletion layer was reduced from 120 min. The depletion layer was maintained fully in CdTe detector. However, the carrier transit time was later than just biasing. This phenomenon indicates that the distribution of electric field was changed (Fig. 6b). Between 120 and 180 min, the distribution of electric field was reducing further from the result of the latest carrier transit time (Fig. 6c). The depletion layer start to reduce in this time and the dead layer was formed from Pt side. After 180 min, the depletion layer was reducing further and the dead layer is wider (Fig. 6d). Therefore the carrier transit distance is shorter than the initial state. As the result, the carrier transit time was earlier than state of 180 min. From this model of including carrier transport properties, the electric field intensity of in side is stronger and it of Pt side is weaker by the carrier trapping. The intensity of In side continue to be stronger and the Pt side intensity is disappeared at the 120 min. The electric field of model estimated from carrier transport properties agree with the electric field of charge accumulation model [1] and field distribution observed by Pockels effect [3].

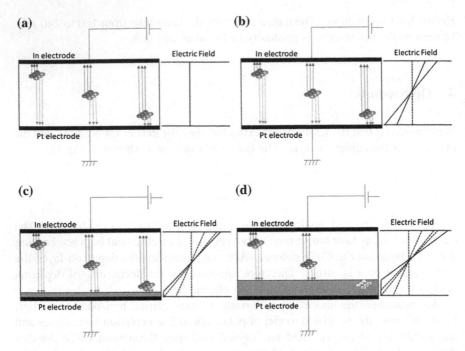

Fig. 6 The estimated model and the distribution of electric field from carrier transport properties from obtained data. **a** is initial state, **b** is the state keeping depletion layer, **c** and **d** is the state after decreasing depletion layer

5 Conclusion

In this study, we estimated the mechanism of polarization including carrier transport properties in CdTe radiation detector with Schottky junction. The measurement system is developed for real time measurement of both pulse height and carrier transit time. The pulse height spectra have shown occurrence of polarization. We estimated the reducing ratio of depletion layer of CdTe detector. The carrier transit time was earlier when the depletion layer started to reduce. We consider the reason that the effect of reducing the depletion layer is bigger than the effect of electric field degradation. Moreover, the carrier transit time have the strong relationship with the reducing depletion layer and the electric field.

References

1. Malm, H.L., Martini, M.: Polarization phenomena in CdTe nuclear radiation detectors. IEEE Trans. Nucl. Sci. **21**, 322–330 (1974)

2. Toyama, H., Higa, A., Yamazato, M., Maehama, T., Ohno, R., Toguchi, M.: Quantitative analysis of polarization phenomena in CdTe radiation detectors. Jpn. J. Appl. Phys. **45**(11), 8842–8847 (2006)
3. Turturici, A.A., Abbene, L., Franc, J., Grill, R., Dědič, V., Principato, F.: Time-dependent electric field in Al/CdTe/Pt detectors. Nucl. Instrum. Methods Phys. Res. Sect. A: Accelerators Spectrometers Detectors Assoc. Equipment **795**, 58–64 (2015)
4. Grill, R., Belas, E., Franc, J., Bugar, M., Uxa, S., Moravec, P., Hoschl, P.: Polarization study of defect structure of CdTe radiation detectors. IEEE Trans. Nucl. Sci. **58**(6), 3172–3181 (2011)

Measurement and Controlling Magnetic Field Strength by Using Hall Effect Sensors with Classical Algorithm

Sławomir Krzysztof Czubaj and Edyta Ładyżyńska-Kozdraś

Abstract In the article a non-contact measurement and controlling a magnetic field on the basis of Hall effect is presented. The structure of a Hall effect sensor was discussed. An example of a simple magnetic field measuring device, which uses the presented sensors, and a model of a magnetic field controlling system with a Hall effect sensor was presented in the paper. A solution system presented in the paper is based on using a ruler of Hall effect sensors to measure a magnetic field induction. The proposed arrangement of the sensors allows to measure the position of a launch trolley of a BSL launcher which uses the phenomenon of magnetic levitation.

Keywords Hall effect sensors · Magnetic field strength · Classical algorithm

1 Introduction

A magnetic field, as opposed to electric field, does not affect a charge at rest. Force appears only when the charge is moving.

Summing up, a magnetic field is a property of space which, tends to rotate or to maintain permanent direction when we place a magnetic needle in it, despite attempts of precipitation from its initial state. Of course a magnetic field can be detected not only by the use of a magnetic needle, such a field influences conductors with electric current (in which charges are moving), as well.

A magnetic field also enables levitation in any medium as well as in a vacuum, because of that for this type of levitation there is no physical contact between the levitating object and the ground. Depending on magnetic field polarisation objects attract or repel each other balancing the force of gravity. Levitation [1] is induced by static or dynamic (changeable in time and/or space) magnetic fields. In case of

S.K. Czubaj (✉) · E. Ładyżyńska-Kozdraś
Faculty of Mechatronics, Warsaw University of Technology, Warsaw, Poland
e-mail: s.czubaj@mchtr.pw.edu.pl

E. Ładyżyńska-Kozdraś
e-mail: e.ladyzynska@mchtr.pw.edu.pl

© Springer International Publishing AG 2017
R. Jabłoński and R. Szewczyk (eds.), *Recent Global Research and Education: Technological Challenges*, Advances in Intelligent Systems and Computing 519, DOI 10.1007/978-3-319-46490-9_67

507

static levitation there is lack of stable static system of magnetic force enabling levitation, that is why additional methods of stabilization must be used. Whereas systems in which diamagnetic materials and superconductors are used stabilization is not required [1]. Magnetic levitation can be induced by systems of permanent magnets or electromagnets. Hybrid systems [2] (an electromagnet interacting with a permanent magnet or ferromagnetic material) are often used as well. Stabilization in permanent magnet systems is achieved by using gyroscope force. A device called Levitron is based on this principle. When using electromagnets an active regulation of electric current in feedback system (with sensors of a levitating object position) is often used.

2 Hall Effect

Hall effect is a basic research method which allows to determine the sign of concentration and mobility of charge carriers in solid bodies. Hall effect has a lot of applications in technology. They are used as electric current sensors, magnetic field detectors, in passive sub-assemblies to measure and control magnetic field strength. Hall effect sensor is a device whose working principle is based on classic Hall effect [3]. Hall effect sensors are made on the basis of semiconducting materials with large mobility of charge carriers and are used to convert magnetic energy into electric energy [1, 4].

3 Construction of the Hall Effect Sensor

To build a Hall effect sensor we use a tile made of semiconducting material, which is then placed in a magnetic field and voltage is applied to it, enforcing the electrons flow in an appropriate direction (perpendicularly to the field line). By measuring the voltage on electrodes placed perpendicularly to a magnetic field line and direction of electrons flow, one can determine in a simple way magnetic field strength, where the sensor is located. In other words—magnetic field and electrons flow through a semiconductor cause changes proportionate to the strength of the voltage change field on the appropriate electrodes. Hall's voltage is too low, about of 30 μv for the field of 1 Gauss, that is why we use a differential amplifier on bipolar junction transistors, which has low level of interference, high output electrical impedance and moderate amplification. A very good solution in this field was offered by Melexis company [5], that is a programmable sensor MLX91208, whose parameters, such as sensitivity, amplification and offset, are programmed by a user and stored in internal memory—EEPROM. Calibration takes place through a special protocol, which involves the administration on one pin modulated voltage. The system is supposed to measure AC and DC voltage within temperature range from −40 to +150 °C and within frequency range up to 200 kHz. The sensor is produced in two versions, which differ in programmed magnetic sensitivity.

An example of a simple magnetic field sensor system is presented in Fig. 1, it is a simple circuit with a switch which turns on the LED diode when a magnetic field is detected.

4 Magnetic Field with Hall Effect Sensor Control Model

An example of controlling a magnetic field can be Levitron model [6], Fig. 2 presents an electronic circuit, in which the position sensor is Hall effect sensor, on the of which voltage increases proportionally to magnetic field strength in which the sensor is placed. The field induced by the electromagnet is regulated by alternating current with adjustable duty cycle. The levitating object must be a permanent magnet, for example neodymium magnet, since something must induce magnetic field which strength will be measured by the sensor, the further the magnet the lower the strength. The figure presented below consists of three fundamental parts: the first is a position sensor, the second—a modulator and the third is actuator.

The SS 495A Hall effect sensor is a magnetic field strength sensor, the Hall voltage induced in the sensor depends on the distance between the levitating object and the electromagnet and a solenoid and is supplied to the PWM MIC 502 BN regulator, which then controls the electromagnet through LMD 18201 system.

When we move the permanent magnet closer to the electromagnet we observe that at some point it starts vibrating and is attracted but only to some extent. The magnet levitates and it continues levitating as long as the system is powered. The levitating magnet can be set in rotary motion on vertical axis and will not lose energy since friction is absent.

Fig. 1 An example of a magnetic field sensor

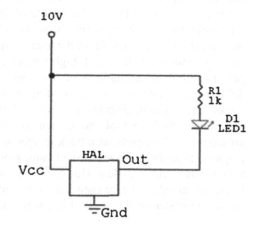

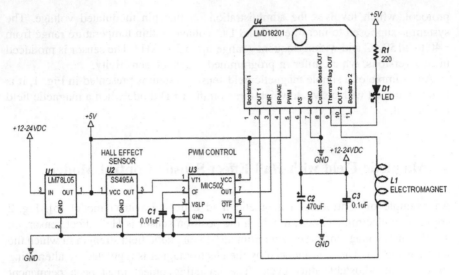

Fig. 2 Magnetic field control system diagram

5 System of Sensors Test

Information gained from a single sensor can be insufficient to determine the position of model tested, a system of sensors which are positioned evenly can provide a lot more information, namely we are provided with a family of characteristics of a magnetic field induction in a displacement function [1].

The proposed sensor arrangement is presented in Fig. 3, it allows to measure the position of the launch trolley [7] of the BSL launcher using magnetic levitation phenomenon. The magnetic launcher system consists of two elements. The first one is a motionless runway made of high-power permanent magnets. The runway consists of two rails with three rows of neodymium magnets generating a magnetic field. The magnets polarization changes across the tracks which makes magnetic field lines form the shape of a flume. The shape of the magnetic field line influences the stability of the levitating system. The second element of the launcher is the levitating launcher trolley. It hovers above the tracks, supported by four levitating posts. These posts are containers filled with high-temperature superconductors and liquid nitrogen. The force interacting between the posts and the tracks depends on the magnetic field strength, which changes depending on the position of the posts relative to the tracks. The magnetic field loses its strength along with changing the distance from the tracks (the height of the launch trolley) and generating a magnetic fume arrangement of magnets on a track results in a field gradient along the tracks. The magnetic field should have constant value along the tracks, thanks to that the launch trolley movement is lossless. However, it is only a theoretical assumption the field is in fact disrupted by the magnets geometry and their imperfect arrangement [7]. The measuring system will consist of linear Hall effect sensors evenly distributed at the base of the container with superconductors [8]. Changing the container position relative to magnetic field will cause a change in output voltage of the individual Hall effect sensors. A microprocessor system reads the state of every sensor and compares it to data stored in the memory, on the basis of the difference between the reading and the container geometry, the relative position of the container is determined on the basis if an algorithm. The voltage signals from the magnetic field sensors will be measured by a data acquisition card and using a computer (PCI or USB bus).

6 The Algorithm for Determination of the Trolley Position

Reading from the sensors (8 to 16 8-bit numbers) will be written in a table, next they will undergo initial processing. First symmetrisation can be carried out. A number, which is an offset of a given sensor, different for every position in the table, is subtracted from every value in the table. Thanks to that the values in the table will correspond to magnetic field polarity. Next will reset of all the values smaller than the maximum value of measuring disruptions determined in the course of research. As a result only two or three other than zero values will remain in the table. This way only two or three values other than zero for those sensors whose signal is dominant will remain in the table. The pre-selection stage will security selection only the sensors, which give dominant signal. In the following stages the algorithm will operate only on these values. The data from the sensors will be compared to the model saved in a form of two-dimensional table. The comparison will be made for each of the values Wk and Wk+1 individually with an appropriate model. The consecutive elements of the table will correspond to the consecutive positions of the trolley xk. For the trolley position w e assume appropriate value xk, for which the difference of values Wk, Wk+1 and values saved in the model is the smallest. As the position choice criterion xk we assume the minimum-quadratic criterion. Ultimately the position of the trolley will be determined by the subtotal of the positions for a sensor and offset resulting from comparison to the formula

$$P_w = P_{sn} + x_w \tag{1}$$

where
P_w ultimate trolley position
P_{sn} position of nth sensor
X_w comparison result

Fig. 3 Sensors arrangement

7 Summary

The model of a system which uses combining sensors presented in the article is just an introduction to a wider research in using Hall effect sensors to measure and control magnetic field. Working out appropriate algorithms processing readings from the sensors can help with building steering systems in the field of automatics and mechanics with the use of microcontrollers. Using non-contact measuring method ensures big permanence of the converter. Hall effect sensor is resistant to dust, dirt and water which clearly makes it better than optical and electromechanical methods if we take into consideration positional sensors.

References

1. Giriat, W., Raułuszkiewicz, J., Hallotrons.: The Use of the Hall Effect In Practice. PWN, Warsaw (1961) (in Polish)
2. Lilienkamp, K.A.: Low-cost magnetic levitation project kits for teaching feedback system design. In: Proceedings of the 2004 American Control Conference, 2004, vol. 2, pp. 1308–1313, July 2 2004–June 30 2004 (2004)
3. Figielski, T.: In the two-dimensional word of electrons, Wiedza i Życie nr 4/1999 (1999) (in Polisch)
4. Kobus, A., Tuszyński, J., Lech, Z.: Techniques of Hall-effect. WNT, Warsaw (1980). (in Polish)
5. Hall effect current sensor aims at electric vehicles [online], data publikacji: 06.12.2013 [acces: 15 maj 2016], http://www.electronicsweekly.com/news/products/sensors/hall-effect-current-sensor-aims-at-electric-vehicles-2013-12
6. Marsden, G.: Levitation! Nuts and Volts Magazine. **24**, 58–61 (Sept 2003)
7. Ładyżyńska-Kozdraś, E., Falkowski, K., Sibilska-Mroziewicz, A.: Physical model of carat of UAV katapult Rusing Meissner effect, Mechanika w Lotnictwie ML-XVI T.I i II/ Sibilski Krzysztof (red.), 2014, Polskie Towarzystwo Mechaniki Teoretycznej i Stosowanej, ISBN 978-83-932107-3-2, ss. 231–242 (in Polish)
8. Sibilska-Mroziewicz, A., Czubaj, S.K., Ładyżyńska-Kozdraś, E. [i in.]: The use of hall effect sensors in magnetic levitation systems. W Appl. Mech. Mater. **817**, s. 271–278 (2016)

The SPM Scanner Head Based on Piezoelectric Unimorph Disc

Krzysztof Tyszka, Mateusz Dawidziuk, Roland Nowak
and Ryszard Jablonski

Abstract The piezoelectric unimorph discs have been recently proposed as a lateral/vertical actuator with nano-scale resolution to be used in a scanning probe microscope scanner head as an alternative for typically used piezo-tube scanners. Here, we reconsider this proposal by investigating deflections of the unimorph piezoelectric disc under voltage bias with a non-contact high resolution optical 3D profiler. The observation shows that high accuracy of X, Y, Z displacement is possible and that unimorph with two opposing electrodes deflects into tilted parabolic shape instead of expected S-shape.

Keywords Scanning probe microscope · SPM · Piezoelectric scanner · Unimorph disk · Optical profiler

1 Introduction

Nowadays the nanotechnology development imposes increasingly stringent demands on metrology [1]. In last 50 years many methods intended for direct surface investigation with nano-lateral-resolution were developed. Among these techniques the great majority is based on Scanning Probe Microscopy (SPM) [2]. The important step towards nano-scale imaging of materials was the invention by G. Binning and H. Rohrer known as Scanning Tunneling Microscope (STM) [3]. The variation of this method allowing investigation of metals, semiconductors, and insulators as well as bio-samples is Atomic Force Microscope (AFM) [4]. The SPMs headed by AFM are the key instruments in today's nanometrology. The SPM devices with different capabilities are used and many AFM-based methods, exploiting different physical phenomena were developed (e.g. Kelvin Probe Force Microscope—KPFM Scanning Capacitance Microscopy—SCM, Scanning Spreading Resistance

K. Tyszka (✉) · M. Dawidziuk · R. Nowak · R. Jablonski
Institute of Metrology and Biomedical Engineering, Faculty of Mechatronics, Warsaw University of Technology, Ul. Sw. Andrzeja Boboli 8, 02-525 Warsaw, Poland
e-mail: k.tyszka@mchtr.pw.edu.pl

© Springer International Publishing AG 2017   513
R. Jabłoński and R. Szewczyk (eds.), *Recent Global Research and Education: Technological Challenges*, Advances in Intelligent Systems and Computing 519, DOI 10.1007/978-3-319-46490-9_68

Microscopy—SSRM, Electrostatic Force Microscopy—EFM, Scanning Microwave Microscopy—SMM). However, it is important to notice, that it was the simple mechanism of STM scanner invented by Binging which started the nanometrology revolution. In STM the atomically sharp tip scans the surface of the conductive sample. During lateral scanning line by line, the biasing of a tip and sample leads to tunneling current through the nanometer scale gap between them. Measured tunneling current is dependent on the tip-sample distance. Therefore, keeping tip-sample distance constant by feedback control application allows detection of tunneling current. Realization of the STM scanning is possible by using piezoelectric scanner tubes allowing lateral tip displacement with nanoresolution. Most of the SPM methods require similar scanning head to work.

Since the evolution of probe microscopes started, the development of more sophisticated SPM tools was initiated. However, with rising need for such tools and SPM being the most basic tool of nanometrology, it is important to design simple, low-cost and more accessible SPM units. One of the proposed solutions for such a design is to use simple piezoelectric unimorph disc instead of piezotubes as an SPM scanner. The unimorph discs are simple to manufacture, less expensive and require lower voltage supply than piezotubes. Recently, basic SPM unit using set of unimorph discs to realize tip displacement was proposed in [5]. Another, older realization described in patent [6] also suggests the use of single unimorph disc with four separated electrodes as SPM scanner.

This work reconsiders the possibility to use a piezoelectric unimorph disc as an SPM scanner allowing nanoresolution displacement of SPM tip in X, Y, Z direction. The unimorph discs in two configurations were investigated: (i) disc with one electrode to realize Z direction displacement, (ii) disc with four electrodes to realize X, Y direction lateral displacement. The non-contact high resolution optical 3D profiler was used to observe the unimorph deflection while different voltage bias was applied to electrodes.

2 Measurement Results

Piezoelectric discs are most often used as sound generators in speakers and buzzers, and are one of the most common piezoelectric devices available. A unimorph disc consists of piezoelectric ceramic as an active layer bonded to inactive layer of a metal disc as it is shown in Fig. 1. The piezoelectric layer has a silver electrode deposited on top of it. When a voltage bias with any polarization is applied to silver and metallic electrode the piezoelectric layer will expand or contract resulting in deflection of a unimorph. This may be used to realize a displacement in X, Y or Z direction. Here, the unimorph with 40 mm diameter of a metal disc and 25 mm diameter of a piezoelectric layer was investigated.

The deflection of the unimorph was observed using the non-contact high resolution optical 3D profiler—Talysurf CCI 600. The working principal of the profiler is based on a type of White Light Interferometry (WLI), the Scanning Broadband

Fig. 1 Piezoelectric
unimorph disc—piezoelectric
ceramic active layer as
positive electrode bonded to
grounded metal disc

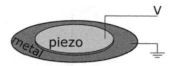

Interferometry (SBI). The surface of a sample is scanned with a light beam with
selectable spectrum, while interference pattern is observed with high resolution
CCD camera. The coherence peak and phase position of interference pattern is
compared for wide spectrum. The result of a measurement is high resolution map of
a sample surface with area of 7 mm^2. The vertical range is 400 μm with guaranteed
resolution of 10 pm.

2.1 Displacement in Z Direction

The results of the measurement of the unimorph disc under changing bias are shown
in Figs. 2 and 3. Three surface maps show how the disc deflects in relation to
no-bias result (Fig. 2-middle). As expected, when the positive bias of 30 V is
applied, disc bulges out (Fig. 2-left). In contrary, when negative bias −30 V is
applied, disc flattens. The dependence of disc profile shape along X axis on bias is
shown in a graph (Fig. 3). It can be seen that dependence can be linearly approximated. This way the slope was calculated to be 0.74 what gives the displacement of
0.74 nm/mV. Such accuracy is sufficient for high resolution Z scanner/actuator.

2.2 Displacement in XY Direction

In Ref. 6 it was proposed to use a unimorph disc with four electrodes as shown in
Fig. 4a. It was expected that when the bias is applied to opposing electrodes the

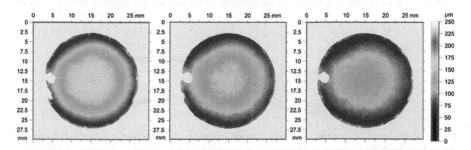

Fig. 2 The result of the observation of the piezoelectric unimorph disc under changing bias—
maps of deflected unimorph with applied bias of 30 V (*left*), 0 V (*middle*) and −30 V (*right*)

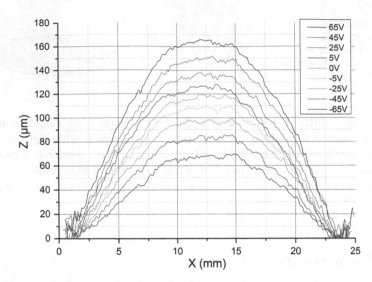

Fig. 3 The result of the observation of the piezoelectric unimorph disc under changing bias. *Line profiles of the unimorph for different biases measured along the disc axis of symmetry*

unimorph deflects into S-shape. This kind of deformation may be used to realize the angular or XY-lateral displacement of the object axially mounted to the unimorph, as shown in Fig. 4b.

The results of the measurement of the unimorph disc under changing bias are shown in Fig. 5. The graph shows how the disc deflects in relation to no-bias result when bias is applied to pair of opposing electrodes while another pair is grounded. The qualification of the obtained shape is not straight forward and does not reflect the expected S-shape. It can be seen that the axis of symmetry of the parabolic profile shifts depends on the bias value and polarization. The symmetry of a profile becomes also distorted resulting in tilted parabolic shape. Although this is far from expected S-shape it can be still used to realize the angular displacement. To estimate the deflection of the unimorph, the slope of an axis via two characteristic points of each profile was calculated (as shown in Fig. 5). Figure 6 shows the dependence of estimated angle of inclination of each axis on applied voltage. Slope

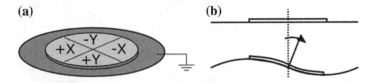

Fig. 4 Piezoelectric unimorph disc with four electrodes, **a** illustration of a piezoelectric ceramic active layer with four separated silver areas to control the tilt in X and Y direction. **b** Illustration presenting the unimorph disc with unbiased electrodes (*top*) and S-shaped unimorph disc with biased pair of opposing electrodes (*bottom*)

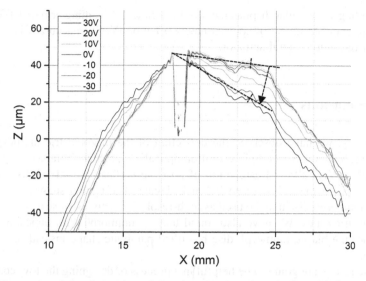

Fig. 5 The results of the observation of the piezoelectric unimorph disc with four electrodes, obtained for different biases. The line profiles are in shape of tilted parabola. To estimate the deflection of the unimorph, the slope of an axis via two characteristic points of each profile was calculated in reference to 0 V result (as marked with *dashed lines* and *arrow*)

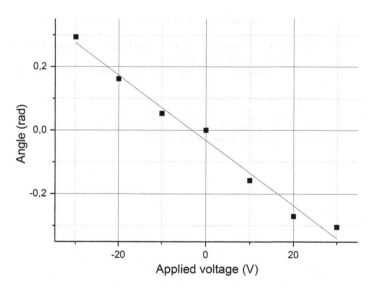

Fig. 6 The dependence of estimated axis angle of inclination on applied voltage, estimated as shown in Fig. 5

calculation gives angular displacement of 0.01 rad/V. Assuming arm of 10 mm this gives lateral displacement of approximately 0.1 nm/mV. Such accuracy is sufficient for high resolution lateral actuator.

3 Conclusions

In summary, we have observed the deflection of the piezoelectric unimorph disc under bias by non-contact high resolution optical 3D profiler. Obtained maps of the unimorph surface allowed estimation of possible vertical and lateral displacement accuracy which may be realized by such device. The results suggest that piezo-electric unimorph disc may be used as high-resolution actuator or scanner instead of piezo-electric tubes. We have also found that for unimorph with opposing pair of electrodes the biased unimorph disc has tilted parabolic shape instead of expected S-shape.

These results are going to be helpful in a process of designing the low-cost AFM scanner head with high-resolution and low supply requirements.

References

1. Wogel, E.W.: Technology and metrology of new electronic materials and devices. Nature 2, 25–32 (2007)
2. Kalinin, S.: Scanning Probe Microscopy—Electrical and Electromechanical Phenomena at the Nanoscale, vol. 2. Springer Science, New York (2007)
3. Binning, G., et al.: Surface studied by scanning tunneling microscopy. Phys. Rev. Lett. 49, 57 (1982)
4. Binning, G., et al.: Atomic force microscope. Phys. Rev. Lett. 56, 930 (1982)
5. Grey, F.: Creativity unleashed. Nat. Nanotechnol. 10, 480 (2015)
6. Alexander, J., Tortonose, M., Nguyen, T.: Atomic force microscope with integrated optics for attachment to optical microscope, Patent US5866902

Printed in the United States
By Bookmasters